Global Environmental Change Science:
Education and Training

NATO ASI Series

Advanced Science Institutes Series

A series presenting the results of activities sponsored by the NATO Science Committee, which aims at the dissemination of advanced scientific and technological knowledge, with a view to strengthening links between scientific communities.

The Series is published by an international board of publishers in conjunction with the NATO Scientific Affairs Division

A Life Sciences **B Physics**	Plenum Publishing Corporation London and New York
C Mathematical and Physical Sciences **D Behavioural and Social Sciences** **E Applied Sciences**	Kluwer Academic Publishers Dordrecht, Boston and London
F Computer and Systems Sciences **G Ecological Sciences** **H Cell Biology** **I Global Environmental Change**	Springer-Verlag Berlin Heidelberg New York London Paris Tokyo Hong Kong Barcelona Budapest

PARTNERSHIP SUB-SERIES

1. Disarmament Technologies	Kluwer Academic Publishers
2. Environment	Springer-Verlag
3. High Technology	Kluwer Academic Publishers
4. Science and Technology Policy	Kluwer Academic Publishers
5. Computer Networking	Kluwer Academic Publishers

The Partnership Sub-Series incorporates activities undertaken in collaboration with NATO's Cooperation Partners, the countries of the CIS and Central and Eastern Europe, in Priority Areas of concern to those countries.

NATO-PCO DATABASE

The electronic index to the NATO ASI Series provides full bibliographical references (with keywords and/or abstracts) to about 50000 contributions from international scientists published in all sections of the NATO ASI Series. Access to the NATO-PCO DATABASE compiled by the NATO Publication Coordination Office is possible in two ways:

- via online FILE 128 (NATO-PCO DATABASE) hosted by ESRIN, Via Galileo Galilei, I-00044 Frascati, Italy.

- via CD-ROM "NATO Science & Technology Disk" with user-friendly retrieval software in English, French and German (© WTV GmbH and DATAWARE Technologies Inc. 1992).

The CD-ROM can be ordered through any member of the Board of Publishers or through NATO-PCO, Overijse, Belgium

Series I: Global Environmental Change, Vol. 29

Global Environmental Change Science: Education and Training

Edited by

David J. Waddington

Department of Chemistry, University of York,
Heslington, York, YO1 5DD, UK

Springer

Published in cooperation with NATO Scientific Affairs Division

Proceedings of the NATO Advanced Research Workshop concerned with education and training in global environmental change science, held at the University of Western Ontario, Canada, June 4 – 8, 1994

ISBN-13: 978-3-642-79772-9 e-ISBN-13: 978-3-642-79770-5
DOI: 10.1007/ 978-3-642-79770-5

CIP data applied for

Softcover reprint of the hardcover 1st edition 1995

Typesetting Camera ready by authors/editors
Printed on acid-free paper
SPIN 10466216 31/3136 - 5 4 3 2 1 0

This work would not have been realised without the original idea and preliminary work of Max Beran.

PREFACE

A workshop was held in June 1994 at the University of Western Ontario, London, concerned with education and training in global environmental change science. It was principally devoted to post-school education from first year university to postgraduate levels.

I have arranged the papers and the results of discussions in the following order:

Section 1: Introduction
These papers set the scene. It begins with one by Beran whose original idea the workshop was. Schneider outlines various ways in which interdisciplinary courses in general and global change science in particular can be set up. Fyfe looks at the wider issues of global change that affect us all. There are then papers which examine major problems (Leal Filho; Harger and Troost) and programmes to meet them (Carter; Fisher).

Section 2: Global Change Science Courses
These papers are concerned with courses concerned with various aspects of global change courses. They include contributions from nine countries at different levels of secondary and university education.

Section 3: Resources for teaching Global Change Science
New resources are needed for effective courses, particularly those in which students can participate actively. There are wide selections both in terms of topics and teaching strategies, most of which are interactive.

Section 4: Workshop discussions
The workshop participants split themselves amongst three groups, one to discuss early undergraduate courses, another with senior undergraduate courses and the third with postgraduate courses. Their discussions are brought together in this section and common themes are identified.

I had nothing but tremendous help from the participants, my colleagues on the Workshop committee, the NATO Scientific and Environmental Affairs Division office in Brussels, those who helped with all the arrangements at the workshop (particularly Marilyn McMahon) and the embattled secretaries at York (Nancy Newton and Sandra Wilmott). Thus, any mistakes in editing the papers and discussions must lie with me.

York, 1994 D J WADDINGTON

TABLE OF CONTENTS

Section 1: Introduction

Section 2: Global Change Science Course

SECTION 1: INTRODUCTION

Education and Global Environmental Change

Stephen Schneider
Institute for International Studies
Stanford University
Stanford CA 94305, USA
and the National Center for Atmospheric Research
Boulder, CO 80307, USA

Joseph Stoltman
Department of Geography
Western Michigan University
Kalamazoo, MI 49009
USA

David Waddington
Department of Chemistry
University of York
York YO1 5DD
UK

Causes of *Global Environmental Change* fall into two major categories: natural variability and anthropogenic contributions, the human component arising largely as a result of vast numbers of people demanding high standards of living and using technology and resources to support their efforts.

An understanding of the processes leading to global change requires detailed knowledge of the natural and social sciences, and a systems approach. This in turn requires cooperation amongst teachers and researchers across the traditional boundaries in education. These boundaries are between subject areas, and because global environmental change is a multidisciplinary problem, a range of potential solutions and an understanding of the complex issues often spring out of the interface of knowledge among the disciplines.

Further, not only must the traditional boundaries of the subjects be broken down, so too must those boundaries between school, undergraduate, and postgraduate education and training.

It will be seen that these two themes are emphasised many times during this book, in terms of the training of global change scientists, in educating school and university students and in providing the public with the knowledge essential to an understanding of the environment. While earth and physical sciences serve as the information and research base, the nature of the issues in global change bridge nearly every component of the curriculum. Thus, the issues in global change require an educational structure that integrates the natural and social sciences, as well as the humanities and professional studies such as business, law, and medicine.

NATO ASI Series, Vol I 29
Global Environmental Change in Science Education and Training
Edited by David Waddington

Why this workshop?
Global change problems are an important and timely focus for research and education and they are indeed being tackled urgently. However, the training brought by most scientists to global change problems consists of first - and often second - degrees in a single scientific discipline, so that the global perspective must be picked up on the job. There are few opportunities for formal training. This problem was recognised during the initial workshop of the NATO Special Programme for the Science in Global Environmental Change (Corell *et al.* 1991), which recognized that the multidisciplinary nature of global change science was a significant obstacle and that NATO had a rôle in assisting with the training and education of the coming generation of global change scientists. The panel in charge of the Special Programme proposed that conversion courses at the masters degree level could be a possible way forward to provide conventionally trained scientists with the necessary global outlook.

A preliminary step prior to holding the workshop was to gather information on existing courses in NATO countries. Enquiries were addressed to research and educational agencies, and centres were visited. From this survey emerged a list of potential participants. However, it was soon apparent that, although there was healthy activity in specific subjects, there were few interdisciplinary systems-oriented scientific research activities or educational programmes. There was, however, an encouraging sign that at least some educational programmes were emerging which attempt to provide students at many levels with a familiarity of the "earth-as-a-system" concept.

These preliminary enquiries also revealed that there was a large measure of interest among educators in having information brought together in a reference volume, which outline global change curricula and resources and experiences of course-providers.

Following detailed correspondence with global science educators over a period of 18 months, it was decided that it was time to bring some people together to share experiences and to chart ways in which global science education at undergraduate and postgraduate levels could be enhanced, and, through this book, to share experiences with others.

The Workshop

Twenty five scientists and science educators from 11 countries agreed to participate in the Workshop. Although the original aim was for the first workshop to identify some of the resources in university postgraduate courses, it was soon apparent that we needed to look at the educational issues below the masters degree level.

During the workshop, it became apparent that in many countries environmental concerns have had little impact on school and first degree science courses. Problems associated with global change education in degrees in single subject disciplines are exacerbated by courses at school and pre-doctoral level, which in many countries are taught in terms of specialisms defined by boundaries set a century ago. This led workshop participants to widen their brief because the training of global change scientists, who have the necessary global perception or requisite combination of knowledge, needs underpinning during first degree courses. With no introduction to the bigger picture of the environment and its relationship to humans, students are not excited about the relevance of the study of science, and we as a society are losing opportunities to further the pursuit of science careers. Getting students interested and committed will in turn contribute to increasing the supply of global change scientists, who have the necessary global perception or requisite combination of knowledge. Further, the lack of a broader science education exacerbates problems associated with a public ill-equipped to face the rapidly changing environment in which we live and is antipathetic to solutions proposed for global change problems. Increasing science literacy through the use of environmental issues will enhance the abilities of the public to understand and participate in environmental policy-making efforts. This broader perspective is reflected throughout the report.

Members of the group had a very wide range of experiences in the education and training of both scientists and non-scientists at university level in various aspects of global change science. They addressed the problems of introducing new and modifying existing courses by having a series of plenary sessions with inputs from participants who outlined the work done in their own institutions and regionally. These are described in Section 2.

The majority of the time was spent in small groups tackling specific issues at three levels of education and training:

(i) senior secondary and freshman university courses

(ii) senior undergraduate university courses

(iii) postgraduate courses

The results of their deliberations are incorporated in Section 3. It was possible to compare experiences, define common objectives for global change educators at many levels, examine the balance between breadth of subjects including various detailed content of specialized topics, identify obstacles to mounting interdisciplinary courses, and, most importantly, fashion a set of recommended solutions to these obstacles.

The differences in outlook of the participants stemming from their different educational systems provided a salutory framework against which we worked, a framework that, all the time, made us think about *flexible* solutions and not solutions that would only apply to one country. There was underlying agreement that there was an urgent need to create an adequately trained talent pool and a global environmentally literate society, which together would provide the foundation for sustainable development.

There was a realisation that representation at a NATO workshop was only from developed countries, and that a vital ingredient was missing: the experience of scientists working in the developing nations. Nevertheless, we had a valuable input from Unesco (Troost *et al.* 1994; Harger and Troost 1994) and a paper from Leal Filho (1994). We did not examine specific ways of educating adults - whether they be policy makers, or members of the general public - either by *formal* or *informal* means, e.g., via the media. We concentrated primarily on the education and training of scientists for research and of those scientists who wish to have a strong element of global change science in their teaching courses at the freshman and senior levels, both in terms of a general education and for development of scientific knowledge and skills.

During the meeting, participants used the same terms with slightly, but still significantly, different meanings. We therefore agreed that we should try to adopt a common terminology so that the reader knew precisely how we were using specific terms. The list of terms was prepared by three participants: Max Beran, Henrietta Lidiard, and Stephen Schneider.

Earth System Science and the Science of Global Environmental Change

The fundamental concept is that of the earth as a dynamic system which has experienced change over all timescales. This is studied as *Earth System Science.* It has been shown that man's activities have the capacity to perturb this global system, so the *Science of Global Environmental Change* is the study of the natural global system focusing on anthropogenic influences and their environmental and societal consequences. This is often shortened to *Global Change Science.*

At present, scientists are focussing on subjects such as climate change, ozone depletion, the loss of biological diversity, degradation of terrestrial and marine environments, and desertification, but the workshop participants were aware that these are but some examples of current and fundamental concern.
There are two terms used in the discussion of earth-systems which need defining. These are *linkages* and *interaction.*

Linkage refers to connections among various sub-elements of a system. For example, the ocean provides a heat source for the atmosphere and thus air and sea are linked. *Interaction* implies linkage in both directions among sub-elements of a system. For example, the ocean provides a heat source for the atmosphere, but, for time scales longer than weeks, atmospheric winds and transparency to solar heat feed back on ocean temperatures. Thus, on these scales, there is an air-sea interaction.

Finally, two terms are used to describe the way global change scientists are educated and trained. These are multidisciplinary and interdisciplinary. We distinguish between these in the following way: *Multidisciplinary* refers to knowledge from multiple disciplines derived from disciplinary methods, practices, and paradigms, independent of problems such as global change. *Interdisciplinary* is used to denote the integration of knowledge from multiple disciplines combined in an original synthesis, in order to explain the behaviour of a complex system or to address a problem of practical significance.

References

Correll RW and Anderson PA (eds) 1991 Global Environmental Change. NATO ASI Series Global Environmental Change, Springer Berlin 1 : 264

Harger JRE and Troost D 1994. This book

Leal Filho W 1994. This book

Troost D, Robinson IS and Blackburn D. This book

Evolutionary Organizational Models for Interdisciplinary Research and Teaching of Global Environmental Change

Stephen H Schneider
Institute for International Studies
Stanford University, California and
National Center for Atmospheric Research
Boulder, Colorado

Environmental science, technology, and policy studies require an integrated, interdisciplinary approach that needs to draw together talents from diverse academic disciplines, and from industry, government, and non-government organizations (NGO's) (mostly non-profit advocacy groups), and must offer incentives to participants to produce team-oriented, problem-driven, interdisciplinary research and teaching. The challenge, of course, is how to encourage this and maintain high quality standards, given the traditional disciplinary and/or departmental structures at most universities, national laboratories, and funding agencies. Some individuals have thus suggested that we need to re-examine and modify these traditional structures to ensure they are not impediments to quality, integrated global environmental systems studies.

Linear versus "Organic Systems" Structure for Academic Institutions

Figure 1 is my "cartoon" of the typical academic research and teaching paradigm, a linear progression of "knowledge flow" (represented by the arrows on the figure) from disciplinary and multidisciplinary (but rarely integrated) natural scientific work fed into social sciences for impact assessments which may spawn isolated efforts at policy analysis, sometimes by interdepartmental institutes. Humanities expertise enters sporadically, usually in the form of an interested scholar or two becoming involved with a team of global change natural or social scientists. By "multidisciplinary, I mean specialists from various disciplines applying their respective disciplinary paradigms and methods in parallel on restricted aspects of some broad systems problem. By "interdisciplinary" I refer to individuals or teams of diverse specialists which learn enough - individually or collectively - about a range of relevant disciplinary methods and content to combine such multidisciplinary knowledge into an original synthesis that helps to raise understanding of systems phenomena or to solve a real problem (Schneider, 1977). Interdisciplinary work is often at the intersection of disciplines; multidisciplinary work is usually centered within several traditional disciplines. Government agencies and private foundations largely fund the

NATO ASI Series, Vol I 29
Global Environmental Change in Science
Education and Training
Edited by David Waddington

spectrum of such research, but, like other "users" of the research who often have to make policy decisions despite the uncertainties, they largely sit on the sidelines once the grant is given and academic research is undertaken.

Teaching is typically concentrated in disciplinary departments, but a broadening of this discipline-oriented approach has long been undertaken at many institutions. There are many examples described in this volume. This approach will be discussed in depth shortly. Outreach of new knowledge from the academic community to the media, the public, and policy makers is sometimes handled institutionally by extension officers, but, as Figure 1 suggests, it is more often provided by mediaskilled entrepreneurs: an unorganized, ad hoc minority the members of which are selfselected "outreach agents" who interpret and "spin" the academic knowledge-base. Such people are typically disciplinary specialists, and, with a few rare exceptions, offer their interpretations and advice from the narrow perspective of their disciplinary paradigms. What is needed, I believe, is to find an organizational vision that leads to institutions which inherently will immerse such specialists in a systems perspective, an experience that could well substantially alter their interpretations.

Figure 2 is a contrasting organic systems paradigm stressing the integrative or "center-like" aspects of global environmental change problems. This figure illustrates the integration and organizational concepts I wish to discuss here. Rather than a linear, disciplinary organization in which knowledge typically flows from physical to biological to social scientists, and eventually to policy assessment teams, Figure 2 offers a circular, problem-oriented, interconnected structure, or an interactive or organic systems approach. The global environmental change issue is subdivided into three primary subsets: (1) the functioning of the global environmental system, (2) disturbance of the environmental system, and (3) the response of the environmental system to global change disturbances. There is feedback among these subsets and a central entity responsible for integration, which is a group also responsible for feeding back to the three sub-problem groups new scientific problems to work on. An input from many traditional disciplines is necessary for describing the functioning of the global environmental system, and likewise for the other two subsets.

Many gaps in disciplinary knowledge will be uncovered by such a problem oriented approach. Thus, rather than competing with disciplinary research

STANDARD ENVIRONMENTAL STUDIES
(Linear Model - Disciplinary to Multi-disciplinary)

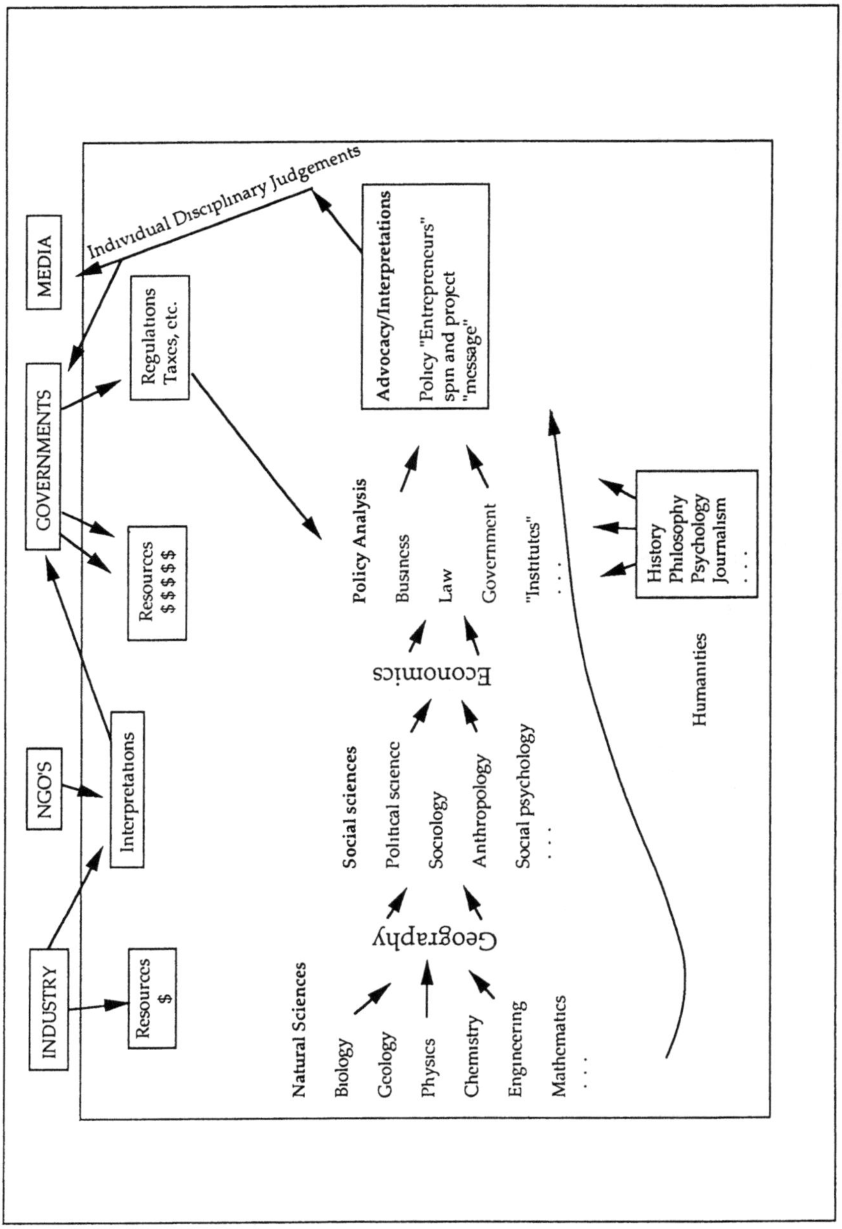

Figure 1

INTEGRATED MODEL OF ENVIRONMENTAL SYSTEMS
(Multi-disciplinary to Inter-disciplinary Research, Teaching & Outreach Connected by Design)

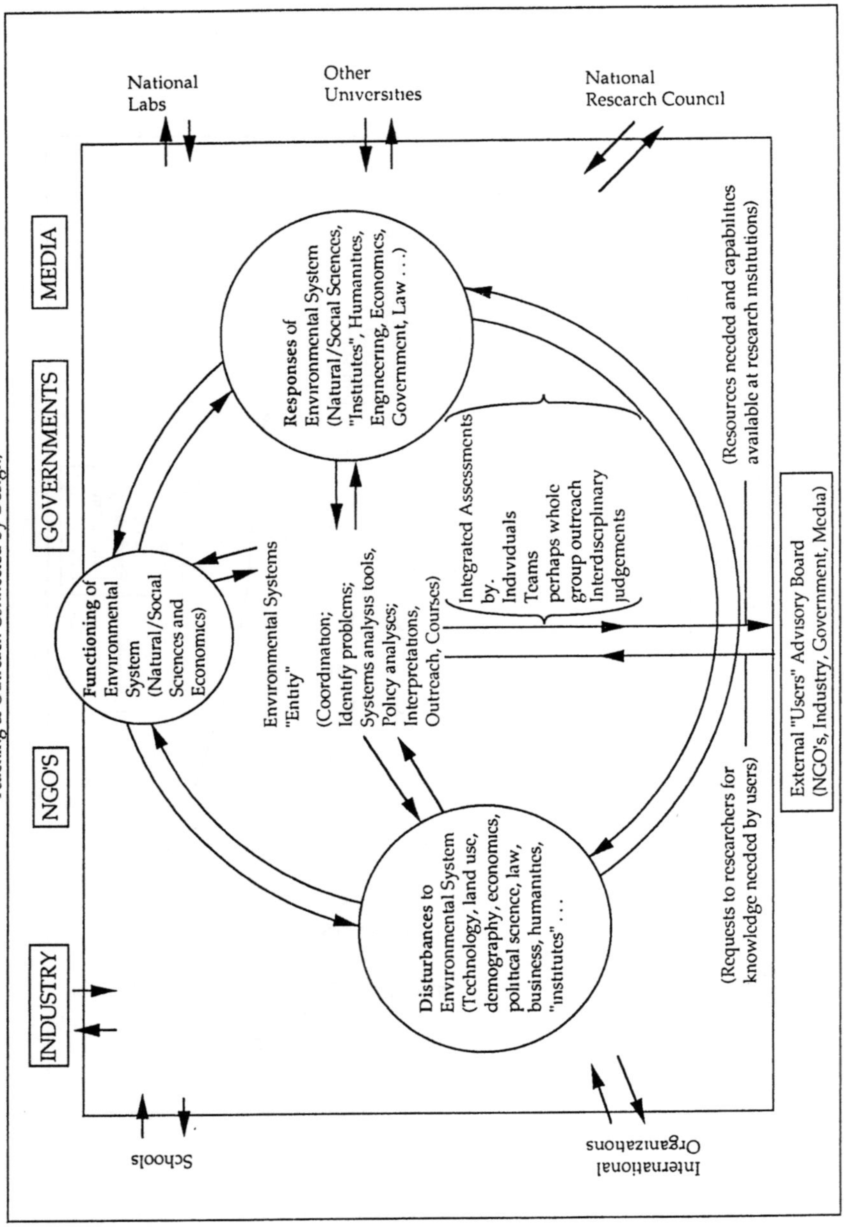

Figure 2

progress, such a systems organization could very well augment it by providing new discipline-oriented tasks for disciplinarians to work on, that are based on overall problem needs and not simply research agendas established in the judgment of disciplinarians as to what is most important for that discipline.

This organic systems structure is designed to unite, evaluate, and reward high quality interdisciplinary research, teaching, and outreach; to coordinate exchanges from multiple disciplines so that an interdisciplinary perspective is facilitated; and to set up formal mechanisms to encourage individuals, teams and (perhaps) whole groups to communicate with the "real world" actors represented on Figure 2 in the "picture frame" margin in the diagram. The three circles reflect a problem-oriented administrative disaggregation of intellectual talent based on environmental systems phenomena, rather than on disciplinary tradition or historic bureaucratic boundaries. Typical disciplinary specialties needed in each sub-problem area are listed in the circles. However, how to organize existing people or groups at an academic or laboratory institution into teams suggested by the circles in Figure 2 is a formidable challenge. In this article discussion will be focused on the intellectual question of whether we would *want* to do this, rather than on the potentially divisive political question of how to effect the transition.

Instead of having practitioners of one discipline pass on their partial view of a systems phenomena to the next discipline, and eventually have the policy implications reported largely ad hoc by disciplinary or special interest-driven policy entrepreneurs (Figure 1), my replacement concept is to evolve a central Environmental Systems Entity (e.g., an integrated department, institute, or some other administrative taxon) to help ensure that actual problems are addressed by having high-quality disciplinary knowledge (available from traditional departments) collected, combined, and focused on problems by the "entity" faculty/staff. Such personnel would include world-leading specialists in the integration process itself. Such scholars may well come from specific disciplines, or be graduates of interdisciplinary programs. Regardless of their origin, they should have demonstrated an ability to seek out specialists with appropriate disciplinary knowledge, and they should help survey and integrate such knowledge in order to enhance understanding of systems phenomena and/or to help solve actual problems.

Evaluating Interdisciplinary Quality

One constraint facing universities considering such a systems approach involves the issue of evaluating interdisciplinary quality. While I could hardly argue with the proposition that researchers should always pursue high quality, it must be recognized that quality evaluation consists of value judgments. Within a university setting, quality generally is defined as any significant quantity of original work that advances the discipline. This definition is perfectly appropriate for disciplinary specialists, but too narrow for evaluation of integration experts. Members of the academic community who define quality under this traditional, narrow definition of disciplinary originality often argue that interdisciplinary programs compromise worth (e.g., Chen, 1981).

I believe that interdisciplinary quality, although substantially different from the disciplinary notion of the concept, is intellectually as viable. Interdisciplinary quality, too, must possess originality. This requirement can be accomplished through an innovative method of combining known materials from various fields rather than within the confines of one field. The materials from various disciplines needed to address a particular interdisciplinary problem must be combined in an original manner that helps to enhance understanding of a systems phenomena or to solve an actual problem. All disciplinary materials used must be up to date and accurate. Furthermore, specialized concepts or jargon must be presented with great clarity to promote cross-disciplinary communication. (These, in fact, have been the criteria for interdisciplinary peer review for the journal *Climatic Change* since its inception [Schneider, 1977]).

Such quality standards can be evaluated, can be enforced, and should be rigorously tested through the peer process - although no individual peer may have the capacity to fully judge the entire nature of the work, nor is that a necessary or a fatal obstacle. The creation of a peer review group for interdisciplinary environmental studies is itself an evolving process, and models of this process will be explored later on in the discussion of Tables 1 and 2. Interdisciplinary work is inherently team-oriented and requires multiple inputs from many quarters. However, over time individuals who become increasingly knowledgeable about the content and paradigms of several disciplines can themselves become interdisciplinary, or in essence, integration experts. These are the kinds of people for whom reward systems are generally not well developed in academe, and for whom funding is often difficult to obtain from traditional sources.

Problem-driven research doesn't exclude disciplinary originality, rather it emphasizes the reverse, requiring more, though problem-focused, disciplinary knowledge. Since that knowledge can be focused on practical problem solving, it should entrain new "customers" for the research products. Thus, this systems model should also provide nontraditional sources of resources to disciplines, that would not likely have been available to curiosity-driven disciplinary researchers from most granting agencies. In short, participation of disciplines in interdisciplinary systems research need not be a threat in the sense of losing out in a zero-sum game, but rather a way of expanding overall support for intellectual activities.

User Interface

In an organic systems model, mechanisms are also built in for interactions with users of research products (for example, environmental activists, industries with environmental connections, elected officials, agency workers, and the public at large) to examine research products and suggest just how these products could - or could not - contribute to improving the knowledge base needed for analysis and ultimate policy choices. In this manner, research relevant to user needs could evolve as an integral component of the center or entity concept. To me, bringing users into research planning is not merely putting "amateurs in charge of disciplines," but rather putting professionals in information use in charge of determining what kinds of information they think they need. Then integration experts at the interface of the user communities and the disciplinary academic and laboratory communities can help to translate what is needed at the policy end into what should be produced at the science end. Quite bluntly, most disciplinary investigators are not qualified by themselves to make that judgment without the input of broader groups, particularly those at the front lines of problem solving. Neither are the problem solvers qualified to dictate the details of what disciplinarians should do, but certainly the former should point out what policy questions are in need of more scientific information, from which integrationists and disciplinarians together could work out focused research agendas that are appropriately multidisciplinary and/or interdisciplinary.

Another vital function that an environmental systems entity could perform would be to establish a place for summer studies, for regional field work, for research fellows to congregate, for media interested in covering global change topics to be put in touch with individuals knowledgeable in various aspects of the issues (rather than the random and/or biased coverage typical of most polarized global change debates), and in which public outreach could be advanced.

Evolution, Not Revolution

In essence, what I am calling for is not the wholesale replacement of the disciplinary paradigm, or its traditions of strict quality review standards for disciplinary research, with something different, but rather the augmentation of standard practices to include problem-driven, interdisciplinary integration as part and parcel of the total academic/laboratory research enterprise. I believe that this evolution toward broader research and teaching paradigms will *not* reduce the quality of disciplinary work. On the contrary, I think that it will expand the opportunities and funding sources, and that evolution of systems organization will not necessarily replace most disciplinary departments, just as biological evolution does not necessarily replace some existing species. Rather, such an evolution will allow analysts and synthesists to address global change problems at the scale they exist (the "problem scale"), a scale not constrained by boundaries and methods that happen to fall into historic entities or disciplines that were created for different purposes.

None of this obviates the need for disciplines. I envision that only a small percentage of overall academic activities would be in the integrationist category, because integration expertise is hard to find, will take time to develop, requires reorganization of training and rewards systems, needs to evolve a peer group, and simply isn't ready to take over a very large fraction of the effort, if quality is to be maintained. Interdisciplinary work cannot function without a high quality base of disciplinary knowledge, but an institution of higher learning cannot efficiently contribute to the solution of pressing environmental problems by retaining a discipline-only, traditional organization and rewards structure. We need both, and there is no time to waste. We simply cannot afford the luxury of waiting for such new structures to emerge bit by bit out of present institutions over the next century, just as specialized disciplines replaced classical training and organizations a century or two ago. We need to be creative and active in accelerating that evolutionary process by perhaps an order of magnitude to address pressing environmental problems whose solutions are already overdue.

Evolution of Global Change Research Organization

Most academic or laboratory research is not organized like the systems model proposed in Fig. 2. However, organizational changes have evolved at academic institutions in an attempt to encourage multidisciplinary and sometimes interdisciplinary activities to take place. Table 1 offers an example of typical organizational issues involved in enhancing global environmental change research activities.

Table 1

Organizational Steps or Issues for Global Environmental Change Research
* Multidisciplinary seminar - to learn who is interested and in what
* Multidisciplinary committees - plan more formal integrative activities and supervise Ph.D. students
* Interdepartmental Institutes (typically not permanent)
- Multidisciplinary faculty with partial appointments in disciplinary departments and institutes
- Senior fellows: non-tenure track but full-time institute appointees
- Development of peer groups capable of evaluating interdisciplinary quality

This NATO volume is dedicated primarily to educational rather than research problems encountered in global change activities. However, research activities typically are given higher priorities than global change educational activities in most of the more prestigious universities, so experience gained in analyzing problems and in finding solutions to interdisciplinary global change research aspects should carry over to dealing with global change educational improvements. Furthermore, the best global change teachers are probably close to evolving research findings.

Returning to the organizational steps in Table 1, the process of using organizational change to encourage global change activities typically begins with the creation of a multidisciplinary seminar. Since the process of establishing networks of multidisciplinary scholars, students, and staff to form multidisciplinary research teams requires a progressive recognition of the problems, practices, and ideas from a variety of disciplines, the formation of multidisciplinary seminars is one of the best ways to build that community of scholars. Talks given by faculty from various departments or from visiting faculty, scientists, decision makers, or people from business and environmental groups from the outside, often serve as a focal point for this activity. Frequently, as interpersonal relationships build around multidisciplinary problems like global change, some subset of faculty, visitors, students, and staff often form the next element on Table 1, multidisciplinary committees. Such groups often plan for more formal integrative activities or can supervise Ph.D. research, evolving to become part of the peer group needed to sustain interdisciplinary quality, as discussed earlier. Such activities could include joint proposal writing, summer studies in multidisciplinary problems, planning for new courses, development of an interdepartmental institute or center, and ultimately, attain a vision for more permanent new departments and/or schools, whose focus would be, in our example, interdisciplinary global environmental change efforts.

Interdepartmental institutes typically are not permanent, hard-money-funded organizational entities at most universities, because they do not have the historical or administrative clout of schools or disciplinary departments. They are typically staffed by multidisciplinary faculty, most of whom may have partial appointments in such institutes, but most of whom have permanence or tenure status in disciplinary departments. Sometimes institutes hire full-time staff, such as senior fellows or research professors, but in many - probably most - current cases such positions are pejorative in the sense that they are not tenure-track faculty slots with full security, privilege, and influence associated with tenure status. In some cases junior faculty (those not yet tenured) are often nervous about participation in such interdepartmental institutes, particularly if their work is interdisciplinary. Such activities can reduce the likelihood of promotion to tenure for junior faculty whose peer evaluation may be made solely on disciplinary standards. Furthermore, they often will be compared to the productivity and disciplinary originality earned by colleagues who may not be splitting their efforts among both disciplinary and interdisciplinary activities.

The tenure issue is explicitly raised next on Table 1. Most tenure-track appointments in academe, or so-called "senior scientist" appointments without term in research laboratories, largely reside in disciplinary departments. Sometimes creative, interim solutions have been attempted across organizational sub-units, such as that at Penn State University, in which the Earth Systems Science program controls the funds for new appointments in global change studies, while the departments retain the legal power to advance or deny a candidate achievement of tenure status. Because of the difficulty in modifying existing legal or traditional appointment practices at academic institutions, such "joint implementation" compromises are often important evolutionary steps of organizational change needed to enhance the tractability and quality of interdisciplinary work.

Another proposed solution to this difficulty was developed in 1969 at Stanford University in its interdepartmental Human Biology program, and is informally known there as "elastic slots." The logic is that research activities and tenure would remain in traditional departments on the grounds that such departments were best qualified to evaluate research quality, but that multi- or interdisciplinary teaching could be carried out in an interdepartmental program or in institute organizations which do not have permanent or tenure granting status. While elastic slots may be a potentially useful interim step, I believe this is not an optimal - or even appropriate - long-term solution, since the highest quality problem solving or interdisciplinary research will require the integration of research, teaching, and outreach in the same organization; and with quality review being applied to both teaching and research, based on how well the candidate performs in the interdisciplinary context, not how well the candidate performs solely in disciplinary research. Rather than seeing departments as best qualified to evaluate research quality, the opposite is more likely: departments are usually not well qualified to judge the interdisciplinary originality or importance of such research activity, and the basic premise behind elastic slots and the assignment of tenure to departments and teaching to programs, is fundamentally flawed with respect to encouraging sustained, high-quality interdisciplinary activities.

The most difficult organizational change for most current institutions would be the creation of new interdisciplinary departments or schools whose prime activity is research, teaching, and outreach on interdisciplinary, systems-oriented problems like global environmental change. To attract the best faculty (especially at early career stages), such an environmental studies entity needs to be backed up with sufficient funding to offer prospective staff the opportunity of permanence on an equal footing with tenure-granting schools and departments. But a new permanent entity is especially hard to create when the overall institution is not growing, let alone when it is undergoing cutbacks. Such new interdisciplinary departments of environmental science and/or management hardly need to add this fiscal obstacle to the many other associated problems of new paradigms, such as evaluating interdisciplinary quality, providing adequate rewards or reprimands based on quality evaluations, and expecting some measure of effective public outreach or technology transfer as part of the job. The suggestion to obtain hard money to support a new interdisciplinary department (which could mean earmarking a fraction of the endowment of a private university for it or becoming a permanent line item in the budget of a large laboratory) is likely to run into a storm of academic opposition from already established organizational sub-units, each of which is trying to hold on to its share of what it perceives as the stable or shrinking pie. Of course, such opposition is usually couched in terms of quality concerns, not competition for scarce resources.

Support for the program, faculty, and staff of a new environmental entity could also be obtained from federal and local grants, as in the case of disciplinary departments and schools, and perhaps could also, particularly in the early phases, get a major boost in support from private foundations. The latter often like to fund new initiatives, particularly for problem-oriented organizations, but do not typically like to be committed to long-term support programs. The creation of new interdisciplinary, permanent organizations would be the ultimate institutional evolutionary step. Most institutions are more realistically described in terms of the activities distributed across the earlier entries on Table 1, but there are some outstanding examples of interdisciplinary environmental studies organization that are already established, (such as the Energy and Resources Group (ERG) at UC Berkeley and the Earth Systems Science Center at Penn State University). At a 1977 symposium on the interdisciplinary process co-organized by the author, the late anthropologist, Margaret Mead, was invited to comment on this question. She characterized three elements she believed were in common to the creation and sustenance of most interdisciplinary ventures: (1) a charismatic leader with sufficient disciplinary credentials to maintain credibility of the enterprise in spite of doubters from disciplines and with sufficient intellectual clout and appeal to procure the next two ingredients; (2) a group of young, dedicated researchers, who are "analogical" rather than "digital" thinkers: or people who at the margins of their intellect like to make connections; and (3) an identified primary funding source for the activities. (Mead's remarks are described in Chen 1981). When pressed by a panelist who was concerned by how difficult these three criteria are to maintain, I recall her replying simply that two out of three of these characteristics needed to be maintained at all times by successful interdisciplinary groups and all three for some of the time, particularly criterion 1 at the outset of the venture. I think her model fits well the ERG and Earth Systems Center examples given above, my former affiliation with the Interdisciplinary Climate Systems Section at the National Center for Atmospheric Research in Boulder, Colorado, and, on a larger scale, the International Institute for Applied Systems Analysis in Laxenberg, Austria.

The Margaret Mead model for interdisciplinary group creation and survival has been closely matched many times and in many types of institutions, but, distressingly, I have also witnessed the withering of such interdisciplinary activities, particularly when the charismatic leader moved on before the unit could become sufficiently established. Thus, such a model implies a burdensome set of hurdles for new interdisciplinary units that most disciplinary groups do not need to address. It is essential for high quality and sustained integrative activities required by such subjects as global environmental change, that support for the activities and staff be institutionalized and put on a par with traditional disciplinary departments. That means having fair access to funding resources, and not being required to maintain indefinitely and continuously either extraordinarily unusual leadership or outside resources as a prerequisite to continued existence. During the transition from initial establishment of interdisciplinary units to more permanent status, it is probably not unreasonable that Mead's model should apply, but eventually a progression through the steps outlined on Table 1 should lead to a more permanent stave .

Successional Model of Organizational Changes to Sustain Global Change

Let us turn now to the question of improving education and training of global environmental students, scientists, professionals, and the public at large. Table 2 summarizes my experience with the teaching of cross-disciplinary subjects at universities and laboratories. The Table suggests a typical progression or successional model for organizational changes in response to the special demands for interdisciplinary teaching. It is a model of a typical successional pattern in which increasing degrees of integrative activities are encouraged and institutionalized. Like a real biological succession, from which this analogy is inspired, sometimes particular examples do not conform with the usual successional sequence (for example, following an intense forest fire ecosystem recovery typically proceeds from grasses to shrubs followed by rapidly colonizing "pioneer" trees and eventually ends up with a final or "climax" state of succession: a mature old growth forest with a stable mix of species). Sometimes standard successional steps are skipped or the order of progression of changes depends on additional factors that range from the scale or intensity of the initial disturbance up to human interventions, such as replanting a certain set of species and eliminating others. The latter could entirely overwhelm the normal successional pattern. Likewise, some educational institutions may have opted for permanent rather than term appointments for an interdepartmental institute's non-faculty staff, contrary to the more usual situation anticipated in

Table 2, in which such permanence for integration experts does not normally occur until the late successional stage when the group becomes a full interdisciplinary department or school. The model in Table 2 is offered simply because this progression seems to capture many typical situations in academe or laboratories, even if individual examples do vary.

Table 2

"Successional" Model of Organizational Steps for Global Environmental Change Teaching

* Multidisciplinary Seminars (get to know the players and rules of their games; typically takes years for individuals or groups to become interdisciplinary)

* Announce Program and Initial curriculum
  - Form multidisciplinary committee to define program goals, participants, and to define core curriculum and electives

* Curriculum Development
  - "Staple-gun" courses (require some half-dozen pre-existing multidisciplinary courses for students to become an "environment major" by displacing similar number of required/elective courses otherwise needed to be a disciplinary major)
  - Tracks (discipline-like course options for upper undergraduates using pre-existing courses)
  - Team-taught, first course (Cameos by major members of staff with interdisciplinary, problem-oriented context emphasized over the disciplinary content
  - New (or modified) core courses (multidisciplinary core knowledge, with content from various disciplines emphasized over integrative, interdisciplinary context)
  - New tracks (add multidisciplinary course options for upper undergraduates)
  - Senior seminar (to reacquaint content-oriented tracked students with integrative interdisciplinary problem context)
  - Co-Terminal Masters allows enough disciplinary content to be equivalent to a B.S. in a major discipline, but with interdisciplinary context of global environmental change problem solving added on)

* Create Environmental Systems Entity with permanence (new department, center or school with tenure-track faculty; research, teaching *and* outreach are combined and their quality evaluated on interdisciplinary systems or problem-oriented criteria)

* PhDs (see Table 1)

* Career change programs
  - A largely multi-disciplinary M.S. program to acquaint already trained disciplinary students and/or practicing professionals with the content of relevant disciplines and the broader context of environmental problem solving
  - Special post-doctoral or mid-career fellowships for senior scientists to retool in an interdisciplinary department, institute, or school.

The successional teaching model in Table 2 is very similar in its basic features to the research steps suggested in Table 1. Both begin with multidisciplinary seminars and progress toward some kind of integrative, systems-oriented environmental science, technology, and policy entity, that could be a center or institute with permanence, a new department, or a school. However, the steps inbetween are somewhat different, with Table 2 chosen to match the typical experience academic institutions have in developing interdisciplinary teaching programs.

Personal and professional relationships often develop at multidisciplinary seminars - seminars that may initially have been organized for building a peer group for research purposes but which also serve to develop relationships necessary for interdisciplinary teaching. At such seminars faculty, students, staff, and other professionals from the region learn enough about each others' interests and abilities to desire further development of a teaching program in global environmental change. The second step often is the formation of a multidisciplinary committee, which tries to define an appropriate curriculum, identify student interest and possible placement of students after completion of courses; and decides who could bring the program both credibility and official status by carrying it through appropriate channels, such as the university management, an academic senate, or governmental officials. Such a committee often writes proposals to help fund or initiate the program and/or tries to attract charismatic leadership either from within the institution or from outside it.

In some cases the multidisciplinary committee might comprise of faculty not only from disparate departments, but from different schools; whereas in other cases faculty might come from different areas of the same department, but be assembled to fashion a new, broad program within that department. Regardless of which model applies, a typical product of such committees is to define a curriculum for a departmental or interdepartmental program related to global environmental change. As the successional model of Table 2 suggests, there are some common stages in the development of such curricula. In the earliest phases after the program is announced, requirements for a core curriculum, required of all students, and/or electives open to upper-class students, are often listed. The courses involved are often already in the institution course catalog and lead to what I have referred to in Table 2 as "staple gun courses", whereby the multidisciplinary program simply "staples" together existing courses in unique combinations not typically open to disciplinary majors. For example, students enrolling in the new environment program either as a major or a minor (depending upon institutional options) might substitute for disciplinary major requirements, several courses from a variety of departments that involve relevant multidisciplinary global change questions. These substitutions would thus displace a similar number of core courses or electives that otherwise might have been required in order to receive a degree in a specific discipline. The obvious advantage of such an approach is that it allows the student and his/her advisor flexibility in putting together an individualized curriculum which more closely matches his or her perception of what constitutes the knowledge base needed to deal with particular aspects of global environmental change, than might be available in a traditional discipline. The disadvantage is that the "stapled" courses themselves are often aimed at discipline majors, and may provide more detailed or peripheral material than is really necessary for the interests of the global environmental change student. For example, an atmospheric science student wanting to learn enough about leaf physiology to understand how to incorporate biophysical processes into a climate model might wish to have some detailed information on leaf physiology, photosynthesis, and perhaps only a few disciplinary lectures would be sufficient for this interdisciplinary purpose. Yet, the staple gun approach would require the student to take an entire semester or more of a botany class to gain the information legitimately. Likewise, a student trying to understand the development process in China and wishing to compare it to the Victorian Industrial Revolution in Europe or North America might register for a history, economics, or political science course that deals with Western industrial development in the late 19th century. Once again, it is possible that the knowledge needed by that student would be sufficient with half a dozen lectures summarizing the highlights of the issues. It may not be efficient to have the student learn in full semester's depth all the labor relations, social and political conditions, innovations in industrial technology, and other details that appropriately belong in the disciplinary courses dealing with that time of Western industrial development.

In view of the demands for multidisciplinary training and the limitations of a four-year bachelor's degree, it may not be practical for a program to strive for full depth in several relevant global change disciplines, and also adequate breadth by exposing the student to a vast array of unmodified physical, biological, and social scientific courses; as well as policy, history, and values studies - all in a standard bachelor's or master's candidacy! This suggests that in the best teaching programs attempts will be made to try to create new core courses and eventually new electives, the latter to be specifically designed to meet the needs of global environmental change students. This could be done by selecting representative samples of material from existing disciplinary courses and combining them, perhaps in an original way, in specially focused global change classes. Of course, such new courses will require staff to teach them, but sometimes difficulty is encountered because the breadth of expertise to teach these "environment" courses is not resident in the department or even in the program. With luck, such expertise is available somewhere across the university, but then getting people to participate can be a problem, especially if there are few rewards for participation, and frequently multiple instructors are needed. This has the advantage of comprehensiveness; the disadvantage of choppiness. One solution is a single course coordinator whose job it is to smooth the edges between jumps, as lecturers dealing with different disciplines follow each other in sequence.

A first course in global environmental change is an example of this issue. First of all, such a course could satisfy the science requirement for non-science majors in a given university. Recent experience at a number of institutions suggests that vast numbers of students are interested in a comprehensive global environmental change course for their general science education requirement. Such a course could also be a springboard for someone in the program planning to take further courses. Cameo appearances by a multidisciplinary set of top faculty typically dominate such a course, but alternatively, visiting scientists, policy leaders, or industrial representatives can be guest lecturers, and together with program staff provide the breadth needed at the introductory level. Such a course stresses the context of global environmental change and its broad multidisciplinary nature over the content that would be more typical in first disciplinary science classes.

I strongly believe that at the outset of learning about global environmental change, students need a sense of how physical, biological, and social sciences all contribute to understanding global change problems. Beyond that, students need to be aware of the policy making process and understand that human values lead to policy choice. Environment programs that have just begun may have no more than one such new course, but more developed programs often attempt to fashion new (or modify existing) courses as part of the environment core. Occasionally, new electives in the global change program may be offered.

The Stanford University Example

Earth Systems Undergraduates. My own experience at Stanford University provides an example. The Earth Systems (ES) program at Stanford was initiated by biologist Jonathan Roughgarden and geologist Gary Ernst. ES is an inter-faculty, inter-school program managed by a committee of some 18 professors from all across the campus. A beginning course, affectionately called the "grabber," is offered as a core requirement for earth systems majors as well as being offered as an option for fulfilling science requirements for all students at Stanford. Its first year saw an enrollment of about 20 students, second year 40, and third year over 100. In 1994, in less than three years, there were nearly 90 majors in this program, attesting to its strong popularity with the students. The Stanford program leaders believe that physical, biological, and social sciences all must be included in a core curriculum taken by all earth systems majors. This core goes beyond the grabber course taught by "cameos" given by half a dozen highly visible professors and a few outside visitors. Three multidisciplinary courses which stress content over context in three physical, biological, and social scientific areas are required. One deals largely with geology (the "geosphere" course), one with biology (the "biosphere" course), and one with economics (the so-called "anthrosphere" course). These courses are cross-listed between Earth Systems Science and the relevant disciplinary departments, and the syllabus has been modified by compromise arrangements with the professors, his/her department, and the earth systems leadership to try to meet the needs both of disciplinary students taking the course and not involved in the earth systems program (who typically request more depth), and earth systems science majors who want to cover more topics in that sub-discipline than might be typical of the major. Stanford has also required calculus, physics, chemistry, economics, and biology background courses from the catalogue (i.e., the "staple gun" approach). The ES requires the students to take those courses (e.g., physics, economics, and biology) that have more mathematical prerequisites, thereby stressing methodological rigor, regardless of whether the subject is economics or ecology. At Stanford, five "tracks" are available to students (Appendix A). These are discipline-like programs for upper undergraduates, that provide depth and content equivalent to roughly two-thirds of the content that might be obtained by a disciplinary major who had not had the benefit of the added breadth associated with the earth systems program. Currently, the tracks at Stanford are "staple gun-like" in that few new courses have been specifically developed for Earth Systems beyond the grabber and the three crosslisted core courses - although a few special topics classes have been offered on land use problems and the management of energy efficiency. However, should the ES program gain permanence and therefore have the capacity to hire tenure-track faculty directly into the unit, it would be desirable to create new courses and even new tracks. (An energy track would be a high priority in the opinion of this author and member of the ES Steering Committee.) However, such events are later stages on the successional model of Table 2, at least so far at Stanford.

Another element to the Stanford program common to a number of other institutions is a senior seminar. Here, the attempt is to reacquaint students with the interdisciplinary context they learned in their first course but moved away from in discipline-like tracks. Such seminars involve outside speakers, faculty members, or, preferably, detailed presentations by the tracked students themselves that are subjected to the comments, criticisms, and perspectives of their multidisciplinary student colleagues.

Senior Honours. Another approach to dealing with the senior level interdisciplinary integration need is a "senior honors" program in environmental science, technology, and policy. Based on recent Stanford experience (Appendix B), this has been run through the Institute for International Studies (IIS), which is an institute consisting largely of a multidisciplinary faculty with half-time, tenured appointments in disciplinary departments and half-time, non-tenured Senior Fellow appointments in the Institute. Institute Fellows, as well as faculty with full-time appointments in departments, make up an interschool faculty committee to oversee the program. (Such a committee helps to advance the process of peer group formation mentioned earlier.) A few full-time institute staff with non-tenure "Fellow" status are also involved. External funding for the honors program was obtained for several years of operation. A competitive application procedure has produced a remarkably diverse and high-quality set of students for the honors program (Appendix B). After the first year's experience, the four faculty involved (with only a dozen or so students each year) recognized that even after three years of multidisciplinary training, Stanford students still needed more hands-on experience with the integration process, particularly for the policy aspects of environmental issues. Thus, the first quarter of the IIS honors program is now devoted to the analysis of practical case studies of environmental science, technology, and policy problems (e.g., acid rain, global warming, etc.). Lectures on risk analysis, cost benefit methods, the regulatory context, and communicating complicated issues into the political process through the media were all included. Students are expected to know the difference between "facts" and "values," what testimony is given, and to be familiar with the nature of the decision making process in industrial, federal agency, and legislative bodies.

The next two quarters are devoted to student presentations on the ideas behind their theses, preparation of interim drafts, and ultimately their final defense before faculty and student colleagues. Simply put, as the year goes on, questions and comments from the faculty at honors sessions are deliberately restrained and the students themselves, with diverse background (Appendix B) begin to take over the job of melding themselves into an interdisciplinary unit capable of penetrating analyses of their colleagues' work.

As one of the four faculty in this experimental program, I can attest to the high degree of satisfaction personally derived from association with these hard working and dedicated students, and the amazingly high quality of their work. (Several of their projects already are being considered for formal publication.) On the other hand, in all candor, it is extremely expensive to have student-to-senior faculty ratios like the one in this program (12 to 4), so this model is simply not feasible for the vast majority of students at most educational institutions. That would have been true at Stanford, too, had not a charismatic leader here secured a generous, but temporary, externally funded gift. Nevertheless, what can be readily generalized from this experience is the important need to bring "tracked" students back together on real world problem-solving exercises, and for them to have multidisciplinary contact with their colleagues that will hopefully lead to some interdisciplinary rapport and understanding.

Co-terminal Masters in Earth Systems. One potential drawback of the successional model in Table 2 or of the Stanford examples presented so far is that by the end of the bachelor's program, an earth systems graduate would typically have had less in-depth training than a disciplinary major graduate, although the environmental change student's training would provide a much broader context in which to place environmental problems. In order to increase the competitiveness of such environment program graduates with disciplinary graduates in job markets, Stanford recently created a "CoTerminal master's degree." This option allows a student to receive both bachelor's and master's degrees in earth eystems in five years. The master's has rigorous requirements (Appendix C) for basic, multidisciplinary methodological training in physical, biological, and social sciences, and math. However, a student is then free to pursue further course work in his or her specialized track. Thus, upon graduating in five years, he or she could have depth in disciplinary content roughly equivalent to a B.S. in a discipline, but the ES student likely would have much greater understanding of the interdisciplinary context of global change problem solving. Preliminary experience suggests that this not only puts the "environment" students on equal footing for disciplinary jobs with bachelor's degree disciplinary majors, but gives them a distinct advantage if they choose to work in a variety of environmental fields.

A Permanent Environmental Unit: The Environmental Educational Climax

On Figure 2 an "entity" was proposed in the center of the diagram to house on a permanent basis the integration experts who serve the purpose of being an interface among scientists from different disciplines, and the outside "user community" which needs emerging global change information for practical decision making. Such an entity could be a new department and/or school, or perhaps even an existing institute or center. The factor that puts it at the "climax end" of the successional model on Table 2 is that it would have permanence, with the capability of granting tenure to deserving and quality reviewed staff (with quality review standards based on interdisciplinary rather than disciplinary criteria, as outlined earlier), and would be an institutional home where research, teaching, and outreach can all be practiced and evaluated as a whole. With regard to outreach, it is my belief that scientists involved in global environmental change studies have the obligation to inform their colleagues and society at large in clear and understandable terms about the emerging state of the art in this field. While not all scientists are equally skilled at public presentation or technology transfer, nor do all desire to be involved in the policy process, at a minimum I believe that students trained in global change science should be given some instruction in techniques for clear communication, the nature of media coverage, the use of metaphor, and the need for public information in this area. Then, they, as well as faculty residing in some global environmental change organizational entity, can choose to what degree they wish to emphasize public outreach in their career; but at least they will be informed of the context and have some skills in this area. At a minimum, I would hope that all global change professionals and students would be receptive to the different outreach styles of their colleagues, including those who invest great efforts at this - provided those outreach activities are of high quality and are responsible representations of often confusing and contradictory results.

At this stage, I might add a personal prejudice: that the best teachers of global environmental change will probably be those who are either themselves researchers immersed in interdisciplinary problems, or who are intimately familiar with research and researchers who pursue this kind of work. For this reason I believe that new organizational entities, which are dedicated to integrative activities that unite research, teaching, and outreach, will have the highest quality global environmental change activities.

Post Graduate Training

Finally, two additional categories appear at the end of Table 2 and deserve brief mention. One is Ph.D. education and the second is career change programs for existing professionals or students who are already trained. With regard to Ph.D.'s in global environmental change, there is legitimate debate among those who believe that global change Ph.D.'s should be problem solvers pursuing interdisciplinary research according to the three *interdisciplinary* quality criteria mentioned earlier; and others who feel it is still in the best interests of students to pursue a more narrowly focused, discipline-like thesis, and then to become increasingly broad in the next phases of career growth 'or perhaps to defer such breadth until after tenure. While there are pro's and con's to both positions, my belief is that institutions should not rigidly adhere to either model, but should make flexible decisions based upon faculty and student interests and abilities, and individuals' willingness to take risks. However, in all cases a Ph.D. candidate in global environmental change needs to recognize that there may be suspicions about his/her capacity to do in-depth work. Thus, regardless of whether the originality of the thesis is in the disciplinary component, the interdisciplinary component, or both, a Ph.D. candidate would be well advised to be particularly careful to produce a substantive, useful, and original piece of work.

As for the retraining of working professionals from industry, government, or individual disciplines, who wish to pursue a career shift into global environmental change areas, a number of institutions have already created master's degree programs for such students. What seems essential in these programs is to have a core set of courses that insure that the student has adequate depth in physical, biological, and social sciences to understand the global change context, while at the same time having the opportunity to pursue special topics in more depth. With regard to the Ph.D. program again, there are many examples in many institutions of individual faculty members who work across departmental or institutional lines doing multi- or occasionally interdisciplinary environmental research. Such faculty also are often willing to fund and train a few graduate students on such projects. However, as laudable as this activity may be, I must ask: How many of these still unusual interdisciplinary Ph.D.'s would be hired as entry level faculty members by a university, whose departments are structured by discipline, given that the disciplinary contributions of these young scholars may not be as weighty as other recent Ph.D.'s who stayed fulltime within the traditional disciplinary boundaries and/or published single-authored, basic research papers rather than team-authored, problem-driven publications?

This raises for me an ethical issue: Should we in academics be training such interdisciplinary doctorates if we wouldn't hire them as colleagues? Furthermore, can we do a truly fine job of integrated interdisciplinary teaching when so few of us have become adequately knowledgeable in three or four environmentally-connected, but still distinct disciplines - disciplines that are needed to do interdisciplinary research that is quality evaluated on interdisciplinary rather than disciplinary criteria?

Advocacy Conclusion

It is my belief that only through the ultimate creation of new, permanent, interdisciplinary environmental organizations within academic institutions can such high quality programs be pursued on more than a piecemeal or temporary basis. It seems unimaginable to this "global changeologist" that universities, or other research and educational institutions claiming a significant share of public resources, could willingly turn down the opportunity to examine both intellectually interesting systems problems and socially important questions, such as understanding, predicting, and managing global environmental change, simply because it demands modification to existing institutional structure or because peer prejudice is in the way. Solutions to the formidable obstacles outlined in this chapter and elsewhere in this volume are given throughout the book. Hopefully, models such as those presented here and in other chapters can help managers, scientists, legislators, and educators to develop or expand research, teaching, and outreach efforts in global environmental change. I do not believe that these models necessarily fit all situations, nor should they be rigidly followed. The essence of the development of global environmental studies entities is flexible institutional change fashioned to meet the specific needs of a wide variety of actors and institutions. But the one thing of which I am certain is that current institutions are not remotely adequate to the task of producing sufficient numbers and quality of students, scientists, managers, and environmentally literate citizens to match the urgent needs of a world undergoing rapid and potentially serious global environmental changes.

Acknowledgments

I appreciate the comments and suggestions of Rosamond Naylor and Donald Kennedy on an early draft of this chapter.

References

Chen Robert S (1981) Interdisciplinary Research and Integration: The case of CO_2 and Climate. Climatic Change Volume 3 No. 4: 429 - 447

Schneider Stephen H (1977) Editorial for the First Issue of Climatic Change and General Guidelines for Reviewers. Climatic Change, Volume 1 No. 1: 3 4

Schneider Stephen H (1992) The Role of the University in Interdisciplinary Global Change Research: Structural Constraints and the Potential for Change, an editorial. Climatic Change Volume 20 No. 1: vii - x

Appendix A

Excerpts from Stanford University's Earth Systems Program Materials 1994.

Faculty

The faculty who teach in the program consist of the most outstanding scientists, scholars, and teachers in the Stanford community. The program is a Stanford contribution to the international consensus that environmental sciences and economics must immediately work together. Leaders from much of the world in politics, industry, and academia, together with a huge number in the general public, have called for reconciling environmental values with economic prosperity while preserving individual freedom. The Earth Systems Program at Stanford is one of the few places in the world where students can study the components of both the natural and social sciences in a coherent, contemporary, and rigorous curriculum of environmental knowledge.

Undergraduate Curriculum

Cognate Courses Required for all Majors:

- Biology (choose one):
  - Biochemistry, genetics, and molecular biology
  - Plant and population biology
  - Principles of ecology
- Chemistry
  - Chemical principles
  - Organic chemistry
- Geological and environmental sciences
  - Planet earth or earth history
- Mathematics
  - Calculus and analytic geometry
- Statistics (choose one)
  - Biostatistics
  - Introduction to statistics methods for earth and environmental scientists
  - Theory of probability
  - Statistics for social scientists
- Physics
  - Mechanics
  - Light and heat
- Economics
  - Elementary Economics
  - Economics Analysis
- Computer Programming
  - Programming Methodology

Courses Required for all Majors:

- Introduction to earth systems or introduction to environmental science and technology
- Geosphere
- Biosphere
- Anthrosphere
- Internship
- Senior seminar

Geosphere Track

- Earth minerals
- The oceans
- Chemistry of the earth
- Environmental and geotechnical geophysics, or general geophysics
- Structural and engineering geology
- Three electives, for example; management of geological hazards; the earth's nonrewable resources; volcanology; natural hazards of man; sedimentary geology and petrography depositional systems; structural and engineering geology ; energy resources; groundwater pollution and: environmental problems in the petroleum industry

Biosphere Track

- Biochemistry, genetics and molecular biology
- Cell and development biology
- Plant and population biology
- Principles of ecology
- Evolutionary genetics
- Plant adaptations, or behavioral ecology
- Three electives, for example: plant adaptation; ecosystems of California; biomechanics of intertidal organisms; biochemical evolution; theoretical population genetics; theoretical ecology seminar, or geophysics

Anthrosphere Track

- Economic analysis
- Economics and public policy
- Economics of development, or Third World population perspectives
- Three electives, for example: world food economy; economic development in Africa; analysis of development projects; economics of legal rules and institutions; International economics; U.S. Agriculture and the Environment; antitrust and regulation; economics of natural resources; environmental issues in international relations

Land Systems Management Track

- Water resources
- Environmental earth sciences I
- Environmental earth sciences II
- Environmental earth sciences III
- Three electives, for example: environmental science and technology, or environmental planning; hydrogeology; introduction to Urban Studies; environmental and geotechnical geophysics

Environmental Technology Track

- Water resources
- Mechanics and fluids
- Introduction to Chemical Engineering
- Two of the following:
  - Environmental science and technology
  - Environmental planning
  - Air quality management
- Three electives, for example: open channel and pipe flow; small scale energy systems; water quality control processes; building energy lab; energy resource; technology and economic change or development and technology in Third World; engineering thermosciences; environmental and geotechnical geophysics

Undergraduate Projects

The program offers students the opportunity to augment their formal coursework with supervised field, laboratory, private sector, or advocacy projects of their choice. In addition to providing experience outside the classroom, the project gives students an opportunity to try out potential career paths prior to making the commitment inherent in accepting a career position or entering into a particular graduate program. The project may be: 1) an internship of the student's choice; 2) directed research under the supervision of a faculty member; or 3) participation in one of several Stanford offcampus programs dealing with earth systems themes (e.g., Stanford in Washington, spring quarter at Hopkins Marine Laboratory, etc.). The project is required of all declared majors in the program, and is normally undertaken between the junior and senior years.

The experience should be a challenging and focused project pertaining to problems in earth systems in the area of the student's declared track, involving a time commitment equivalent to 9 units of academic credit.

Each student is expected to submit a written report on the earth systems project to the academic coordinator. This paper should include the objective of the project, the student's role and function while engaged in the project, and the relationship of the project to the student's track within the Earth Systems Program. This could be followed by a techniques/methods section and a presentation of any data acquired. Finally, the student should include a section addressing the implications of the work and possible directions for future work. The report should be 10-15 pages in length. In many cases, reports or documents prepared as part of an internship are themselves suitable earth systems project reports.

Senior Seminar

Each student is expected to participate in the Senior Seminar. As part of the seminar, the student will make a presentation based on the earth systems project report, and lead a discussion of the implications of the work. The presentation should include a historical perspective, the goals of the project, the student's responsibilities, and the results of the work. The student should be prepared to discuss the work with the rest of the class and with students from different tracks, and should be able to place their work within the larger earth systems focus.

Appendix B

Excerpts from Materials for the Goldman Interschool Honors Program in Environmental Science, Technology, and Policy at the Institute for International Studies, Stanford University

Environmental studies at Stanford have expanded rapidly in both breadth and depth during the past 5 years in response to widespread faculty interest and increasing student demand for further research and teaching in the field. Almost 50 faculty members and permanent research staff from the Schools of Humanities and Sciences, Earth Sciences, Engineering, Law, Business, and Medicine are pursuing research directly related to environmental science, policy, and technology. Among them, at least 25 are involved in curriculum development and programs that integrate teaching and research. These efforts are helping to serve three important functions: to provide environmental literacy to all undergraduates; to offer specialized study for some undergraduates; and to develop distinguished research and graduate training programs in environmental science, technology, and policy that span several disciplines.

The Institute for International Studies (IIS) is taking a leadership role in promoting both the interschool and international policy dimensions of Stanford's environmental studies program. The importance of these efforts is underscored by the facts that: 1) environmental problems, such as ozone depletion, loss of biodiversity, water contamination, and acid rain, are not restricted by political boundaries; and 2) the solution to these problems requires he combined efforts of scientists, engineers, economists, and policy specialists. No single discipline is capable of addressing the full scope of any major environmental problem.

The IIS's strategy is to facilitate interschool cooperation, enhance interdisciplinary research opportunities, integrate research and teaching activities, and develop research and internship programs for students in the environmental field both in the U.S. and abroad. During the past 5 years, the IIS's efforts have been very successful in improving communication among faculty in different disciplines and reducing the departmental barriers that often prevent effective interdisciplinary work.

In September 1993, the IIS initiated the Goldman Interschool Honors Program in Environmental Science, Technology, and Policy for undergraduates throughout the university. The program serves as a focal point for uniting widespread student and faculty interest in the environmental field, and contributes substantively to the development of an integrated academic program among departments with courses directly related to the global environment. The goal of the honors program is to facilitate communication among students in different departments and to prepare them for the type of interdisciplinary interactions that they are likely to encounter as professionals following graduation.

The program currently draws four to five exceptional students from each of the three undergraduate schools (Earth Sciences, Humanities and Sciences, and Engineering) into a small group seminar. The seminar meets throughout the year; the first phase consists of faculty lectures and problem sets, and the second phase consists of student presentations of their work. The seminar is led by Lawrence Goulder, Donald Kennedy, Rosamond Naylor, and Stephen Schneider, all of whom have appointments at the IIS. An interschool faculty committee, consisting of 10 senior faculty members from Earth Sciences, Engineering, and Humanities and Sciences, selects the students each year and certifies their theses.

The focus of the program is the completion of a thesis by each student that reflects innovative and individually motivated research. Ideally, each thesis demonstrates intellectual depth in the student's primary field (e.g., economics, biology, earth sciences, or engineering) and some breadth to place the thesis in a scientific, technological, and policy-related context. The theses are oriented around scientific, technological, and policy-related aspects of leading environmental issues of international scope and relevance, such as biodiversity loss, greenhouse gas emissions, toxic wastes, deforestation, degradation of productive agricultural lands, transboundary water pollution, and population growth.

Appendix C

Further excerpts from the Earth Systems Program Materials at Stanford University pertaining to the coterminal bachelor's and master's degrees

The Stanford coterminal degree plan enables an undergraduate to embark on an integrated program of study leading to the master's degree before requirements for the bachelor's degree have been completed. An undergraduate majoring in earth systems may apply to work simultaneously toward bachelor's and master's degrees. The master's degree in earth systems provides the student with enhanced tools to evaluate the primary literature of the discipline most closely associated with the student's track, and allows an increased specialization through additional course work that may include nine credits of thesis research. Finally, integration of earth systems concepts is furthered by participation in the spring seminar, where the master's student is expected to play a leadership role.

Students should apply through the Earth Systems office. They are required to complete an application which includes a statement of purpose; a Stanford transcript; two letters of recommendation, one of which should be from a faculty member of the program; and a list of courses in which they intend to enroll to fulfill degree requirements. Students may be admitted as early as their eighth quarter at Stanford, or after earning 105 units, but no later than their eleventh quarter. Students may either 1) complete 180 units required for the B.S. degree, or 2) complete a total of 15 quarters during which the requirements of the degrees are fulfilled concurrently.

The student has the option of receiving the B.S. degree upon completion of that degree's requirements, or of receiving two degrees concurrently at the end of the master's program.

Four levels of requirements must be fulfilled to receive a M.S. degree:

1) all requirements of the B.S. degree
2) an enhanced set of cognate courses or their equivalent
3) further course work (and/or thesis research), all of which should be at the 100-level or above, and including 18 units at the 200-level or above, leading to further focus within the student's track
4) participation in the spring seminar

The program must consist of a minimum of 36 units of course work and/or thesis research, at least 18 of which must be at the 200-level or above. A more detailed description of the coterminal master's degree may be obtained from the Earth Systems' office.

1. Cognates

The following subject areas should be mastered, if this has not already been accomplished at the B.S. level. Suggested courses are indicated, but others may be substituted with the advisor's approval.

Courses in differential equations, linear algebra, statistics, optimization theory, ecological systems, thermodynamics, geochemical cycling (given by the department of Mathematics, Statistics, Economics, Biology, Chemistry and Earth Sciences).

2. Specialization Within Track

The student must devise a program of study that shows a level of specialization appropriate to the master's level, as determined in consultation with the advisor. At least 18 units must be at the 200 level or above. The program should demonstrate further specialization and focus within the student's undergraduate track.

The program should be developed prior to application to the master's program, and will be part of the application to the Earth Systems Committee. Nine units may be in the form of research, culminating in the preparation of a master's thesis. A thesis is not required for the degree.

3. Integration

Master's students must take part in the spring seminar (ESys 290), and will have additional responsibilities appropriate to the master's level (thesis presentation, modeling problems, etc.), 2 units.

GEOSPHERE TRACK

Required courses:

Chemistry of the earth (4 units); the oceans: an Introduction to the marine environment (3 units); sedimentary geology and petrography (4 units); hydrogeology (5 units) or hydrology (3 units).

Other recommended courses:

Surface waters; low temperature aqueous geochemistry (3 units); sedimentary basins (3 units); sedimentary petrography (2 units); introduction to micropaleontology (5 units); energy resources (3 units); management of Earth Resources (3 units); role of fluids in geologic processes (3 units); introduction to groundwater solute transport (3 units); environmental aspects of petroleum engineering (3 units); plate tectonics (3 units); tectonophysics (3 units).

BIOSPHERE TRACK

Recommended courses:

Genetics (3 units); general botany (3 units); plant adaptations (4 units); ecosystems of California (4 units); systemics and ecology of vascular plants (5 units); plant genetics (3 units); biomechanics of intertidal organisms sum (6 units); plant physiology (4 units); conservation biology (4 units); biology of insects (4 units); biology of birds (3 units); biochemical evolution (3 units); advanced plant systemics (2 units); mathematical models in population biology (3 units); behavioral ecology; theoretical population genetics (3 units).

ATMOSPHERE TRACK

Required courses:

Price allocation theory (5 units) or economics analysis (3 units); economics of the environment (5 units); economics of natural resources (4 units).

Recommended courses:

Population perspectives in the third world (5 units); industrialization, growth, and economic development (5 units); international trade theory (5 units); public finance and taxation (5 units); introduction to political economy (5 units); essentials of political theory (5 units); world politics and the global economy (5 units); advanced ecological anthropology (5 units); public policy design (3 units); environmental law and processes; environmental strategy.

Education and Training of Global Environmental Change Scientists

Max Beran
TIGER Programme
Institute of Hydrology
Wallingford, UK

Introduction

As preface to an analysis of education and training for Global Environmental Change (GEC) scientists I wish to pose four questions:

1. What are we teaching?
2. Who are we teaching?
3. Why are we teaching it?
4. How do we teach it?

These questions can be asked "in the round" as an aid to setting the scope of the Workshop, or they can be posed quite specifically to help to design an individual teaching programme.

The first question concerns the definition of the science taught. It is the one to which I will devote most space for two reasons:

1) a wide "between-scientist" divergence of view and
2) a "within-scientist" tendency towards an unnecessarily narrow view of what constitutes Global Environmental Change (GEC).

The other three questions bear on the relative emphasis to be given to the different components of GEC and to the style of teaching. A relatively small space is devoted to them because I am not a professional educator, but one who would have benefited from a taught course. My other qualification for entering into this field is that I conducted a small survey through the NATO region on what training facilities are already in place. My comments should therefore be seen as an amateur's *aide memoire*.

What are we teaching?

NATO's intention in mounting this workshop within its Special Programme on the Science of Global Environmental Change, was to promote a discussion on whether there is a need to teach the science of Global Environmental Change and, if there is, what should be taught and how.

NATO ASI Series, Vol I 29
Global Environmental Change in Science
Education and Training
Edited by David Waddington

It is rather revealing that, despite a decade or more of practical progress in the science of GEC, there is still no agreement on what is meant by Global Environmental Change. One explanation is that the GEC issue is multifaceted, and its science has been addressed through multidisciplinary teams that are often distributed through multilaboratory consortia. With all of these "multi's" it is perhaps unsurprising that there is still an absence of consensus arising from the different biases that are carried to GEC by the specialists who work in the field. True consensus will probably not be achieved until the *multi*disciplinary has been replaced by *inter*disciplinary actions.

Another contributory factor is that GEC is an issue, as well as a science. This introduces additional groups into the community of interested parties: those concerned with actions, ethics, and policies. Currently it is the "greenhouse effect" which has the public's main attention, and this has led some to simply equate Global Environmental Change with Global Climate Change. While it is certainly arguable that at the time of writing climate change is the single most worrying global scale threat it is actually but one of a set of major global environmental issues. Equating GEC with any one issue is an error that should be avoided in the workshop.

One might then equate GEC with the sum of the global environmental issues. But as well as leaving the definition unresolved, this too lacks generality, since it would logically confine attention only to those issues that have global impact - mainly those which have a negative influence on the atmosphere.

Conscious that criticism is easy, I wish to suggest another definition for Global Environmental Change. The following approach to a definition is a personal construct not found in textbooks but it is one that I have found helpful when explaining the uniqueness and the unity of the science, and useful in setting bounds to what might otherwise develop into an amorphous and arbitrary set of large-scale questions.

Two analogies help in constructing this definition of GEC:

a) Family tree
b) Venn diagram.

The family tree analogy allows us to trace a genealogical descent from Global Change, the "grandfather", down to GEC, the "grandchild". We will need to have clear a distinction between Global Change and Global Environmental Change. Traditionally the term *Global Change* has referred to a branch of earth science which deals with the "deep palaeo". To most people *Global Environmental Change* comprises a set of issues such as climate change, ozone depletion, biodiversity reduction, desertification, eutrophication and so on.

Scientists who study Global Change address major questions relating to how long the earth has had an atmosphere, or a magnetic field, or a salty ocean; and why there are ice ages, soils and mountains. Perhaps the biggest question of all for Global Change scientists is how - despite tectonic upheavals, an ever brightening sun, and bombardment by planetesimals - life persists. In attempting to answer these questions, two valuable concepts have emerged from a marriage between Global Change and a branch of engineering mathematics termed Systems Science:

(a) the earth as a system, and

(b) negative feedbacks that keep the system operational.

The combination of these two concepts is called *Earth System Science.* The characteristics of the hydro-, litho-, atmo- and bio- spheres, which make up the skeleton of the earth system, the role of biogeochemical fluxes in linking these system components, the crucial importance of negative feedbacks in maintaining the stability of the whole, and the way that the very largest (the earth) appears to be under the ultimate control of the very smallest (photons, radicals, microbes) are all Earth System Science topics of great relevance to GEC. However, the object of this article is to try to generate a shared understanding of the boundaries of the subject, not to list syllabus topics.

Returning to the family tree analogy, Earth System Science (the father science) now begets a child - Global Environmental Change science. This comes about through an alliance between Earth System Science and man-made environmental changes such as land-use changes and industrial and agricultural emissions. This relationship is possible because these man-made changes appear to attack critical points along the chain of feedbacks studied under Earth System Science - *eg* the radiation balance - and so risk destabilizing those mechanisms that have been responsible for past stability.

In the Venn diagram analogy Global Environmental Change is perceived as the intersection of two sets: one containing Global Changes (GC, *sensu* geologist), and the second containing Environmental Changes (EC, *sensu* "environmental" activist) on GEC = GC $\cap$ EC.

What is helpful about this line of thought is that it avoids making special cases for changes that are less than global in impact. The latter are normally justified by reinterpreting "global" as meaning merely "big", "important", or "widely felt". Indeed most examples of GEC have to be geographically circumscribed, for example, desertification and marine pollution. Conversely, every widely experienced environmental change is not a global change, as illustrated by the following anecdote. According to contacts in the UK Environment Ministry, the two environmental problems of greatest public concern (as measured by the size of the ministerial postbag) are dog-dirt in public places and the playing of loud music. Neither can be called global on my definition (no earth system process involved), and nor would become so, either even if the whole globe shared these British preoccupations!

To express the idea briefly: *an Environmental Change becomes a Global Environmental Change if it has a significant effect on an earth system process.*

A feature common to many Earth System feedbacks is that they involve the earth's radiation budget. Industry, agriculture, land-use change, and the sheer force of our numbers have the capacity to alter the composition of the atmosphere and the properties of the earth's surface, all known points of sensitivity in the feedback systems that global change science tells us have kept the earth functioning properly for 3 to 4 thousand million years. A small intensification of the chlorine cycle and we let extra UV radiation through the atmosphere; a small nudge to the carbon cycle and we run the risk of causing climates to change; a larger nudge to the sulphur cycle and we degrade soils and hardly anybody in the northern hemisphere knows what a really clear sky looks like; a bigger nudge to the nitrogen cycle and we enrich biota but impoverish species; and with an even larger nudge to the phophorus cycle, we run the risk of doing the same to remote oceans.

I have spelt out a list of global environmental changes and the biogeochemical cycles that are implicated. It is worth reiterating that not all of the changes are global in the sense of being experienced worldwide. For example, not all places on earth consist of land that might become desert, nor are all species being impoverished because of biota enrichment. Nevertheless they are examples of GEC because they arise from interventions in the global control systems. It may be argued that this approach is no less arbitrary than either one based on whether changes are global, or on an agreed litany of big issues. I maintain that this approach does include the big issues and moreover has the considerable pedagogic advantage of permiting a variety of apparently disparate topics to be addressed inside a single framework.

Another advantage of this approach to defining GEC is that it distinguishes cause from effect. Social scientists in particular are inclined to include forcing factors like land-use change, over-population, and exploitation of natural resources in the same list as climate change and desertification. Widely felt impacts of GEC, such as increased flood risk due to sea level rise, are also sometimes bracketed with the change. For a scientific analysis it is more natural to separate the causes and the consequences from the basic phenomena.

While it is true that this scheme gives primacy to Earth System Science and biogeochemical cycles, it does not equate GEC science with Earth System Science. Only processes that operate in the human time frame are relevant. The rock cycle has controlled atmospheric and ocean properties on the multimillion year time scale but has no relevance to the current greenhouse problem. So whatever local problems they cause, mining and quarrying do not cause global environmental change. In general, biogeochemical exchanges with the biosphere and atmosphere are most relevant for GEC.

In summary I have proposed a way of thinking about GEC that links it to recognized bodies of science. It should therefore give the student the feeling that it is a science with a coherent structure. There are many ancillary topics relating to the causes and consequences of GEC that I have not addressed here. This is not because I think forcings and consequences are unimportant or that they have no place in a taught course, it is because they are somewhat independent of the way in which the core science is structured. Some thoughts on individual topics are included as Appendix A.

Who, why, and how are we teaching GEC?

Having a broad encompassing framework for GEC does not of itself determine the shape of a curriculum or the detailed contents of a syllabus. This can emerge only after we consider who is receiving the training and for what purpose. The different age groups/levels are

- senior school (school years 12 and 13 in the UK)
- freshman undergraduate
- senior undergraduate
- masters and other post-graduate
- career.

The different motivations include

- awareness of environmental issues
- learning science
- training for a research career
- training for a professional career.

There is a correlation between the two lists in that much of the interest at school level will be for more general citizenship purposes, and at the upper levels the need for information on GEC by someone in a senior position is likely to be tied to considerations of policy and advice rather than to advancing the science.

However, there are many points of overlap which weaken the correlation. Particularly important at school level will be a strong motivation for using GEC education as a means of making scientific concepts more relevant. This same motivation may well apply at freshman level where GEC science may be taught, even though it will not become the student's main subject at senior undergraduate level. The senior undergraduate teaching programmes will be beneficial to the taught element of masters and other post-graduate work, particularly in informing students more broadly beyond the confines of their research or dissertation topics.

GEC is by its nature and parentage multidisciplinary. A paradigm for higher education is that there must be discipline at its heart. I have not heard this paradigm critically analysed, but I assume it is based on the need for training in thinking, not just in providing facts. We presumably will not spend time in the workshop in re-examining accepted wisdom, but we need to acknowledge that it does give rise to a built-in tension. The obvious way to resolve this tension is through some educational system consisting of a major subject to give depth and a set of minor subjects, which convey extra GEC information. But are there other structures? Does it mean that there should never be an undergraduate degree in "Global Environment Changeology"?

In my opinion GEC requires a certain way of thinking that may be difficult to acquire unless it is taught formally. The body of facts may be no more particular than other subjects, but the need for appropriate compromise is pervasive. Is it any more than the remark attributed to Aristotle that an instructed mind should be satisfied with the requisite degree of precision demanded by the task in hand and should not seek more. In this regard GEC may be thought to resemble engineering, but the part played by principles appears much greater when dealing with the earth system. While other applied sciences can rely on empirical findings, GEC science continually needs to return to fundamentals like the periodic table, the gas laws and cellular processes.

The workshop should discuss special needs and attitudes relevant to those who will become involved in giving advice. There has been something of a paradigm shift in this area of recent years, partly brought about by some spectacular failures in the sphere of development. A story told by Joe Farman (of Ozone hole fame) comes to mind. When asked about the risk of an ozone hole over Europe, Joe said that the rate of depletion was slow and confined to the time of year when radiation was in any case low. He also said that people who lived in sunnier regions were currently more exposed to UV-B than a European would be even if the hole became worse. He suggested that the questioner could stop worrying. This drew the sharp rejoinder that it was Joe's place to give the facts; it was the questioner who would decide whether or not she would stop worrying!

This story helps make several points: the need to work with people and to give them credit for knowing their minds, and when requirement for skill in advocacy. In part such empathetic skills are innate, but much depends on having confidence in our knowledge. With such a broad subject area as GEC this does present special difficulties therefore building confidence should be one of the particular aims of GEC training. One way to achieve this is to develop a capacity to forecast an approximate result and refrain from giving extreme answers or implausible opinions. Environmental problem-solving provides many opportunities for this and course material should capitalise on these.

Teacher's perspective

I close with a few remarks concerning teachers as elicited from discussions with course providers and designers.

We need to distinguish multi- from interdisciplinarity. GEC science is much affected by both, although the latter is the ideal and should be the *aim* of training courses for scientists. Skills in team work are required for both. Teaching and learning are not the same thing, and teaching can get in the way of learning. We touch here on good pedagogical practice, which is not the particular preserve of GEC science. However the subject has the capacity to energize young trainees into exploring issues for themselves, especially if they have access to data sets and "toy" versions of models of global systems. Appendix B provides a list of centres of GEC educational interest.

Attitudes to texts and teaching resources tend to polarise members of the teaching profession. Some take the lofty view that the good student is best served by being obliged to make personal discoveries, while others seek out computer-assisted learning, video, and distance learning aids.

Higher education institutions tend not to reward cross-discipline endeavours. Structures are needed at the institutional level to assist career development of individuals who choose to work across discipline boundaries. This includes recognition and credit for joint supervision of doctoral research and publication in multidisciplinary journals.

Appendix A - Topics for taught courses

The lists below are intended as an aide memoire without special pretensions for completeness and certainly without prescription as to the depth appropriate to an individual course or student.

Basic science concepts
Physics - radiation, absorption, diffusion, transport
Chemistry - reactions, kinetics, pH, eH, "chemistry of life"
Biology - levels of organization, cell processes, dynamics
Geography - geomorphology, biogeography, human geography
Mathematics - chaos, complexity, systems theory, model building

Earth system science concepts
Reservoirs and fluxes
Conservation laws, rate laws
Feedbacks, linear and non-linear systems
Earth history - solar system, geological eras, evolution and radiation
Biosphere - plant distributions, trophic relations, exchanges
Lithosphere - erosion and weathering
Atmosphere - layers, circulation, chemistry, radiation
Hydrosphere - ocean circulation, freshwaters, phase change, transport
Cryosphere - ice caps, glaciers, radiation budget

Biogeochemistry
Weathering processes, humus formation, carbon stores
Biochemistry and cellular processes
Cycles of water, C, N, S, P, etc.

Causes of change
Industrial emissions
Agricultural emissions
Land use change, deforestation
Population and resource use

Global Environmental Changes

Greenhouse effect

Ozone depletion

Global climate change and warming

Reduction of biological diversity

Desertification and soil degradation

Eutrophication

Marine pollution

Acidification

Socio-economic consequences and responses

Sustainable development - Rio declaration, Agenda 21, minority issues

Conventions, laws - climate, biodiversity, sea, poles, desertification

Control measures - financial, legal, and educational

International policy and research structures

Tools and techniques

Modelling

Sensors and instrument systems, satellite remote sensing

Monitoring and networks

Data banks

Special skills

Advocacy

Empathy

Global thinking

Dealing with uncertainty

Interdisciplinarity and team building

Appendix B - Contacts for Global Environmental Change Teaching Initiatives

Head of Environmental Education and Training
Commission of the European Communities
DG-XI, Brussels

Education in Global Change
ICSU Committee on Teaching of Science
Professor D J Waddington
Department of Chemistry
University of York, Heslington
York YO1 5DD, UK

Education in Atmospheric Chemistry and
Global Change, IGAC Project Convenor
Professor J W Winchester
Department of Oceanography
Florida State University
Tallahassee, FL 32306-3048, USA

Dr N Azimi
UN Institute for Training and Research
Environment Training Programme
Palace des Nations
CH-1211 Geneva 10
Switzerland

Dr Donald Johnson
Dir, Earth Systems Science Education Prog
Space Science and Engineering Center
1225 West Drayton St
Madison, WI 53705, USA

Global Change Education Diskette Project
National Geophysical Data Center
NOAA, E/GC1, Dept 891
325 Broadway
Boulder, Colo 80303, USA

IGBP-START
1825 K St NW
Washington DC 20006, USA

Secretariat
European Association for Environmental
Management Education
viale Ippodromo, 9
I-21100 Varese, Italy

Institut Laplace
10-12 Ave de l'Europe
78140 Vélizy Villacoublay
France

Education and Communication on
Environment and Development (ECO-ED)
720 Bathurst St, Suite 500
Toronto M5S 2R4
Canada

Dr D Troost
Training and Education in Marine Sciences
Unesco
1 rue Miollis
75732 Paris Cedex 15, France

Vice Rector, United Nations University
15-1 Shibuya 2 chome
Shibuya-ku
Tokyo 150
Japan

The Director
Stockholm Environment Institute
Box 2142
S-103 14
Stockholm
Sweden

Education and Training Division
World Meteorological Organization
Geneva 20
Switzerland

Dr J J Nijman
EERO Training Centre
PO Box 182
6700 AD Wageningen
Netherlands

The Editor
Journal of Environmental Education Research
Carfax Publishing Co
PO Box 25
Abingdon OX14 3UE, UK

Carbon Dioxide Information Analysis Center
Oak Ridge National Laboratory
PO Box 2008
Oak Ridge, Tenn 37831-6335, USA

Int Environmental Education Programme
Unesco
7 Place de Fontenoy
75700 Paris
France

Sustainable Development
Towards Global Change partnerships

W S Fyfe
Deptartment of Earth Sciences
University of Western Ontario
London, Ontario
Canada, N6A 5B7

Introduction

The twentieth century has been truly remarkable in the development of natural science. The giants of the early 1900s laid the foundations for the description of matter and energy, and the developments of the past few decades have led to our present age of observation. Given support we can "see" almost everything from a galaxy to a genetic defect. The same developments have led to the age of information and modern computers. And via the new communication systems, particularly television, we are in the age of expectation. Television, which influences the daily life of almost every human of every age, shows more and more the differences between rich and poor - those who starve to death and those who drive a Buick. Today, about 40 nations have deep internal social conflict. In the 1940's the great British writer, Aldous Huxley said we were driven by a creed of greed, violence, and incomprehension. We still worship wealth and violence, but at least the incomprehension is no longer excusable.

The science and resulting technologies of this century have supported the phenomenal growth of human population. In 1750, less than one billion; 1850, about 1.2 billion; 1950, 2.5 billion; now we add almost 100 million per year, and the World Resources Institute in their brilliant reports (1994) project over 10 billion by 2050. Population in Africa grows annually by 2.93%; Asia, 1.78%; the world, 1.68%.

The technologies which have supported this phenomenal growth are changing the most basic elements of our life support system. We have perturbed the chemistry of the atmosphere and the emission of gases such as carbon dioxide is increasing. Soil degradation is catastrophic. As Postal (1992) shows, at least forty nations have an acute shortage of water. There are clear signs of unusual climatic conditions, record temperatures, floods.....

NATO ASI Series, Vol I 29
Global Environmental Change in Science
Education and Training
Edited by David Waddington

King and Schneider (1991) of the Club of Rome stress that in many of our strategies for development, for the use of knowledge, we have failed. In large part this failure can be attributed to the recent systems promoting specialization. It is interesting to reflect on how this has arisen. First, industry needs specialists, the best material for a semi-conductor, the best alloy for a jet engine....the best chemical to kill a mosquito.... The agencies which fund research look for the specialist, and they support each other. The ultimate insult is to tell the scientist that his work lacks focus! And in this century a vast amount of the support for basic research has been derived from the need to have the best weapons.

Further, the demand for specialists for our modern technologies has led to a huge expansion of the university systems. Early this century universities were small. The student of chemistry or physics probably knew the history professor and drank coffee with students from the arts. Today, this is rare. The great concern of the modern student is, "Can I compete to obtain a job?" University departments and their students have retreated into their cocoons of specialization in order to survive and compete.

Thus, while there have been amazing advances, the global situation is causing vast concerns. The Stockholm (1972) and Rio (1992) Earth Summits demonstrate these concerns, but they have been failures in terms of effective action.

The ultimate objective of science must be to improve the human condition for all people. Since the famous report of the World Commission on the Environment and Development (the Bruntland Report, 1987), a term increasingly used is 'sustainable' development. Many argue what this means. For me it is simple: will I leave the planet in better condition for all, all species who support me, and who will follow me? It is unfortunate that the Bruntland report is very anthropocentric, and lip service only is paid to the crucial question for the importance of biodiversity.

The Future

To support well the growing human population, there will be, there must be, vast development. Urgent problems are concerned with areas such as:

- energy for 10 billion
- water supply
- soil conservation
- transport systems

- waste disposal
- mineral resources

All on a planet where the 'environment' never was, never will be, in a steady state (e.g., Stine, 1994). This implies that for security in the future there must be reserves and technology to distribute such reserves.

In the past, strategies for development have come mainly from specialists with the support of economists. I agree with King and Schneider that many of these developments have been truly non-sustainable. It is clear that if the situation is to improve, we require new systems. I remember well the words of the late Roger Revelle, the director of the Scripps Institute of Oceanography. He said the needs were "urgent and hard". Urgent for obvious reasons. Hard, because they involve cooperation between the specialists and across disciplines.

The Need for Teams

Environmental problems are complex. The earth environmental system and its mass-energy exchanges involve interactions between:

- the sun
- the ocean-atmosphere transport systems
- the biosphere
- the geospheres
- and now anthropogenic technologies

These all fluctuate on a wide range of time scales from seconds to millennia.

It requires specialists to study any aspect of these complex forcing interactions. I don't think that many of us have the time or ability to be an expert simultaneously in solar physics, the response of a plant to changes in atmospheric carbon dioxide, and the impact of a new defoliant on the albedo. This implies that to arrive at the best systems for the needed new developments requires development of improved systems for effective cooperation.

I very much like the words of Jay Forrester, a systems analyst quoted by King and Schneider: "In any complex system, attack, however apparently intelligent, on a single element or symptom generally leads to a deterioration of the system as a whole". Examples abound: If we pump groundwater for agriculture at a rate greater than replacement, the system will crash. If we burn all fossil carbon; i.e., oil, gas, and coal, as rapidly as possible, we will have a greenhouse.

In terms of needed developments and th teams needed, every case will have very specific needs. There is no doubt that our standard of living is related to the availability of energy, and today over 90% of the world's energy comes from oil, gas, and coal. There is also no doubt that standards of education are related to energy, and in those fortunate lands where there is universal access to quality education, the population problem is coming under control (see data in World Resources 1994). For example energy consumption per capita (1991, giga-joules) is 320 - U.S.A., 187 - Germany, 9 - India, 12 - Africa, 23 - China. While some waste, most have great need.

Imagine that we faced the challenge of providing increasing availability of energy for the developing world (I hate that term!) by a factor of ten. Who would be on our team? For certain we must consider the sustainable energy alternatives:

- solar of all types (photoelectric, wind, biomass......)
- geothermal potential
- nuclear potential
- fossil fuel potential

We would need experts in all aspects of food-fibre production, water resources, solar technologies, climate experts and statisticians, biodiversity experts, nuclear technologists......... And teams, local and global, must be involved in the planning *from the start* and be proactive not reactive.

The Technological Opportunity

I think that there is little doubt. Those who recognize the great environmental problems and act, will lead the technological structures of the next century. The crisis areas, some of which are listed above, all require new systems on a grand scale. There is a vast opportunity for the innovative practical mind. The specialists we train in universities must be kept in contact with the environmentalists, industrialists, and economists. But as always, there are opportunities at more simple technological levels (e.g., low tillage agriculture for soil conservation, or the work we are doing in Canada and India, using sewage and coal ash to greatly enhance the growth rate of trees and remineralize degraded soils). For a vast array of our present environmental waste problems, there are simple solutions of easily proven economic advantage. The key to action is to assemble the groups involved and engage them in realistic dialogue.

The Role of Education

If we are to improve the present situation and systems, we must also improve our educational systems that ultimately drive our political systems. I very much like the words increasingly used: "an objective of universal literacy, numeracy, and science". It is strange that on our planet so many do not know why we have so much oxygen in our atmosphere, that if oceans are warmer there will be more rain somewhere, or that frogs can control mosquitoes. The average citizen has no idea what happens to waste products. Planetary science has something for every student of every age, of every ability. The challenge is to choose the best examples to illustrate general principles. I have had great success with elementary school children in discussions of the economics of saving energy and stopping waste production in general. Children are fascinated to learn that ECO-logy and ECO-nomy have a common Greek root, Oikos-house, no conflict!

At university all students must have time to learn a little about the general topic of the Earth System and their life support systems. For example, I think all students should read the reports of organizations like the U.N. population council, the Worldwatch Institute, and the World Resources Institute which includes wonderful data tables (also available in diskette form). But perhaps most of all we must develop systems that allow students from different disciplines to meet, discuss, and debate, formally and informally. We must have specialists, but we need specialists who can communicate with other specialists and who can overcome the jargon of their specializations. A great advantage we also have at the university level is that normally there is a significant range of national and ethnic diversity.

We have always been concerned with the problems of illiteracy. I suggest that today the most serious problems we face result from the total lack of knowledge, ignorance about why we are here and how we have evolved. This ignorance is vast, from the Pope to our leading politicians.

At this university we have started an experiment. At the graduate level a student of astronomy, physics, chemistry, and so on, whose work has application to the development of the environment, can elect to graduate Ph.D., M.Sc., Discipline - Environment. All these students must attend a full course on the planetary environment - from solar physics to convection on Earth. This course is given by faculty from about 12 disciplines. Then all students must attend their own course, where they present the topic of their research and describe why it is important for the improvement of the environment. It is a start, but we have a long way to go.

We need more specialists. All must read and write, add and subtract. But all must learn to respect their ultimate home. And hopefully, whether by positive action or catastrophe, the population is reduced, and the billion or so humans remaining will have a joyous planetary experience.

I like the words from a green U.S. company, Ben and Jerry's Ice Cream: "What we cannot do alone, we can do together".

Or is Richard Dawkins (1991) correct: "Like every plant, ant, worm, we are simply bio-robots."

References

Dawkins R (1991). A new world view, in, from Gaia to selfish genes. Ed., C Barlow, MIT press, Cambridge, Mass 195-212.
King A and Schneider B (1991). The first global revolution. Simon and Schuster, London, 397 pp.
Postal S (1992). The last oasis-facing water scarcity. WW Norton and Co. New York, 239 pp.
Stine S (1994). Extreme and persistent drought in California and Patagonia during mediaeval time. Nature 369: 54 6-549.
World Commission on Environment and Development (The Bruntland Report) (1987). Our common future. Oxford University Press, 383 pp.
World Resources Institute. (1994). World Resources 1994-95, a guide to the global environment. Oxford University Press, 400 pp.

Environmental education and training of global scientists for sustainability

J R E Harger
UNESCO/ROSTSEA
Jln. M.H. Thamrin 14
Jakarta 10012, Indonesia

Dirk Troost
UNESCO
1, rue Miollis
75732 Paris, France

Introduction

An element in modern thinking concerning environmental education as defined by "environmental educators" concerns the notion that one may divide the field into: 1) education about the environment and 2) education for the environment. It is proposed by some that "environmental education" is not "environmental science education". This seems to mean that environmental education (education for the environment) is exactly what one might call "environmental indoctrination" or even "environmental ideology". Although this may be acceptable to some extent, it seems to be entirely removed from the environment and apparently deals with 'values', and 'ethics'. However, the way in which these values are arrived at is not stated. Likewise the problem of rationalizing an ethic is simply avoided. In using this approach, nobody gets to go into the environment to actually check up on the real situation, since all answers are apparently known, and the idea that students should be encouraged to verify information for themselves is not well supported.

It is important to stress that "environmental-global change education" should start as early as possible within the education experience, and this should take place at least in high school. This form of education concerns assessment of the real world with the intent of using the resulting information to obtain 'reality checks'. It involves transmission of the viewpoint that calls everything into question and seeks answers, concerning the processes governing sustainability of the biosphere, from the earth. It deals with the methodologies involved in testing and verifying elements of local, regional, and global environmental change. Values and so forth flow from the information deduced and adduced in relation to the question of "environmentally sound and sustainable development". The element to consider in all contexts is sustainability. It is, however, important to ask what exactly this term means in the context of scientific practice, with respect to the information that science brings to light, in respect to social practice, in terms of environmental processes, and so forth. The

NATO ASI Series, Vol I 29
Global Environmental Change in Science
Education and Training
Edited by David Waddington

term "development" should not be regarded as daunting since it can at the limit refer to qualitative factors. This approach can provide a realistic basis for elaboration and definition of the values the educators are looking for. The practices and attitudes that are of value to society are thus those which are in line with sustainability.

The question underlying local, regional, and global sustainability and its definition and analytical relation to all that is learned about the environment should be the mainstay of all training for environmental global change research and understanding. This is important because comprehension of the term "sustainable", in the context of the principles governing maintenance of the biosphere, actually provides the basis for values and ethics in the modern world. Behavior that promotes sustainability, or at least that does not promote the opposite trend, should be regarded in a positive light and seen as ethical. Positive value thus would lie in the reinforcement of sustainable practice in all respects. The mode by which science interacts with this form of analysis is thus of some importance, since it is scientific practice that can be used to frame, analyze, and test the validity of claims that will be made towards sustainability in everyday affairs.

This notion is most important because presently there are rather significant distortions of the work carried out under the Global Environment Facility (GEF) research programs, that in some ways amount to focal points on one side or the other of the north-south (developed-undeveloped) split. For instance, work on methane production associated with rice growing appears to be undertaken in an effort to actually show that the developing world is at fault in generating significant contributions to greenhouse gas sources as the result of depending on rice as a staple. Nevermind that most rice fields are now located on what were previously methane-producing wetlands, that rice merely vents methane that has formed from the previous year's green fertilizer, that the carbon released goes up mainly as CO_2 and will be reclaimed from the atmosphere as rice growth proceeds, that the slightly positive methane/CO_2 ratio that might develop is a small fraction of the carbon/global warming term incurred by the fossil fuel based agricultural effort, and so forth.

Field work philosophy

The educational programs required to combat environmental change must also be anticipated in the form of concretely focused studies for graduating students.

Problems associated with environmental management options will increasingly require the attention of an educated population to render solutions compatible with the emerging environmental ethic. Educational institutions will face new pressures to provide adequate training in this regard, and the best way of ensuring an appropriate response is to expose students to field situations requiring their attention to observation and detail at an early stage. Indeed, examples drawn from environmental studies can be used to illustrate almost all facets of a modern scientific education, and the increasing use of the phrase "environmental education across the curriculum" is an indication of the extent to which serious environmental studies are now required at all educational levels and, most importantly, in high schools.

There is widespread agreement that the principle of *ecologically* or *environmentally sound and sustainable development* should be adhered to in both developed and developing countries. An important step in ensuring that this will be undertaken on a broad front involves engaging high school students with the realities of environmental analysis, so they may contribute to defining exactly what future generations will mean by this (at present) uncertain phrase. High school science faculties will increasingly be called upon to design interlocking programs around the major areas of environmental studies, although at first this may only involve token adjustments, such as the inclusion in chemistry courses of behavioral characteristics of common salts in solution to promote understanding of saline versus fresh water environments and so forth. Yet, at a later time it may be possible to imagine a far wider array of interlocking subject areas within standard and classical curricula, which will be pointed towards an increasingly sound and scientific understanding of environmental processes, and eventually lead to the development of specific curricula on the subject.

Field studies objectives

Properly planned field studies can provide the opportunity for team collection of environmental data at sufficient densities so as to require the use of microcomputers for analysis and reporting. These same data sets over the long term may then be saved to provide a systematic time sequence for the detection of secular change. Field studies should form a key part of an interdisciplinary curriculum that can prepare present and future generations with an understanding of marine, terrestrial, atmospheric, and associated systems, as they relate to ecologically sound and sustainable development in a continually changing world. A field studies curriculum should bring together elements within the existing syllabus to: (1) make advanced knowledge of marine terrestrial and associated systems accessible to teachers and students through the analysis of existing problems; (2) catalyze interaction among high schools and established research and education systems to ensure that activities significantly contribute towards required educational goals and are accepted as credits towards qualification in different programs where appropriate; (3) encourage the collection, recording, and dissemination of information relating to the environment as the result of the accumulation of class data over the years; (4) encourage implementation of standardized monitoring and testing protocols to enable consistent utilization of data over a wide front; (5) progressively establish educated communities in all countries of the world to understand factors associated with environmental change and the resultant impacts, thereby encouraging global ecologically sound and sustainable development in a changing world; (6) at a broad level encourage sharing of knowledge, resources, and facilities, and selectively enhance facilities to maximize the capability of all participating institutions.

Ecologically or environmentally sound and sustainable development

A major concern in environmental thinking and practice currently deals with the area known as "environmentally sound and sustainable development". Implementation of well thought-out national ecologically sound and sustainable management plans would be a major asset in the development of an overall response to environmental change. A systematic approach to this problem will permit the allocation of resources to upgrade public analytical thinking about the oceans and the terrestrial environment and their role in global change. It will also contribute to significantly raise the level of awareness for environmental problems in all segments of the population at both national and global levels. Recent attempts have been made by many international organizations, including the UN system, to direct the global community in adoption of a policy of environmentally sound and sustainable development (ESSD). In a crude sense, we may define ESSD as development which permits further development without terminal closure of favorable options. In pursuit of this end, the general approach towards defining the potential for progression of civilization throughout the world must be addressed through ecologically sound and sustainable development. This in turn means that only human actions which permit continued development, with the environment as the final arbitrator, can be considered as constituting an acceptable strategy for social activity at all levels. The challenge at this stage is to elaborate useful models which will permit specification of such action. We know what we should like the phrase "ecologically sound and sustainable development" to mean, but we currently have no clear idea as to how this can be attained in practice, nor indeed how it can be measured objectively.

A major task for environmental education will be to lay the groundwork to permit definition of what a community will understand by this phrase through early experience with the subject matter at a sufficiently detailed level. Eventually this will require an increasingly heavy commitment to gather and process environmental statistics and to define and quantitatively evaluate development options. Suitably educated students will be able to participate in this exercise by contributing informed opinion, as required in the future. One of the most difficult areas to deal with will be the assessment of the effects of energy injections. On the one hand, energy enrichment to human social systems is clearly used to promote increases in carrying capacity and the elaboration of useful structures. On the other, the act of energy capture and subsequent energy degradation promotes vast and as yet unquantified negative environmental impacts and instabilities, among which the increasing concentration of atmospheric CO_2 stands out as a significant warning signal.

Organization and structure

The field exercises should be proposed to highlight and analyze environmental problems relating to the ramifications of change. Although it can be anticipated that not all influences associated with global change will be deleterious, the exercises should focus on raising student awareness of the environmental issues faced by modern society. The suggested method is to take a particular local problem or a set of problems and to analyze these in detail in order to (1) make students familiar with the particular circumstances through their own efforts (data collecting and analysis) and (2) help them gain an understanding which will relate to the whole globe.

Lack of understanding about the influences operating in the modern environment continues to be widespread. Societies in many countries have evolved through substantial periods of dependence on economies operating under apparently balanced ecological conditions. For many, the effects induced by global industrialization and global economic processes are new phenomena, ones which often cannot be understood by the direct transfer of practices which were formerly adequate to deal with local causes and effects. In the face of advances in technology and changes of scale owing to the operation of the global economy, traditional practices, such as those involved in fishing and ocean management, often prove inadequate to guide further development. For instance, in the Pacific Ocean traditional management practices have evolved successfully over long periods of time to ensure that local stocks of living resources are not fished out. The use of modern fossil-fuel-based enterprises which key into the global economy however, often result in severe over-fishing, as efforts are made to capitalize on local resources and to repay loans for fossil-fuel-powered equipment.

Environmental studies and traditional approaches

In order to be realistic, the interdisciplinary nature of environmental studies does not have to be emphasized to the extent that it cannot be accommodated into existing curricula. On the contrary, topics selected should be relevant to existing education requirements and should be related to the socio-economic needs of the community that will receive the graduating student. Any subject taught using a field-work approach should have an existing knowledge base and a theoretical underpinning. If the subject is to be used to generate effective environmental insight, it must also be taught using a practical approach involving learning by doing, problem solving, and decision making. An important aim should not be so much to orient the subject for merely passing exams, important as that may be, but to conduct it in such a way as to contribute to the fulfillment of the needs of learners in terms of generating insight into the current environmental problem and enabling those concerned to realize their own role in the issue. In other words, the educational implications of environmental studies, particularly environmental field work, as part of the school education experience must be clearly understood and subscribed to before field studies are undertaken. It is the purpose of this paper to advocate a field work experience focused in such a way that the major objectives within existing curricula can be achieved more effectively than would be the case if no field experience was offered. In addition, the information gained should be of direct use in the construction of strategies for obtaining successful examination results.

Environmental ethics

As human population pressures increase throughout the world and the influence of industrialization spreads from formerly restricted areas in ever widening circles powered by fossil fuel energy accumulated from ancient ecosystems, the factors mitigating against continuity of undisturbed natural communities capable of maintaining a buffered environment intensify. With the possible exception of remote communities in the abyssal depths of the oceans or perhaps under the Antarctic and Arctic ice-caps, many ecologists now feel that there are few, if any, undisturbed ecological communities remaining on the face of the earth.

Traditionally, ecological field-trips have been marked by a heavy emphasis on sample collection and preservation for later identification in a laboratory setting. For the most part such activities encouraged the destruction of organisms in the name of science with the majority of the formerly living animals and plants left to rot in forgotten corners. In some places on the shores of continental North America successive class field trips caused discernible damage. Even today an active search is maintained for comparatively rare shells, particularly cowries which are already considered to be under threat. The modern wilderness and eco-tourism slogan of "take only pictures, leave only footprints" has surely found its way into the consciousness of most high school students, so that it is important to design environmental studies programs having the minimum of impact upon the communities studied.

As with the Jains of India who believe that the conscious or even the unconscious harm exerted on the least of living creation detracts from their own quest for perfection to the extent that they wear no leather, silk, or even fibers obtained from living plants, preferring instead to go naked in the world and to studiously brush aside insects that might otherwise be harmed by their passage; so too should the execution of environmental field studies be conducted with the utmost in circumspection with respect to the object of study. Insofar as it is possible, the activities planned in the program developed within this context should be designed to exhibit the minimum of impact. Experimental actions against larger and rarer life forms, that result in destruction of the individuals concerned, are expressly frowned upon.

In spite of this injunction it nevertheless remains true that some form of impact will be involved with even the most carefully planned field course. Where coral reefs are examined using snorkeling techniques, stray fins will surely break some of the most fragile forms, but even here, careful preparation can hold the damage to a level that is far below that which would accompany a heavy storm. In wading across a reef flat it is certain that some organisms will be adversely affected. Yet again the teacher should devote time and attention to these details for it is only by arguing for the good of the whole by virtue of knowledge gained that the sacrifice of the few can be justified. Clearly some ecosystems such as coral reefs adjacent to urban population centers or in association with massive tourist developments are more at risk than, say, the plankton blooms that develop in response to sewage discharges (nutrient enrichment) around the outskirts of the ever growing human city ecosystems. Considerations such as these can be used to develop guidelines concerning the extent to which sampling mortality in the name of ecosystem knowledge can be justified.

In other words, the whole concept of environmental field study programs must be exercised in an environmentally sound and sustainable manner. We may approach this problem best through consideration of the extent by which it may be possible to conceive a form of natural self-sacrifice, which nature may be prepared to offer in exchange for providing the basis for the development of human understanding that will permit elaboration of survival strategies for the biosphere as a whole and thus for humankind as well. By this viewpoint, humanity depends on nature, and at this stage in the evolution of our planet, nature depends, as never before, on humanity.

Expected output and accomplishments

Improvement in the quality and quantity of environmental education and related analytical capabilities of graduating high school students will enable institutions such as universities, government agencies, and private firms to carry out their missions more effectively by having access to informed personnel. Although the final outcome of a course in environmental studies should lead to an integrated comprehension of the major factors shaping the world and the extent to which mankind may interfere with them, nonetheless this overall perspective will normally have to be reached through a variety of course options that may be offered at a particular high school. The broad questions set in examinations often require students to interpret local conditions in the formulation of answers. Often a considerable array of material can be assessed if students have been exposed to a concentrated field program where they have been forced to analyze information they have themselves collected. Questions relating to the implementation of particular sampling protocols and to ecological principals such as niche theory, the pyramid of numbers, etc., can be covered in a meaningful way in an environmental field studies program. Other general biological principles may also be addressed, including identification and classification, food-webs and food-chains, energy flow patterns, and population studies. In certain instances, the study of management practices may be considered.

The aim of the field trip is to provide an integrated exposure to environmental analysis, to enable the students to obtain unequivocal evidence concerning the effect of at least one or more environmental problems, and at the same time to provide the necessary exposure to practical examples of facts, processes, and interactions that are specified within existing syllabus plans. Since the fundamental rationale for the activity involves the use of environmental problem areas as the skeleton upon which the individual syllabus concerns are to be related, and since time and costs are limiting factors, it is essential to select a program which will accomplish both objectives.

Report production

The field experience and the associated data collection exercises form the main component of the overall activity being designed, in part to force attention of students towards structures under study. The main part of the learning component is, however, focused on the reflection required to produce a comprehensive written report. In addition to the simple description of the exercises involved and the overall account of the activity, the report should contain a detailed analysis of the data collected. The activities should be designed for implementation by study groups involving from 4-6 students (not more). The amount of data generated from a class of 30-40 can be quite significant. For instance, a coral reef transect exercise in the tropical marine environment can readily generate 12 transect readings for each student group. Each of the transect readings may involve positional information for the life-forms encountered, as well as density and coverage measurements. This form and density of information can only be sensibly analyzed with the help of a microcomputer and in addition can form the basis for studies of secular change.

An Environmental Education Framework Based on Ecological Sciences

Within the concept of "Environmentally Sound and Sustainable Development", an important overall issue is development without excessive outside interference. Relevant specific issues are (1) loss of biodiversity, (2) land degradation, (3) degraded ecosystems, (4) socio-cultural gaps, (5) socio-economic imbalance and problems associated with ensuring social justice, (6) conflict between traditional values and foreign life-styles, and (7) use of biodiversity. The issues underlying the individual geographical components of an environmental education system are, (1) requirements for the development of ESSD plans, (2) preservation and maintenance of biodiversity, (3) control of global warming through the requirement for the elaboration of carbon sinks (coral reefs, inland and mangrove forests, etc.), (4) assessment of external pressure in the form of exploitation of mineral, timber, and other natural resources, as well as eco- and cultural tourism etc., (5) relationships with national parks, (6) requirements for ecological repatriation of damaged ecosystems, and (7) requirement for development of effective local area management.

A preliminary approach towards customized courses to address environment and sustainable development might be envisaged to cover the following 3 areas:

Area I

SOCIO-ECONOMIC DEVELOPMENT

1. Development models/theories
2. Indicators for ESSD
3. Development agencies and activities (Stockholm, Bruntland, UNCED, post-UNCED Agenda 21)
4. Natural resource accounting
5. Environmental impact assessment and cost/benefit analysis
6. International relations

Area II

BIOSPHERE PROCESSES AND BIODIVERSITY

1. Evolution of species, ecosystems, and biomes; function of biodiversity
2. Ecosystem function in relation to global climate stability
3. Rates/causes of ecosystem changes
4. Conservation of nature: notations from different cultural backgrounds
5. Values associated with protected areas (PA's) as bio-diversity centers, conservation, management, etc.
6. Protected area management and development linkages
7. PA's and Post-UNCED, Agenda 21

Area III

GLOBAL CHANGE

1. Pollution, pollution as externality of socio-economic development
2. Greenhouse warming, ozone depletion and other manmade environmental changes
3. Global change and options for future

As an example, the following outline covers the major features to be considered in an ecology course, which are also the appropriate elements of environmental studies. Particular field exercises can be slotted into the framework and given the emphasis necessary to address traditional subject areas. It should be noted that unless some thought is given to this, teachers may have some difficulty in determining the relevance of specific exercises that may suggest themselves when facing field situations.

Structural outline for ecology and environmental studies

1. The natural environment and interactions among organisms
    - 1.1 Living space
        - 1.1.1 Effects of physical factors (determining where organisms can live)
        - 1.1.2 Effects of biological interactions (determining where organisms may live)
            - 1.1.2.1 Effects of the same kinds of organisms (intraspecific competition)
            - 1.1.2.2 Effects of other kinds of organisms (interspecific competition), predation
            - 1.1.2.4 Evolutionary adjustments: mutualism, commensalism, etc.
    - 1.2 Mortality factors, physical, biological
2. Biological processes and the physical environment
    - 2.1 Role of energy in community maintenance, photosynthesis (primary production)
        - 2.1.1 Active transfer of biologically fixed energy, conditions governing stability
            - 2.1.1.1 Predation as an energy transfer device
            - 2.1.1.2 Food webs
            - 2.1.1.3 Pyramid of numbers
    - 2.2 Role of nutrients and essential elements in community maintenance
        - 2.2.1 Identification of major elements and nutrient components
        - 2.2.2 Origins
        - 2.2.3 Recycling
    - 2.3 Energy-mediated injections into natural systems
        - 2.3.1 Nutrients
        - 2.3.2 Elements (carbon, etc.)
        - 2.3.3 Complex chemicals (chloroflurocarbons, organic pesticides, etc.)
3. The biosphere
    - 3.1 Conditions associated with sustainability
    - 3.2 The Gia hypothesis, energy use, feedback control
    - 3.3 Biodiversity and long-term changes
4. Human impact on the environment
    - 4.1 Assessment of biological communities
    - 4.2 Monitoring of physical components (pollution)
    - 4.3 Reconstruction of damaged ecosystems: why do it ?
    - 4.4 Human population: growth and impact
    - 4.5 Human ecology and the concept of sustainability

The Role of Environmental Education in the Training of Scientists on Global Environmental Changes in Developing Countries

W Leal Filho
European Research Centre on Environmental Education and Development
University of Bradford
West Yorkshire BD7 1DP, UK

Introduction

Although over half a century has passed since Haeckel launched the concept of ecology, ecosystems, and environmental balance, only in the last two decades has the problem of global environmental changes become an issue to be seriously discussed. New scientific evidence shows that there is indeed a strong sign that various alterations (for example in the planet's climate) are taking place, which are bound to lead to major changes in habitats and in the ecological dynamics of various plant and animal species, as well as in human behaviour.

A number of on-going research programmes have been used to try to understand the origins, structure, and dynamics of global environmental changes. These vary from research aimed at detecting increases in the temperatures of various parts of the planet (which may possibly be linked to the so-called 'greenhouse effect'), to the preparation of models for predicting the evolution and eventual advancement of desert areas in regions such as northern Africa. A fact that is clear from the body of research already existent, and that has been registered by UNDP (1991) and IUCN (1991) in two of the most significant publications in the field of environment and development in recent years, is that the current models of economic growth, adopted by the various countries which now suffer intensively from the effect of climatic variations, are not the most suitable ones. In addition, it is clear that alternative ways of improving a country's economic performance should be found that are preferably based on the principle of sustainable development.

However, due to the magnitude of the problem and its various variables, many research topics are yet to be covered comprehensively. Leal Filho (1992) listed some research needs seen in terms of global environmental changes in developing nations, with a special emphasis on climate changes, and concluded that many items are yet to be fully investigated.

NATO ASI Series, Vol I 29
Global Environmental Change in Science
Education and Training
Edited by David Waddington

Among today's research needs, mention also ought to be made of investigations of the ways through which national policies implemented in developing nations tackle the problem of global environmental changes. Such a need, as acknowledged by Magalhaes and Glantz (1992), is illustrated by the fact that, although various studies have dealt with policy on global environmental changes in industrialized nations, there are few registers of comparative studies which have focused on the political behaviour and the social responses from developing nations to these changes. This is especially conspicuous in economically powerful developing countries, such as Mexico and Brazil, in countries considered overpopulated like India, and in countries which severely suffer the effects of global changes, such as Bangladesh and the Philippines.

From Discussion to Action

Despite the research needs discussed above, a steady evolution has been seen in the process of discussion on the item "global environmental changes". For example, being aware of the increasing evidence of climatic variations, the World Meteorological Organization (WMO) set up its International Climatic Programme in 1976. This development was followed by a conference in 1979 about the world's climate, the World Climate Conference. In that event, the concept of global warming, or the greenhouse effect and the role of CO_2 in exacerbating it, was one of the most discussed items. A few years later in 1985 a conference in Villach, Austria, approved a set of research studies on CO_2 (by then clearly acknowledged as a major factor associated with climatic variations), which included quantitative and qualitative analysis of CO_2 concentration in the atmosphere, and its effects on the climate and on ecosystems. The Villach Conference set in motion further research studies.

More recently, the United Nations General Assembly at its 1990 session set up the Intergovernmental Negotiating Committee for a Framework Convention on Climate Change (INC/FCCC), which was given a mandate to "draft a framework convention and any related legal instruments it considered necessary" (IUCC 1993). Since then, negotiators from all over the world have been working on the text of such a convention. In May 1992, they adopted the United Nations Framework Convention on Climate Changes which was signed by more than 150 countries.

According to the latest ratification status report, over 50 countries ratified the convention and a meeting of all ratifying countries, dubbed as the "Climate

Summit" is expected to be held in Berlin in March 1995 to discuss its implementation. Once this process is completed, the INC/FCCC will be disbanded and the responsibility of the fulfillment of the convention's recommendations and implementation will rest with the Conference of the Parties, a body to be created, which will be composed of representatives of the countries which ratified the convention.

The Need for Training

In consideration of the need for education, training, and awareness as tools for the implementation of the convention and as instruments for effectively dealing with global environmental changes as a whole, the United Nations Framework Convention on Climate Changes has made provisions for their incorporation as part of its structure.

Article 6 of the convention (Education, Training, and Public Awareness) requires the parties to promote and facilitate at the national, sub-regional, and regional levels the following measures:

1. the development and implementation of educational and public awareness programmes on climate change and its effects
2. public access to information on climate changes and its effects
3. public participation in addressing climate change and its effects, and in developing adequate responses
4. training of scientific, technical, and managerial personnel
5. the development and exchange of educational and public awareness materials on climate change
6. the development and implementation of education and training programmes, including the strengthening of national institutions and the exchange or secondment of personnel to train experts in this field, in particular for developing countries

Due to the relevance of training especially designed to correct the deficiencies in the knowledge of the relationship between environmental policies and environmental changes (at the local and global level) found in developing nations, some considerations of the role of environmental education, as a tool for understanding global environmental change as a whole and climate change in particular, should be made. Environmental education, as a process aimed at raising awareness on environmental issues (Leal Filho 1993, Unesco 1987) may catalyse action aimed at reversing the lack of information seen today among people in developing nations, especially among policy-makers.

The systematic provision of information on environmental matters, one of the applications of environmental education, is also very useful in the sense that the body of knowledge gathered from training may be used as reference points in helping developing countries to prepare more reliable national policies, which would allow them to properly deal with global environmental changes.

An analysis of some on-going national environmental policies related to global environmental changes was undertaken using a sample of six developing countries. In the analysis, the sample countries' effectiveness under pre-determined criteria was evaluated, and a comparison of the basic lines of their policies was made with a view of suggesting ways of improvements. The study, which consisted of a survey involving a sample of 6 developing countries and done over the period 1992-1993, aimed to:

- investigate the existence of policies associated with global environmental changes in individual countries
- assess the effectiveness of the existing policies in individual countries on the basis of their responses to environmental problems (including disasters) and on their preparedness to deal with global environmental changes
- suggest, based on the principles of environmental education, some items that should be considered in the setting up and/or review of national environmental policies, aimed at improving a country's likelihood of effectively dealing with sudden environmental and climatic variations and handling long-term environmental changes

In terms of the sample used, two representatives of the three largest geographical areas of the planet, where developing nations are concentrated, were chosen. In Latin America, Brazil and Mexico were selected on the basis of their size and economic influence in the region. In Africa, focus was given to Morocco and Kenya, the former having a large desert and the latter suffering an advanced process of desertification, which makes an investigation into their level of response to global environmental change a promising issue. In terms of Asia, Bangladesh was selected due to the fact that it is one of the region's most populated countries and because it has been victimized by various environmental problems related to global changes. The second Asian country, namely the Philippines, was chosen on the basis of its historical vulnerability to environmental disasters, which makes it an interesting case study for assessing the existence and scope of policies oriented towards environmental variations and long-term environmental changes, adopted by a developing country. Language barriers were probably minimal due to the investigator's fluency in English, Portuguese, French, and Spanish. Data collection consisted of a combination of one empirical and two theoretical methods. In terms of the theoretical methods, a historical investigation of the scientific evidence for environmental changes in the selected countries was undertaken. This was combined with a structural analysis of national environmental policies. In empirical terms, an evaluation of the countries' infra-structure for dealing with the local effects of global environmental changes was undertaken. The establishment of relevant collaboration schemes with persons and institutions, nationally and internationally,.was paramount in the execution of the survey. Table 1 shows the criteria used in the survey and Table 2 shows the results obtained.

Table 1 - Criteria used in the Study

Assessing National Environmental Policies, with a Special Reference to Global Environmental Changes, in Developing Countries

a) preliminary procedures:

- diagnostic of a country's economic and political situation
- main environmental resources found and commercialized
- significance of the trade of environmental resources to the economy
- social problems related to the use of environmental resources

b) policy study

1. existence of National Environmental Policy (NEP)
2. date of implementation of NEP
3. organization(s) responsible for the NEP
4. integration between organizations
5. existence of training programmes
6. status of personnel (i.e., well trained, reasonably trained, poorly trained)
7. prospects for improvements

Table 2 - Emphasis given to Training on Global Environmental Changes in the Context of National Environmental Policies in 6 Developing Countries

Country	National Environmental Policy	Emphasis on training
Brazil	Yes	3
Mexico	Yes	3
Kenya	Yes	4
Morocco	Yes	3
Bangladesh	Yes	2
Philippines	Yes	3

Scale of emphasis 1. strong, 2. considerable, 3. moderate, 4. negligible, 5. none

Conclusion

Although a larger sample would be needed for drawing more definitive conclusions, the survey illustrates the fact that the integration of environmental education and training, in the context of national environmental policies in the sampled countries, can play a relevant role in enabling them, as developing nations, to effectively deal with global environmental changes. Although training needs are many (see Appendix 1), such programmes are seen to have a strong potential for improving a country's ability to deal with environmental phenomena, especially those which may have strong national implications.

A special emphasis should also be made on the need for further training in the methods through which climatologic data may be disseminated in a "easy-to-understand" language, thus enabling people, mainly the non-expert public, to understand them. In addition, ways through which sets of especially relevant information might be made more widely available should be further investigated and disseminated as part of training programmes.

The provision for systematic training may provide a country with the means for understanding the reasons underlying global environmental changes and the associated problems, which may guide the search for long-term solutions.

References

Information Unit on Climate Change (IUCC) (1993) Framework Convention on Climate Change. IUCC, Geneva

IUCN / World Conservation Union (1991) Caring for the Earth. IUCN, Gland.

Leal Filho WDS (1992) Research Needs and Research Agenda. I "International Conference on Impacts of Climatic Variations and Sustainable Development in Semi-Arid Regions". Fortaleza, Ceará, Brazil. January 27 to February 2

Leal Filho WDS (ed) (1993) Environmental Education in the Commonwealth. Commonwealth of Learning, Vancouver

Magalhaes AB and Glantz MH (1992) Socioeconomic Impacts of Climatic Variations and Policy Responses in Brazil. Esquel, Brasilia

United Nations Development Programme / UNDP (1991) Human Development Report. Oxford University Press, Oxford

Unesco (1987) Report on the Intergovernmental Conference on Environmental Education, Moscow, 1987. Unesco, Paris

Appendix 1 - Training Needs in Global Environmental Changes (GECs)

1. Training is needed on how GECs affect the environment at a local, regional, and global level. In particular, attention should be paid to semi-arid regions and to threatened ecosystems, such as the tropical forests.

2. Training on the recognition of the phenomena that lead to GECs, in the framework of which the elements which create them should be investigated and ways through which their effects may be minimized, should be sought. This particular topic may include the principle of sustainable development, as a means of combining environmental conservation with economic growth.

3. Training is needed on the methods through which climatologic data may be disseminated in a "easy-to-understand" language, thus enabling people, mainly the non-expert public, to understand them. In addition, ways through which sets of especially relevant information may be made more widely available, should be sought.

4. Training is needed on how to make data on GECs available to decision-makers in specific countries, in a easy-to-understand way, thus guiding them to plan their economic and social policies, and thus stimulating sustainable economic development.

5. Training on the methods of recognition of the environmental problems associated with GECs, using data and inputs from individual and comparative research should be instituted. These may identify the extent to which problems are interrelated. The greenhouse effect, the depletion of the ozone layer, acid rain, and certain methods of energy production and fuel use should be looked at in detail and any eventual links with GEC's should be pointed out, thus guiding the adoption of proper and effective corrective measures.

6. Training aimed at catalysing the effectiveness of current international cooperation programmes on issues related to GECs is needed. This would address the deficiencies existent today in terms of information exchange, technology transfer, and transfer of resources for research, among industrialized and developing nations, with a view to improving and facilitating the access and proper exchange of valuable data.

7. Training is needed on the understanding of the phenomena related to GEC's, such as storms, inundations, cyclones, droughts, and frosts. Data on their origins, impacts, and effects should be provided during training, thus preparing trainees to identify eventual common patterns.

8. Training is needed on the measures aimed at the prevention and/or minimization of environmental damage caused by GEC's. This particular topic should be reinforced to provide suggestions of actions that should be taken at a short- and long-term level when a certain event (e.g., a rise in sea level) or series of events, occur. This is especially relevant to developing countries which often lack the means to cope with sudden changes.

NASA/USRA Cooperative University-based Earth System Science Education Program

George W Fisher
Morton K Blaustein Department of Earth and Planetary Sciences
The Johns Hopkins University
Baltimore MD 21218, USA

In an effort to strengthen the teaching of earth system science at the university level, the Universities Space Research Association (USRA) and NASA have developed a cooperative program called Earth System Science Education (ESSE) at 22 universities in the United States (Table 1). Begun in 1991 and scheduled to last through 1996, the program is intended to strengthen interdisciplinary earth science education in areas relevant to NASA's Mission to Planet Earth, to promote cooperation between NASA and university faculty, and to attract students and faculty to earth system science.

The core of the program is a pair of undergraduate courses that each participating university agreed to develop: a general survey course accessible at the freshman level, and an advanced course for senior science majors. The advanced courses were expected to involve faculty from at least two departments related to earth system science and were intended to provide a basic grounding in system concepts and models. Most universities responded by developing courses in which students from different backgrounds worked together in interdisciplinary teams to model components of the earth system. The course content tended to reflect the strengths of each university and so varied substantially from institution to institution. In return, the ESSE program provided modest financial support (for teaching assistants, a small amount of faculty release time, instructional materials, and so on) for each university, plus some additional travel funds to encourage the use of lecturers from other ESSE institutions, both to broaden the range of expertise available at each university and to promote collaboration among ESSE faculty.

In order to facilitate the program, ESSE sponsored a variety of meetings for representatives from participating universities. Steering meetings, attended by two representatives from each university, were held twice yearly to compare strategies, successes, and problems, and to share teaching materials. Given the difficulties of developing interdisciplinary programs on many campuses, these meetings proved very useful by providing a variety of program models several of which are presented in the Appendix to this volume. Two workshops devoted to exploring specific topics were also held. One was devoted to Stella[1] (a graphical

Table 1: Universities participating in the ESSE program

University of Alabama
University of Alaska - Fairbanks
University of Arizona
University of California - Los Angeles
University of California - Santa Barbara
University of Florida
University of Iowa
Johns Hopkins University
University of Minnesota
University of New Hampshire
New York University
Northwestern University
Ohio State University
Pennsylvania State University
Princeton University
Purdue University
Rice University
Rutgers University
Stanford University
Utah State University
Washington University
University of Wisconsin

differential equation solver for the analysis of dynamic systems), and the other to Geoscope (a CD-Rom-based hypermedia encyclopedia of global and regional earth science data sets developed by the Canadian Space Agency for ISY '92). Both workshops were highly successful, and facilitated the spread and development of computer-based material for teaching earth system science. These materials have proved so useful that USRA is now in the process of establishing a server to make the materials available over Internet to interested faculty at non-ESSE universities.

We have the impression that ESSE has been very worthwhile for the 22 universities participating in the program. It has clearly enabled those universities to develop undergraduate programs in earth system science education more rapidly than they would have done without the ESSE program. It has promoted the development of instructional materials and the use of

1 Stella is a trademark of High Performance Systems, Inc., 45 Lyme Road, Hanover NH 03755.

programs like Stella and Geoscope, and it has enabled those universities to develop or enhance links with the NASA community.

After participating in the ESSE program for three years, I have three observations about teaching earth system science at the university level:

- A variety of curricular and programmatic models work well as vehicles for teaching earth system science, and it would be a mistake to attempt to impose a standard earth system science curriculum on the community. Universities have different strengths and different areas of faculty interest, and each should be encouraged to develop programs which build upon its own strengths and areas of expertise.

  There is an urgent need for development of curricular materials which are responsive to the diversity of both subject matter and depth of presentation in earth system science programs. Textbooks covering the full range of earth system science topics may be less useful than short books dealing with specific topics, which may be combined by an instructor in order to meet the needs of an intended course. Some publishers are now offering to combine chapters, which they have published, to form books tailored to a specific course. This style of publication seems particularly well suited to earth system science courses. Hypertext materials available in CD-Rom or on-line via Mosaic or other Internet browsers provide an interesting alternative for those with access to the necessary technology.

- Most college professors were trained in individual disciplines, and so there is a special need for short courses in earth system topics designed for instructors, who would like to be able to teach in disciplines in which they have little background.
- There is a need to involve the solid earth science community more fully in earth system science. The geologic record provides a valuable record of the way the earth system has behaved in the past, and solid earth scientists are expert in several areas relevant to earth systems: geochemistry, rock-water interaction, sedimentation, and so on. One of the most neglected areas is that of soil development. Soils are the interface between the solid earth and the rest of the earth system, and it will be very hard to link the solid earth to the fluid envelopes and the biosphere without understanding soils better than we do now.

Global Change Education: The Evolving U.S. Federal Plan

Lynne M Carter, co-chair Education and Communications Working Group
Graduate School of Oceanography
University of Rhode Island
Narragansett, RI, USA 02882

The scientific study of global change has advanced our abilities to understand and predict environmental causes and effects, both natural and anthropogenic. The U.S. approach to the complicated science issue of global change comes through the Subcommittee on Global Change Research (SGCR), a subcommittee of the Committee on Environment and Natural Resources (CENR) representing 18 federal agencies. The members of the SGCR believed that the long term scientific study of global change required informed and educated participants. The SGCR established an Education and Communications Working Group (formerly the Education Task Group) to develop an education plan with a primary focus on the need for future scientists. However, as the questions about global change cause, effect, and policy options have evolved, so has the need for education opportunities. Preparing future scientists is referred to as a *pipeline* issue when the level of final participation is as active scientists and engineers. The first global change education program was featured in *Our Changing Planet* (1991). While identification of the need for a broad base of scientific literacy and an informed citizenry and decision-makers was briefly mentioned in the Future Program Needs section of the report, the short-term identified focus was

> To inspire and train the scientific cadre required to carry out the U.S. Global Change Research Program

It was envisioned that the Education Program would take the form of postdoctoral fellowships, graduate student fellowships, undergraduate research experience, high school summer internships, research training group awards, and summer institutes and training workshops. All of these activities would occur through the National Science Foundation (NSF) grant process and be managed by a governing board representing the SGCR global change agencies.

A number of agency activities in global change education were also underway. They included mostly graduate programs through the Department of Energy, National Aeronautics and Space Administration (NASA), Department of Interior, National Science Foundation (NSF), and National Oceanic Atmospheric Administration (NOAA).

NATO ASI Series, Vol I 29
Global Environmental Change in Science
Education and Training
Edited by David Waddington

The support mechanisms that were envisioned for the global change education initiatives were twofold: agency programs that were underway, and CENR initiative funding. As it turned out, only activities that were agency funded actually occurred.

All of the on-going and planned Education and Communications Working Group and agency global change education opportunities were intended to complement the science priorities of the SGCR. Those priorities were in climate and hydrologic systems, biogeochemical dynamics, ecological systems and dynamics, earth systems history, human interactions, solid earth processes, and solar influences.

The FY '92 Plan

The education plan as outlined in *Our Changing Planet: (1992)* identified the same pipeline issues and agency programs already underway and related to global change as in the *FY '91 Plan* (1991). The FY '92 program clearly identified the education planning effort as *in progress,* with a small amount of funding available for pilot projects in both the FY 1991 and 1992 years. Future education programs were expected to

> further enhance programs in the areas of research experiences, educator training, and environmental education and outreach activities at all levels.

The additional focus of environmental education and outreach at all levels was reflective of the growing interest in global change, environmental, and science education for all levels of the public.

The issue of broadening the base for global change education to include those before and following a collegiate experience leads to global change education for non-scientists but also back to the pipeline issues. Many have discussed various projections concerning the imminent national crisis in the supply and demand for scientists and engineers. Some of this shortfall is being addressed by an increasing number of foreign-born students, but that is not likely to increase enough to make up future supply needs. Although there may be enough qualified students to avoid this problem, they are not going on to college from high school (Atkinson 1990). Special focus education, aid, and grant programs for the pre-college level could encourage continued education of those specially qualified in science, math, and engineering. We cannot expect to improve the projected shortfall of scientists significantly without a continuing stream of students.

The FY '93 Plan

Global Change issues provide a relevant theme for scientific inquiry and for the policy issues that are becoming important elements in an increasingly technical society. In recognition of the relevance of global change and of a federal role in science and mathematics education in general, a broadened committee scope was developed. The Working Group long-term global change education strategy now included the need for *multidisciplinary studies, transgenerational opportunities, long-term commitments to change,* and *coordination with other relevant national and international education programs.* As a result, any global change long-term education plan needed to include opportunities for all education levels: precollege and early college, upper-level college and graduate school, and informal continuing/public education (Task Group on Education Workshop 1993).

The Education and Communications Working Group also developed a series of short-term recommendations. These short-term recommendations were meant to close the gap between what was presently being offered in global change education and the long-term strategy. The short-term recommendations for FY '93 were to: 1) focus highest priority on enhanced educational efforts for K-14/16 and on public education (since most of the present effort was for upper level science major college and graduate students); 2) utilize in-place education mechanisms in new and innovative ways (since it was unlikely that the education system could be totally reinvented); and 3) provide a central clearinghouse on global change education programs and information (since such programs are many, varied, and difficult to locate). The support mechanisms envisioned for global change education initiatives were threefold: agency global change focused initiatives; Working Group initiatives; and portions of agency environmental education and other math and science funding that was focused on global change issues (Task Group on Education Workshop 1993).

FY '94 and '95 Activities

In support of the long-term strategy and short-term recommendations developed, the Education and Communications Working Group proposed the first national global change education initiative entitled *Project Earthlink* (Our Changing Planet 1995). *Project Earthlink* will span FY '94 through FY '96. A major public kick-off will occur on the 25th Anniversary of Earth Day, April 1995. The mission of *Project Earthlink* is to: build national capacity among educators for developing global change education programs; facilitate community access to global change information; and increase scientific and technologic literacy of targeted audiences. The message is, "Our individual and collective actions change the global environment" (Project Earthlink 1994). The scientific understanding promulgated is based on IPCC and other scientific assessments (IPCC 1992).

Project Earthlink activities are aimed at six audiences: community leaders, informal educators, teachers and students, journalists, students and leaders from other countries, and the general public. Activities to reach these groups include (but are not limited to) workshops for journalists, teachers, informal educators, and community leaders; satellite video conferences; a national conference for educators, scientists, and policy makers; demonstrations of *information super-highway* access to global change data sets; and student science fair projects on global change (Project Earthlink 1994). A variety of special materials appropriate for the different audiences are being developed. For example, an educator resource guide covering natural variability, the greenhouse effect, sea level rise, ozone depletion, ecosystem response, health effects, and decision-making under uncertainty has been developed and is awaiting printing.

Other celebratory and educational activities will occur in conjunction with the 25th Anniversary of Earth Day in April 1995. An example is the year-long educational project, *Building a New World,* that will culminate in the construction of a 45-foot diameter globe by school children from the U.S., Canada, and Mexico on the Washington D.C. Mall (Project Earthlink 1994). This will also coordinate with the National Science Foundation's Science and Technology Week for 1995.

There has been a traditional education perspective to this Working Group. Since its name change to include Communications, the Working Group has redefined its vision, mission, and goals. The proposed vision is that *people of all ages will have a broad-based understanding of global change enabling informed choices leading to a sustainable environment.* The proposed missions are: *to promote understanding of global change issues leading to informed decision making, and to promote long-term formal and life-long informal learning programs in global change education.* The Working Group will continue to *oversee agency activities, support interagency projects, encourage additional agency activity in global change education, and provide federal support, mandate, and encouragement for education toward a sustainable environment* (Education and

Communication Working Group 1994). The Working Group, and correspondingly the US federal global environmental change education efforts, will continue to evolve in programs, focus, and activities, as the issues surrounding global environmental change become better understood.

References

Atkinson R (1990) Supply and Demand for Scientists and Engineers: A National Crisis in the Making, Science 248 : 427.

Education and Communications Working Group (1994) Minutes for May 17, 1994. US Global Change Research Program, US Department of Agriculture, Global Change Office, 1621 N Kent St., Rm 61 LL, Arlington, VA 22209, USA.

IPCC (1992) Climate Change. Intergovernmental Panel on Climate Change. WMO/UNEP, Cambridge University Press.

Our Changing Planet (1990). Our Changing Planet: The FY 1991 Research Plan. The U.S. Global Change Research Program, A Report by the Committee on Earth and Environmental Sciences, Global Change Research Information Offices, 1825 K. St. NW, Suite 805, Washington DC 20006, USA.

Our Changing Planet (1992): the FY 1992 Research Plan. The U.S. Global Change Research Program, A Report by the Committee on Earth and Environmental Sciences, Global Change Research Information Offices, 1825 K. St. NW, Suite 805, Washington DC 20006, USA.

Our Changing Planet (1995) Our Changing Planet: the FY 1995 U.S. Global Change Research Program, A Report by the Committee on Environment and Natural Resources Research of the National Science and Technology Council, Global Change Research Information Offices, 1825 K. St. NW, Suite 805, Washington DC 20006, USA.

Project Earthlink (1994) Private communication.

Task Force on Education Workshop (1993). Task Group on Education Workshop Report: Framework for Long-term Planning and Short-term Program Recommendations FY '93. Subcommittee on Global Change Research, US Global Change Research Program, US Department of Agriculture, Global Change Office, 1621 N Kent St., Rm 61 LL, Arlington, VA 22209, USA.

Visiting Lecture Schedule, Spring 1991

Science Issues
Overview of Global Environmental Change: The Science and Social Science Issues
William W. Kellogg, Senior Scientist (retired), NCAR

Atmospheric Chemistry: Causes and Effects
Sherwood Rowland, University of California, Irvine

The Role of the Oceans in Global Environmental Change
Berrien Moore III, University of New Hampshire

Climate and Society
What is the Role for the Press in Global Environmental Change?
Charles Alexander, Senior Editor, Time Magazine

Population and Global Environmental Change
William Clark, Kennedy School of Government, Harvard University

Energy: Its Use and Effects
Irving Mintzer, Center for Global Change, University of Maryland

Agriculture, Forestry, and Food Security
Gary Evans, U.S. Department of Agriculture

Tropical Forests
Pedro Sanchez, North Carolina State University

Potential Health Effects of Global Environmental Change
James Fouts, National Institute of Environmental Health Sciences

Policy Options and Mechanisms
Mitigation Options and Economic Consequences
Alan Miller, Center for Global Change, University of Maryland

National Legislation and Other Options
Claudine Schneider, Congresswoman (R), Rhode Island

International Treaties and Activities
Alan Hecht, Environmental Protection Agency

SECTION 2: GLOBAL CHANGE SCIENCE COURSES

Teaching teachers about global change science: Project COPE

Richard C J Somerville
Scripps Institution of Oceanograph
University of California, San Diego
La Jolla, California 92093-0224, U.S.A.

Project COPE is the name of a unique science teacher training program which took place in 1991 and 1992 at the University of California, San Diego (UCSD). The acronym 'COPE' stands for Change on Planet Earth, and the basic concept of this novel effort was to teach school teachers about the interdisciplinary science of global change. The core of the project was a series of lectures that I gave to teachers who teach science in elementary and middle-junior high schools. The grades taught by these teachers were kindergarten through eighth grade (K-8) and the ages of their students were approximately 5 through 15 years. Project COPE was sponsored by the U.S. Department of Education and administered through UCSD Extension, a part of my university which specializes in outreach activities and which offers many kinds of both conventional and non-traditional instruction for a wide variety of students.

The teachers came to the UCSD campus for one Saturday each month during 1991, except during July and August. On each of the ten Saturday mornings, I lectured to them for 3 hours, while afternoons were devoted to laboratory sessions arranged by grade level, and designed and run by science teaching experts. The laboratory sessions provided the teachers with a series of experiments, demonstrations, and other activities, which could be used in their classrooms. The lectures were intended to inform and stimulate the teachers, rather than to provide them with specific material which they could directly use with their own students. The lectures covered some of the most prominent global problems of contemporary earth. Each science, including the ozone hole, the greenhouse effect, acid precipitation, air pollution, and the loss of biodiversity. These topics are interesting and important to a broad spectrum of people today, and they are also active areas of scientific research.

The essential purpose of Project COPE was to augment the content of the curriculum and enhance teachers' instructional strategies. The idea behind the project was very simple: that earth science is alive and exciting today, and that informing teachers about current research on topics of great public interest

NATO ASI Series, Vol I 29
Global Environmental Change in Science
Education and Training
Edited by David Waddington

would help them convey the fascination and power of science to their students. I am convinced that the effort to increase science literacy in the general population must begin while children are very young. Science is not dull, and it need not be perceived as difficult. There is a great deal of evidence that many young children acquire a strongly negative impression about science in the early stages of their education, in part because their teachers are often inadequately equipped by their own training in science and mathematics. Project COPE was an effort to light one small candle, rather than curse that particular darkness in our education system.

How did some 60 kindergarten through eighth grade teachers come to assemble once a month on the UCSD campus to hear a university professor talk about global change? This unusual sounding project actually evolved quite naturally. UCSD Extension is a highly entrepreneurial part of the university, and it has a long history of organizing highly successful short courses for teachers. I had taught for several years in a series of summer science teacher training institutes offered through UCSD Extension. These institutes were largely inspired and organized by Paul Saltman, a biologist who is a colleague of mine on the UCSD faculty and a master teacher himself. The concept of a more elaborate science teacher training program, based on current scientific research on earth science topics of widespread general interest, appealed greatly to me. In a series of discussions in 1989, Dr. P.A. Moore, an education specialist with UCSD Extension, worked with me to plan an ambitious effort, partly modeled on these summer institutes, but stretching over a full two years and focussing on global change science. She found a possible source of financial support in the U.S. Department of Education, and together with Professor Saltman, we submitted our proposal through UCSD Extension in February 1990. It was approved in June of that year.

In the next few hectic months, we planned all the details of the program and advertised it widely in the school systems of San Diego and Imperial counties, which are two large counties in southern California near the Mexican border. Many teachers responded, and it was difficult to narrow the field to about 60. This limit on the number of participants was due mainly to the amount of money in the grant awarded by the Department of Education. Those funds were used mainly to buy materials and to pay the staff who administered the program and ran the laboratory sessions.

In teaching in the COPE Project, I quickly developed a deep respect for the teachers who were my students. They were all accomplished professionals in their own field, and they brought a rich spectrum of real world classroom experience to our program. Their energy and dedication were impressive. The great majority of them, however, were not especially well informed about many aspects of science, despite their enthusiasm for it. Nearly all of them also shared a general distaste for mathematics. In some of them, a real fear of mathematics was sadly apparent. I must also say that, taken as a group, these teachers were well aware that their own background in science and mathematics was considerably weaker than they would have preferred. This was not the fault of the teachers, After all, they were themselves products of the American system of higher education. It is a sad fact that this system produces university graduates who, more often than not, are literally unable to give a more-or-less correct answer to the question, "Why is summer warmer than winter?" As F. Sherwood Rowland has noted (Rowland, 1993), the preferred wrong answer is that the earth is closer to the sun in summer. I have given this basic science literacy question as a litmus test to UCSD undergraduate students, and I tried it on the Project COPE participants as well. The result is always the same: most people just do not know. If anyone doubts the need for improved science education, that fact should help make the case.

We met on campus on the second Saturday of every month that year, except for July and August. During the second year of the project, there were no formal lectures. There were more field trips, however, and emphasis was placed on the role of the participating teachers in staff development within their schools and districts.

Project COPE was intended to enhance the earth science knowledge base and teaching strategies of K-8 educators and to assist teachers in conducting in-service programs on contemporary earth science issues in their schools and school districts. The focus of Project COPE was to trace the scientific investigation of patterns of change in the global environment.

To increase the likelihood that school teachers would derive practical benefit from project COPE, the instructional content and laboratory activities of the project were based on the Science Framework for California Public Schools, also known as the California Science Framework, or simply the Framework. This document, adopted by the state in 1989, helps to unify systematically many aspects of teaching science. The Framework "calls for a new dynamic in science learning." Project COPE was consciously and explicitly designed to be consistent with this framework.

The Framework is detailed in a 220 page book bristling with pedagogical detail of interest primarily to professional educators. One feature strongly emphasized in the Framework seemed particularly appropriate as a guide to the Project COPE lectures. This feature was called "The Major Themes of Science."

In the words of the Framework, "Themes are the big ideas of science, larger than facts and concepts: they link the theoretical structures of the various scientific disciplines. Themes are a way to integrate the overarching concepts of science into a curriculum, much as theories encompass and connect the basic data and evidence of science."

The Framework recognizes six major' themes:

- energy
- evolution
- patterns of change
- scale and structure
- stability
- systems and interaction

Quoting from the Framework again, "The use of conceptual themes will not by itself solve all the problems of science education. However. the incorporation of themes in thoughtful and judicious ways should improve the integration of facts and ideas, the interrelationships of theories and disciplines, and the quality of instructional materials." These themes proved to be very helpful in selecting and organizing the material for the lectures, and in allowing the teachers to relate global change topics to the other subjects they taught.

Teachers who participated in Project COPE received five quarter units of UCSD credit, a stipend to help cover their expenses, and a considerable amount of documentation. We relied especially on *The Next One Hundred Years* (Weiner 1990), which became the unofficial textbook for Project COPE, and on two special issues of *Scientific American* - "Managing Planet Earth" (September 1989) and "Energy for Planet Earth" (September 1990). Jonathan Weiner's book is an excellent popular account of many aspects of global change science, and it is especially good at conveying a picture of research as an exciting activity carried on by people who are often interesting human beings! Weiner paints an engaging picture of many of the scientists who have pioneered this field, such as Charles David Keeling, whose careful measurements of atmospheric carbon dioxide concentration have rightfully become world famous as a foundation for the realization that human activities can affect our planet on a global scale.

The topics covered in the COPE lectures were varied. We covered two in considerable detail: the depletion of stratospheric ozone (dramatized by the ozone hole over Antarctica), and the potential for global climate change due to human activities, particularly by an increase in the greenhouse effect. Through these two topics, we explored many aspects of the climate system that are critical to determining the state of the environment of our planet. We then moved on to briefly cover several other topics which included acid precipitation, the global aspects of air pollution, and, finally, threats to biodiversity.

The lectures were taped, and the transcribed tapes served as the raw material for a book which I wrote after the program ended. This book *Change on Planet Earth,* was printed in 1993 at UCSD using desktop publishing technology and was distributed to the participants in the program. The book is simply a reworking of about 30 hours of my lectures. It is in no sense a complete account of Project COPE. It contains nothing of the laboratory sessions or the field activities, and it omits entirely any account of the work the participants did in their own classrooms or with their colleagues in their schools. In that respect, it leaves out evidence for how well Project COPE succeeded in its primary goal of helping teachers to improve science teaching.

Even as a record of the lectures, this book is incomplete. The transcripts of the tape recordings managed to capture most, but not all, of what I actually said. Then a skilled and patient editor, Phil Lauder, worked closely with me to translate the rhythms and conventions of spoken English into something approximating readable text. In this painstaking job, we tried to improve organization and clarity while retaining something of the informality of the lectures.

However, some special aspects of the lectures could not be fitted into this format. One was the large collection of slides, overhead transparencies, and handouts that were part of every lecture. To help compensate for this omission, I added an extensive annotated bibliography to the book to help guide readers into the rich literature of this subject.

I also decided to use the Project COPE experience as the basis for a new book on global change science aimed at the general reader. For this purpose, over a hundred copies of *Change on Planet Earth* were distributed to scientists and educators for review, many of whom sent back extensive comments and suggestions. These critiques are being incorporated in the new book. The title is taken from a poem by Elizabeth Bishop which is especially relevant to global change, *The Forgiving Air*.

Personally, I greatly enjoyed the experience of Project COPE. For me, the most rewarding aspect of the program was the interaction with the COPE participants. Self-selected and highly motivated, they enthusiastically asked questions, challenged one another and me, held spirited discussions before and after our sessions (and sometimes during them), and brought back valuable lessons from their classroom experiences. Their energy and commitment to their profession were impressive. Project COPE was an immensely rewarding experience to me, because the participants made it so.

It was a somewhat difficult experience, too, because at the outset of the course the participants established two rules for me to follow. I was told to avoid scrupulously the technical jargon that is the natural dialect of a working scientist, and I was warned not to employ any mathematics whatsoever. Forcing a scientist to lecture under these restrictions is a little like asking a cat to bark. I managed to break both of these rules. Technical terms arrived in our classroom uninvited and were introduced politely, one at a time, with proper definitions, and the participants soon became adept in the jargon in spite of their initial fear of it. Then, somehow, after a lot of numbers, one little algebraic question managed to sneak into the lectures, and I proudly included it in *Change on Planet Earth*.

The equation, I=PAT, is due to Paul Ehrlich and his collaborators, who use it as a simple model to illustrate the fact that the impact (I) of human beings on the environment is related to the size of the population (P), the affluence (A) of the people, and the technology (T) which they use. Obviously, it is a highly idealized description of reality, but it is pedagogically useful in many ways. Nevertheless, a substantial number of the COPE participants found this equation an obstacle rather than an aid to understanding. This aversion to even the simplest mathematics shocked me, and I still wonder how well a person can appreciate, say, a computer simulation of the circulation of the global atmosphere and ocean, if that same person is truly perplexed by I=PAT. Nevertheless, I may leave the equation out of *The Forgiving Air*. In his best-seller, *A Brief History of Time*, Stephen Hawking remarks that he had been advised that every equation he put in would halve the sales of the book, and that may well be so. Nevertheless, Hawking did dare to include one equation, $E=mc^2$, and I take that as a good omen for the future of science literacy.

References

Rowland FS (1993) President's Lecture: The Need for Scientific Communication with the Public. Science 260: 1571-1576

Somerville RCJ (1995) The Forgiving Air. University of California Press

Weiner J (1990) The Next One Hundred Years. Bantam Books, New York

What is Environmental Chemistry?

J Buffle and M Filella
Department of Inorganic, Analytical and Applied Chemistry
University of Geneva
Quai Ernest Ansermet 30
CH-1211 Geneva 4, Switzerland

> If you ask a chemist to find for you what a dynamo is, the first thing he would do is to dissolve it in hydrochloric acid *(Albert Szent-Gyorgyi, 1893-1986; biochemist, Nobel prize for Medicine 1937).*

Introduction

Not all scientific disciplines have reached the same stage in their development. Environmental chemistry today parallels the emergence in the nineteenth century of chemistry as a rational discipline and even more closely that of biochemistry (formerly called biological chemistry) at the beginning of this century. Achievement of scientific knowledge includes several steps: collection and classification of data, interpretation and generalization, and formulation of new, more rigorous, and quantitative concepts. Both chemistry and biochemistry have passed through long descriptive periods in which important discoveries led to the deduction of the properties of elements and compounds. Following this came a quantitative formulation of modern concepts which allowed predictions to be made on compound reactivity. In chemistry, this descriptive phase has probably reached its pinnacle, but in biochemistry there is still a great deal to be done. This lag is due to the greater complexity of the biochemical systems studied, particularly those of supramolecular organized structures.

Environmental chemistry deals with even more complicated systems than biochemistry, and it is a much younger discipline. Therefore, despite the development of modern sophisticated analytical techniques, it is still in its infancy, i.e., at the experimental data collection and classification stages. For instance consider sediments, soil aggregates, or humic compounds which are key environmental components, but whose rigourous chemical description is extremely difficult. This enormous difficulty does not mean that rigorous description of environmental systems is impossible. Biochemistry has succeeded in describing in detail the primary to quaternary structures of proteins, polysaccharides, or DNA, and even more complicated structures like biological

NATO ASI Series, Vol I 29
Global Environmental Change in Science
Education and Training
Edited by David Waddington

membranes, even though such achievements were considered to be impossible not so very long ago. Undoubtedly, environmental chemistry will succeed in describing the components and structures that play key roles in environmental processes, but to achieve this, as much time and effort as have been put into other disciplines will be required, and new sophisticated technologies will need to be developed to overcome the problems posed by the enormous complexity of environmental systems.

At present, environmental chemistry faces a peculiar and difficult situation. As mentioned above, classical chemistry (i.e., chemistry of small molecules of molecular weight < 1000 Dalton) has largely overcome its descriptive phase so that most chemists are trained to think in terms of predictive quantitative concepts, while the description of environmental systems and their supramolecular structures still needs semi-empirical approaches. It is therefore understandable that some classical chemists look with scepticism and sometimes with disdain on the work of environmental chemists. They should realize however that the relatively high predictive level of classical chemistry is limited to homogeneous systems that involve a restricted number of small molecules. This is at odds with most situations of the real world (industrial, life, and environmental processes), which include physically and/or chemically heterogeneous systems, and where supramolecular structures play an important role.

Like biochemistry, environmental chemistry should rather be considered by classical chemists as a challenge and an opportunity to make further progress by applying existing concepts and developing new ones for real systems.

The purpose of this paper is to discuss the specificity of environmental chemistry as compared to other environmental sciences and to classical chemistry.

The place of environmental chemistry in environmental sciences

Not only are environmental sciences in their infancy, but during the last decade their evolution has shown a leaning towards empirical studies. Indeed, because of the urgency to solve pollution problems, most administrations and many research foundations favour the funding of strategic, short-term, and therefore empirical studies. There is no doubt that urgent problems must be solved, but the authorities should not forget that for the long-term, basic research results in the most useful and the biggest pay-off. But for the results of long-term fundamental studies in biochemistry and microbiology, it would not now be possible to combat tuberculosis, diabetes, cancer, and many other illnesses. By favouring short-term empirical studies in the environment, thorough-minded scientists are discouraged from entering this field, thus hampering environmental science from developing into a sound science. This may have dramatic consequences for future generations which will be confronted by more serious pollution problems than ours, but which will have no rigorous scientific discipline to solve them.

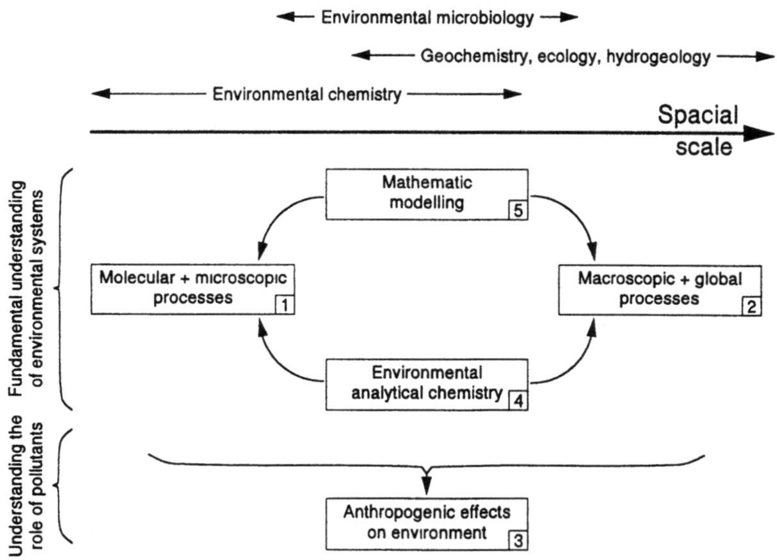

Figure 1. Different aspects covered by environmental natural sciences

Environmental chemistry should not be confused with pollutant analysis and control. On the contrary, environmental chemistry is a multifaceted discipline. A simplified but useful representation (Fig. 1) distinguishes among three aspects of all environmental natural sciences: the fundamental understanding of the functioning of unperturbed environmental systems (boxes 1 and 2), and the understanding of anthropogenic perturbations on these systems (box 3). It is essential to make the distinction between aspects 1 and 2 on the one hand and 3 on the other: anthropogenic perturbations cannot be understood and controlled without a preliminary understanding of the behaviour of unperturbed ecosystems (see below the analogy with medicine, Fig. 2). It is also useful to discriminate between environmental "microprocesses" (occurring on small scales of nm to mm) and "macroprocesses" (large scales of mm to km) corresponding to whole ecosystems. Although the border between these two scale ranges is arbitrary, the nature of the corresponding processes, the techniques used to study them, and, consequently, the corresponding conceptual approaches are very different. Teaching environmental sciences should cover all aspects of Figure 1, linking each particular discipline more closely to one or other of the aspects shown.

As far as *environmental processes* are concerned, environmental chemistry is closer to microprocess studies (Figure 1). Figure 1 shows that such studies should be done in close association with microbiology and geochemistry. In order to understand chemical microprocesses however, environmental chemistry also has to develop *analytical chemical methods* (box 4) needed for such studies. These methods are required not only by chemists, but also by many other environmental scientists since chemistry influences many geological and biological processes. Analytical chemistry has thus a key role to play in almost all aspects of Figure 1. Because of this, the general idea has developed in the scientific community that the only role of environmental chemistry is to develop chemical analysis methods for environmental quality control. Although unquestionably environmental analytical chemistry is a crucial issue for all environmental scientists, environmental chemistry should not be confined to chemical analysis. The parallel development of analytical chemistry and process studies is as vital for solving environmental problems as for combating illness in the medical domain. Figure 2 shows the obvious analogy between environmental chemistry (study of processes), monitoring, and management on the one hand, and biochemistry, medical analysis, and health control, on the other.

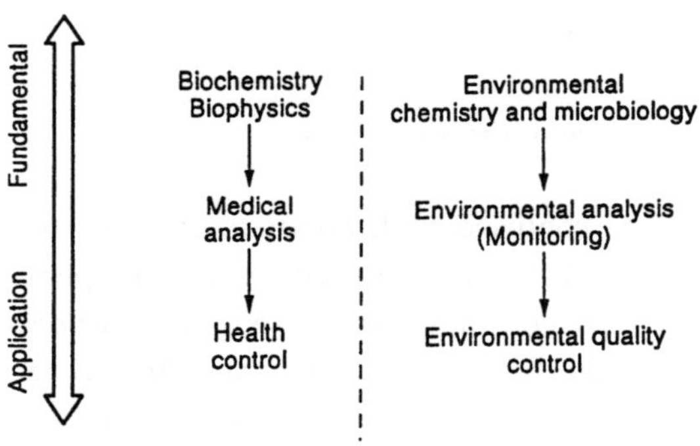

Figure 2. Analogy between environmental and medical sciences.

Specificity of environmental chemistry as a chemical discipline

The aspects related to environmental analytical chemistry will not be discussed here. Analytical chemistry - development of new analytical methods based on new physico-chemical principles - is a well-established branch of chemistry. Let us just say that, as already discussed, environmental sciences greatly need new methods applicable to diverse environmental systems.

As far as physico-chemical processes are concerned, work conducted over the last 20-30 years by Stumm, Schindler, and others has clearly shown that global geochemical cycles of elements are ultimately controlled by the reactivity and physico-chemical properties of compounds at the molecular level, and that, without a good knowledge of the behaviour of molecules of chemical compounds, no prediction will be possible at the global level. For instance, the cycling of phosphorus in lakes (with possible concomitant lake eutrophication) depends on many physico-chemical mechanisms, such as transfer of phosphate through biological membranes; oxidation of organophosphorus compounds; phosphate adsorption on iron hydroxide; nucleation of iron phosphate; oxidation of iron(II) to iron(III); calcium phosphate crystal formation, growth, and dissolution, which itself may depend on adsorption of surface active organic compounds and so on. Each of these steps is intricately related to other physico-chemical processes in a manner analogous to the metabolic cycles in biochemistry (e.g., Krebs cycle for glucide metabolism in organisms). The primary role of environmental chemistry is to elucidate the most important of these steps, in particular those which play a rate-limiting role in the overall cycle of the element of interest. In contrast to the usual systems studied in classical chemistry, the systems in environmental chemistry have the striking feature that many properties of the systems are closely related to their chemical and physical heterogeneity. The study of physico-chemical processes therefore requires using new conceptual approaches that combine classical chemistry concepts with new ones, in particular reactivity probability concepts and supramolecular structural physical models. The main differences between environmental chemistry and classical chemistry approaches are discussed below.

a. The *chemical heterogeneity* of environmental systems is a first major difficulty for classical chemists who are usually trained to think in terms of reactivity of individual molecules. The classical approach for studying the complex environmental mixtures, regardless of their complexity, consists of characterizing all the components of the mixture. Even if one casts aside the enormous efforts required to solve this problem in this manner, the solution to the questions posed are not often obtained. Indeed, this approach ignores that heterogeneity by itself is a property of the environmental systems of interest. Fulvic and humic compounds of soil and waters provide a good example of such a case. Even after "purification", fulvic and humic samples are composed of a large number of similar but not identical compounds. As an ensemble, they act as a complexing system for metal ions (and small organic molecules), with buffering capacities spanning an extremely wide concentration range of metal ion. As such, they play a key role in maintaining acceptable concentrations of vital and toxic metal ions for organisms in soils and waters. Alone, none of the individual species will impart this property. Hence, they should be thought of as a single chemical system, rather than as a mixture of individual species. This buffering action due to chemical diversity (or chemical heterogeneity) can be seen as being equivalent to the increased stability of ecological niches when the number of biological species increase (biodiversity). It has been shown that the overall properties of heterogeneous chemical systems may be rigourously related to the properties of individual molecules by introducing the concept of reaction probability. Descriptions of such chemical systems should then be based on rigourous statistical concepts, applicable not only to a large number of identical molecules (as is usually done in classical thermodynamics), but also to a system consisting of a large number of similar but not identical molecules.

b. It is now well established that the most important reactions in environmental systems occur in solid-liquid (soil, sediment, water) and gas-liquid (soil, ocean, fog droplet surface) *interfaces*. The electric field and solvent structure are very different at these interfaces (over a thickness of tens of nm) than in the bulk solution and thus may drastically modify the nature and rates of chemical reactions. A first glimpse at many chemistry curricula for chemistry students shows that surface reactions (adsorption isotherm, interface catalysis, etc.) are barely (or not at all) taught.

c. The previously mentioned interfaces are the simplest examples of *environmental microstructures*. All environmental systems are physical structures at all levels, from nm to km range. Examples of large structures are soil horizons; lake, ocean, or atmosphere stratification; and geological layers. Microstructures are those with which environmental chemistry is more concerned. Aggregates between large polysaccharide macromolecules, clays and microorganisms in soils, water, and sediments are examples of microstructures in the range nm to mm. They play a key role in the circulation of oxygen, water, and nutrients, and therefore in the maintenance of life. Sediment-water interfaces in lakes and oceans are other examples. They are stratified at the μm to mm level by successive layers of varying chemical composition, which controls the overall exchanges of chemicals between the whole of the water body and the sediment. Similarly, the gas exchanges between the ocean and the atmosphere are predominantly controlled by a diffusion layer of 30-80 μm whose chemical composition is very different from the bulk solution. In all cases, the functioning of living environmental macro and micro systems is very much independent of their structure (see d). Indeed, living systems (organisms or ecosystems) need pH, redox, or other chemical gradients which can only develop with physical heterogeneity. A good characterization of environmental microstructures is therefore a key task of environmental chemists (as characterization of cell structure is a key aspect in biochemistry). This implies knowledge of colloid chemistry, macromolecule physical chemistry, and basic biophysical concepts. This is unfortunately at odds with the classical training of chemists which is usually restricted to homogeneous solution chemistry (see the introductory quotation by Szent-Györgyi).

d. The physical structures existing for all scales of environmental systems and the corresponding chemical gradients imply the existence of *fluxes of compounds* within the system of interest (e.g., soil aggregate, sediment-water interface) or between one system and the neighbouring systems. Environmental systems (micro and macrosystems) are therefore dynamic systems. Thus, understanding environmental processes necessarily implies coupling kinetics of chemical reactions with physical transport processes, which often occur by molecular diffusion in microsystems. This is an additional lacuna in the training of most chemistry students. Not only is the coupling of chemical and diffusion equations rarely taught, but often more emphasis is laid on chemical equilibria as compared to chemical kinetics.

It is essential to realize that properties (a) to (d) above, in combination with classical chemical reactions, are the key factors which assure the *homeostasis* of environmental systems. Therefore, they also correspond to sensitive properties of environmental systems, which may be particularly affected by anthropogenic activities. These concepts should be taught in detail to all environmental scientists.

Teaching and research in environmental chemistry

Role of modelling

In recent years, the use of numerical simulation models as a means to evaluate large scale or complex natural processes has increased sharply. In some cases, the predictions generated by these models are considered as a basis for public policy decisions (e.g., global circulation models to predict the earth's climate). There is an over-emphasis on the predictive value of modelling, often neglecting the fact that all models are oversimplified compared to natural processes, many natural processes are still unknown, and, above all, that the primary value of models is heuristic. On the other hand, it is unquestionable that models have enormous importance as educational tools. They allow in particular i) classification of information, ii) checking working hypotheses with experimental data, iii) formulation of new hypotheses by eliminating unimportant factors, and iv) sensitivity analysis.

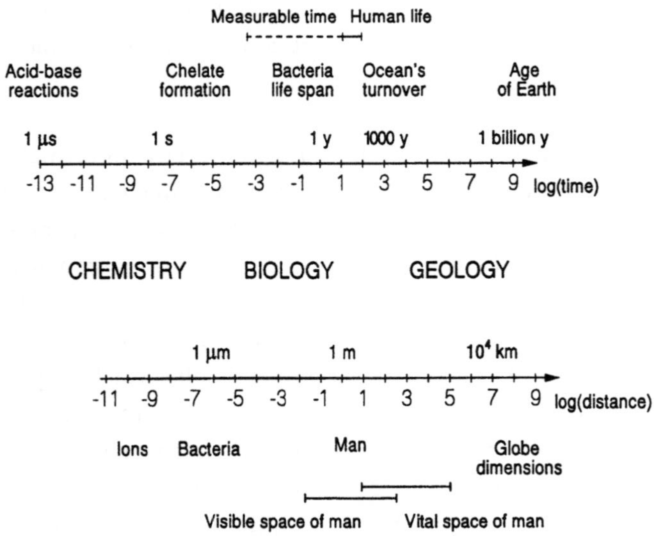

Figure 3 Time and space domains covered by environmental sciences

These aforementioned considerations apply to all disciplines. But as a learning tool, modelling is particularly important in environmental chemistry. Probably it is the only tool available for quick evaluation of the influence of factors affecting molecular or microscopic processes in the macroscopic evolution of ecosystems. As explained later, the time and space domains of chemical processes are much farther away from those occurring on the macroscopic or global scale (Figure 3). It is thus extremely difficult, both for students and researchers, to assess the actual importance of chemical processes on ecosystem evolution, and in this respect modelling is an extremely useful tool for linking microscopic and macroscopic processes.

Experimental work

As mentioned earlier, expertise in any scientific discipline can be acquired only by working experimentally in that field for at least some period of time. This is even more important in the environmental sciences because of the complexity of environmental systems and the multidisciplinarity of the problems involved. As far as environmental chemistry is concerned, the student will have to deal directly or indirectly with three experimental aspects:

- Study of physico-chemical processes *in situ* (in the field): planning and performing well-controlled experiments in the field is much more difficult than in the laboratory, because of the large number of factors to be considered and controlled in order to allow sound interpretation of data.

- Study of physico-chemical processes in the laboratory with model compounds simulating natural systems: studies on natural systems allow one to discriminate between predominant and secondary factors affecting a particular macroscopic process. However, results of *in situ* measurements cannot provide precise values of physico-chemical parameters. These can only be determined in the laboratory by using model experimental systems that mimic natural ones.

- Development in the laboratory and application *in situ* of analytical methods specifically designed for studying environmental processes: measuring the right parameters is a key issue in order to understand natural processes. This is far from trivial in complicated natural systems.

Teaching environmental concepts

The previous section discusses the basic physico-chemical concepts which are currently missing in many chemistry curricula and are a prerequisite for understanding environmental chemical processes. Teachers in the classroom will be faced with two additional difficulties of a general type - other than those specifically related to the principles - while teaching these concepts to the students. The first difficulty is related to the time and space scales covered by the environmental sciences (Figure 3). It is not always easy for students to grasp the idea of how chemical processes occurring at the nm to mm level can affect ecosystem evolutions over centuries or even geological times. Another problem is our "sensitive" time window. Every-one understands the meaning of time flow, if the time ranges from fraction of a second to millenia. Scales such as these are met in environment macroprocesses (often biological, geological, and physical processes), where students do not have problems visualizing them. However, the time-space windows in physico-chemical processes are quite different from the latter ones, and it is difficult to help students understand their relevance in environmental chemistry (i.e., the link between chemistry and environmental processes). The situation becomes worse if one compares the "sensitive" space window of man (fraction of nm to tens of km) to the size of plankton (10^{-3} - 10^{-6} m), colloidal particles (10^{-8} - 10^{-6} m), or simple molecules (10^{-9} m). These entities differ in size by orders of magnitude, and therefore behave in a completely different manner. The importance of these differences is not obvious to the students who are new to these concepts.

A second difficulty of environmental chemistry is that many different concepts must be integrated in order to understand the behaviour of any particular environmental system. The principles of the various individual chemical reactions (acid-base, redox, complexation, precipitation, hydrolysis, etc.), that may occur in this system, are often easy to understand. It is, however, much more difficult for students to decide which processes to ignore and which ones to consider, i.e., which ones play a major role in the evolution of the environmental system of interest, in particular when non-chemical processes have also to be taken into account. In most chemistry courses, students learn to understand single processes in depth, but they are not trained to think in terms of combining many processes together. At the general teaching level this may be one of the most important impacts of environmental chemistry on chemistry students. Even if they do not pursue their studies in the environmental sciences, learning how to combine quantitatively many different factors affecting a given process will be very fruitful for any real world application.

Conclusion

The learning of a science involves more than the acquisition of new knowledge by a blank mind. Instead it involves a substantial restructuring of pre-existing knowledge, in which new knowledge must compete with students' previous knowledge and familiar ways of thinking. This is particularly true for chemistry students learning environmental chemistry and environmental sciences.

It should also be stated clearly to students that any curriculum of one to three years in environmental chemistry or environmental sciences is inadequate to make them specialists of environmental processes. The basic requirement for a medical degree is five years, and specialisation in a particular field of medicine is required: that is, additional years must be spent before one can claim to be a specialist. Environmental systems are certainly at least as complicated as the human body. There is therefore no hope of producing environmental scientists with expertise within 1-3 years. The only thing we can do is to give students the basic concepts necessary to learn so that they can progress further by themselves.

CHEMISTRY, RESOURCES AND THE ENVIRONMENT
A course for chemists interested in environmental issues including global change phenomena

C Anastasi
Department of Chemistry
University of York
Heslington
York YO1 5DD, UK

Introduction

The interdisciplinary nature of Global Change Science brings problems and challenges to those developing courses in this area. There is a strong feeling that students should be exposed to the interaction between disciplines as much as possible so that a global perspective is gained. As noted elsewhere in this book, the absence of a department in an institution dedicated to Global Change Science usually means the relevant expertise required for such a course is in institutions having departments whose focus and application of resources lie in different directions. A way forward to course development is the "staple gun" approach with its inherent strengths and weaknesses, examples of which are presented elsewhere in this book, (Schneider 1994). In many ways it is a revolutionary approach to the problem of delivering Global Change courses.

An alternative, more modest way forward for institutions that wish to educate students in Global Change issues, but whose resources are limited, is the subject of this article. It essentially champions an evolutionary approach to teaching the subject, and builds on the knowledge of the two most important groups of people: the students and the academic staff. For the students, the material presented builds on a firm and expanding knowledge of their chosen major discipline: chemistry, physics, biology and others. For the academics, Global Change issues are discussed within the context of their own expertise. The common ground for both students and academics lies in the major discipline.

The course in the Department of Chemistry at the University of York is an example of this model. In describing the course in detail, the strengths and weaknesses of the approach will be highlighted. It is a course that has been run successfully for about fifteen years and has been described as an environment course. What will become obvious is that it covers many of the elements needed to educate students in Global Change issues.

NATO ASI Series, Vol I 29
Global Environmental Change in Science
Education and Training
Edited by David Waddington

Key aims of the course and departmental strategy

There are three key aims of the course:

- To illustrate the challenges facing society as a result of the increasing demand for natural resources
- To show the essential role of chemistry in meeting these challenges
- To highlight the impacts of human activities on the global environment

Additionally, the course material lends itself readily to the development of other, transferable skills including communication (both written and oral), team work, problem solving and development of analytical skills beyond those found in the major disciplines.

The strategy adopted by the Department was to harness the in-house expertise available and to create a course that is concerned with *illustrating* environmental principles rather than one that seeks to cover *all aspects* of the subject. This immediately removes the problems associated with inter-departmental barriers, both technical and behavioural, that have been highlighted elsewhere in this book. Also, the resources needed are considerably reduced and the integration of material is much easier. However, the course presented is invariably less broad than may be ideal, and the student body the course services has a narrower background than might be desired.

Course Framework which Incorporates Environmental Option

The Department of Chemistry at York provides a three year degree dedicated to students wishing to specialise in chemistry and related subjects. The environmental option is taught side by side with the larger chemistry course. The objectives of the former are:

- The application of 'core' chemistry in an environmental context
- The development of environmental ideas within a secure and continuing chemistry framework
- An increasing emphasis on environmental content as the course progresses

The table below summarises the course structure. All students take the same 'core' chemistry in the first year; this part of the course provides students from different schools and curricula with the same knowledge base and prepares them for more advanced material in the second and third years.

Part I	Year 1	All students take the same (core) chemistry course.	
Part II	Year 2	Two-thirds of the course (the core) is the same for everyone.	One-third consists of two optional units.
	Year 3	One-third of the course is core chemistry.	Two-thirds consists of four optional units, a literature review and research project.

In the second year, two-thirds of the course is the same for everyone and concentrates on core chemistry and one-third of the course is made up of optional units chosen by students to suit their interests. In the third and final year of the degree programme, only one third of the course is core chemistry common to all students, while two thirds of the year is spent on optional units. During their last year students also carry out Literature Reviews and Research Projects, which are usually related to the areas covered in their optional units.

Students find the course attractive because essentially they are able to create one-third of their course through their choice of a portfolio of optional units. Some students wish to concentrate on mainstream chemistry, and so choose the appropriate optional units that allow them do this. Other students wish to study environment, industry or life science issues, and so choose the optional units that form the respective courses: *Chemistry Resources and the Environment* (described in detail below): *Chemistry, Management and Industry:* and *Chemistry, Life Sciences and Pharmaceuticals.* There is also an opportunity for students to take a selection of units from each of the course options. In this way it is possible to raise the awareness in environmental issues of a large fraction of the undergraduate school..

Environment for All

As indicated above, the unit options are available to students in their second and third year. However, there is an attempt to increase student awareness about environmental issues through workshop activities in both the first and second years. The activities are diverse in nature and are aimed at developing different skills.

An example here is the annual *Happening*. This is a workshop in three parts which involves all of the first year school. Small groups of 3 or 4 students are formed, and each group then formulates responses to specific technical questions. The first part of the workshop aims to educate students about the industrial production of chlorofluorocarbons, with particular emphasis on why these compounds were developed in the first place, and highlighting their strengths and weaknesses. In the second part of the workshop the fate of chlorofluorocarbons in the atmosphere is discussed so that students are then able to understand the impact of these compounds on stratospheric ozone and global climate change. The students are then sufficiently well informed to be able to contribute to the third part of the workshop, a debate in which two student teams present the case for and against a motion such as: *"This house believes that all man-made halogen containing organic compounds should be banned"*.

This activity allows students the freedom to learn from their tutors and their peers within relatively secure technical boundaries, and in a generally informal environment. The debate provides the vehicle for using the knowledge gained in a decision making way. It also allows students, probably for the first time, an opportunity to explore the interaction between the certainties of science, and the uncertainties of emotion often engendered by environmental issues. In a broader context the workshop is an excellent vehicle for students to discuss the fundamental question of whether technology is able to provide solutions or 'fixes' to environmental issues, an aspect that is of increasing importance in Global Change Science.

The Environmental Option: 'Chemistry, Resources and the Environment

As indicated earlier, this option begins with two units in the second year, followed by four more units in the third year. The four units that form the basic environmental option along with the main aspects covered in each unit are given below:

- *Earth Sciences:* Geosphere; Rocks; Hydrosphere; Biogeochemistry
- *Energy Studies:* Energy in the environment; Fossil fuels; Electricity generation; Renewable energy sources
- *Land and Water:* Water resources; Crop production; Fuels from biomass
- *Chemistry in the Atmosphere:* Chemistry in the atmosphere; Measurement and control of air pollutants; Atmospheric composition and climate

It is clear from this description that many of the fundamental components of Global Change Science, as defined elsewhere in this book, are covered in these units, and in an integrated manner.

Two other units have a significant content relevant to those taking the environment option and complement the material given above:

- *Raw Materials, Processes and Products:* Resource utilisation; Carbon based minerals; Mineral economics
- *Managerial Economics:* Market economics, Project appraisal and risk analysis; Location of industry and the environment

In these last two units the students must rely and expand on their knowledge of chemistry as they are exposed to issues that are largely the concern of industry, yet also impact on the environment. This allows students to gain a more complete insight into the driving forces associated with the consumption of resources and their impact in global change issues.

Students with a keen interest in environmental issues will tend to take all of the optional units. Other students may take a smaller selection, and although these students have a less broad perspective of issues and lose a little in terms of the integration, they at least have been exposed to some aspects of the subject.

A breakdown of one of the units, *Chemistry in the Atmosphere,* is shown overleaf a nd serves to illustrate the breadth and depth of the material involved. The description also gives an indication of the kinds of activities the student is exposed to that enhance other skills, both analytical and behavioural.

The lectures are approximately 1 hour long. Each tutorial involves 9 hours work of which 8 hours are for completion of set work and private study, and 1 hour is spent in small group discussion with a tutor. The workshops have a time allocation of 16 hours, again spent in small groups. An example of the kinds of workshop activities is given elsewhere in this book (*The Truth About Catalytic Convertors*. A typical unit then, has at least 70 hours of formal study associated with it.

CHEMISTRY IN THE ATMOSPHERE UNIT

The chemistry of our atmosphere is an important environmental issue that has received considerable attention in recent years. In this unit the interplay between the structure of the atmosphere and the chemistry is explored, and the role of laboratory studies, ambient measurements, pollutant fluxes and modelling are discussed. Control of air pollution through legislation, and the costs involved, are also important issues addressed in this unit.
The Greenhouse Effect, Acid Rain and Photochemical Air Pollution receive individual attention and are used to investigate possible approaches to air pollution in the future.

MAJOR AREAS COVERED:

Chemistry in the Atmosphere:

- Structure in the atmosphere
- Stratospheric and tropospheric chemistry
- Laboratory measurements of gas phase processes
- Liquid and solid phase chemistry in the atmosphere

Measurement and Control of Air Pollutants:

- Inventories of air pollutants
- Measurements in the atmosphere
- Modelling of air pollutants
- Pollution abatement strategies
- Legislation and policy
- Economics of air pollution

Atmospheric Composition and Climate:

- Radiation, atmospheric composition and temperature
- Climate change
- Planetary atmospheres

LECTURES: 27 over three 9 week terms (this includes some lectures on specialist subject matter given by experts from government research institutes)

TUTORIALS: 1 per term

PRACTICAL WORK: A selection of workshop activities dealing with paper comprehension, problem solving and data analysis

SITE VISIT: Visit to an urban air pollution monitoring site

As indicated earlier, an important feature of this general approach to Global Science Education from the students point of view is that they are able to use aspects of their core chemistry to support and build on in the environmental option. In the unit *Chemistry in the Atmosphere* summarised above, links to the 'core' chemistry include:

- Gas phase chemical kinetics including reaction rates, thermochemistry and mechanistic chemistry
- Liquid and heterogeneous phase chemistry
- Spectroscopy and analytical methods

There are obvious links in this unit to global change issues and examples include:

- Stratospheric ozone
- The emission, chemical transformation and measurement of trace greenhouse gases in the atmosphere
- Trace gas contribution to radiation balance in the atmosphere

The assessment of the units is in two parts. The first contribution is a continuous assessment based on the various workshop activities, including *The Truth About Catalytic Convertors. The* major assessment is through the a formal examination of the unit at the end of the course. Examples of the kinds of questions set in the climate change area for students taking the *Chemistry in the Atmosphere* unit are given below.

SAMPLE EXAMINATION QUESTION FOR THE "CHEMISTRY IN THE ATMOSPHERE" UNIT

The information summarised below is often presented in discussions of the Greenhouse Effect:

Trace Gas	*Lifetime/years*	*Global Warming Potential (GWP) relative to CO_2 (per molecule basis)*
CO_2	uncertain	1
CH_4	10	21
N_2O	150	206
CCl_3F(CFC-11)	60	12400
CCl_2F_2(CFC-12)	130	15800
$CHClF_2$(HCFC-22)	15	10700
C_2HF_5(HFC-125)	28	13400

(a) Explain why the lifetime of CO_2 is uncertain, and why calculating GWPs of trace gases relative to this species is unsatisfactory. (5 marks)

(b) For each of the gases in this table, with the exception of CO_2, write down the key reactions that determine its lifetime in the atmosphere. (7 marks)

(c) Explain the differences in GWPs shown in the table above. (3 marks)

(d) Explain why the GWP of atmospheric O_3 is difficult to include in such comparisons (5 marks)

SAMPLE EXAMINATION QUESTION FOR THE "CHEMISTRY IN THE ATMOSPHERE" UNIT

Methane is a key component of the natural and polluted atmosphere. Its concentration in the troposphere is steadily rising, as shown in the figure below.

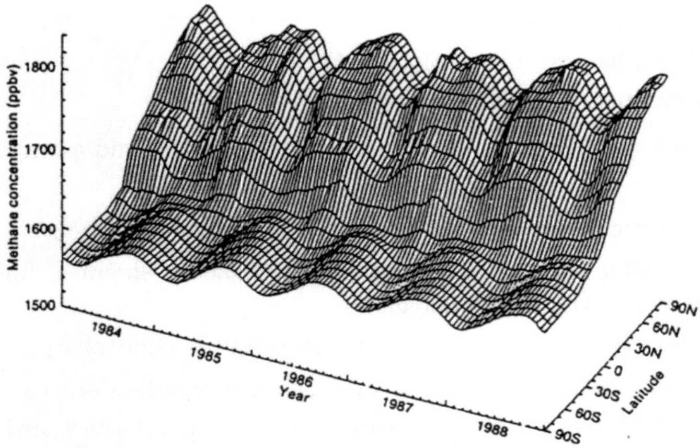

The global distribution, seasonality, and trend of methane in the troposphere

(a) State the significant sources of methane, indicating the approximate contribution each makes to the total concentration in the troposphere. (6 marks)

(b) Explain the latitudinal and seasonal variations depicted in the figure above. (6 marks)

(c) Describe the fate of methane in terms of the reactions it undergoes in (i) the troposphere and (ii) the stratosphere. (4 marks)

(d) Describe briefly the importance of methane to the greenhouse effect. (4 marks)

The examination paper is two hours long and students will be expected to attempt three such questions out of four. Each unit contributes 6% to the overall degree mark, so that in total, the six optional units that make up the environment course contribute 36% towards the final degree. Also, the literature review and research project described below contribute a further 12% to the degree. This is of considerable benefit to students since they are being assessed in a variety of ways on issues and content that they have largely chosen.

Literature reviews and research projects

Literature Reviews and Research Projects are final year undergraduate activities and are attempts to take the student into the more speculative areas of the subject. The objective of the Literature Review article is to collect into one document in a coherent way, information which has been obtained from a variety of published sources. The time allocation for the study is approximately 35 hours and the final report does not normally exceed 2500 words. Examples in the climate change area demonstrate the breadth possible:

- Absorption of infrared radiation by trace gases in the atmosphere
- Aircraft emissions and their contribution to atmospheric chemistry
- Aerosols in the atmosphere
- Feedback mechanisms in climate change

The Research Project is intended to provide the student with an opportunity to learn how to plan, carry out and report a piece of original work, and thus to develop further analytical skills. The time allocation for the study in this case is approximately 100 hours and the final report does not exceed 4500 words. Once again examples in the climate change area include:

- Free troposphere chemistry
- Absorption bands and overlap effects in infrared radiation modelling of the greenhouse effect
- Latitudinal distribution of methane and hydroxyl radicals
- Future population growth, fossil fuel usage, and the greenhouse effect

The project description for the *Free Troposphere Chemistry* research project is given next to illustrate the range of activities involved. It is common practice for the student to be immersed into an established research group with all the hidden benefits that entails

FREE TROPOSPHERE CHEMISTRY

An undergraduate research project in the climate change area

PROJECT AIMS:

- To develop and validate a model that describes the chemistry of trace gases in the troposphere
- To use the model to assess the impact of increasing concentrations of greenhouse gases on the production of tropospheric ozone and the cleansing potential of the hydroxyl radical

STUDENT ACTIVITIES INCLUDE:

- Establish extent of knowledge in this area
- Develop chemical models for the tropospheric degradation of methane, hydrochlorofluorocarbons, and carbon monoxide for incorporating in computer models
- Establish fluxes of key pollutants
- Conduct a suitable variational analysis using computer based models
- Write a formal report on the findings of the research

In summary, the Literature Review and Research Project activities allow the student to develop research skills, to demonstrate initiative, and to take responsibility for the programme of work and its implementation. The actual review and project report also allow the student to demonstrate writing skills with particular emphasis on content, delivery, and balance. A bonus is that the student's research work can lead to a full research project, which the students themselves can develop in postgraduate programmes.

Future Development of Courses for Global Change Science

There is little doubt that the course described above has been, and continues to be successful. The Department of Chemistry at York takes approximately 115 undergraduate students each year. Approximately one quarter will take all of the environmental units and, as indicated earlier, many more students will take one or more of the units summarised above.

The course also continues to evolve. This is essential not only because our knowledge and understanding of environmental issues is increasing, but also because the character of the student intake to undergraduate degrees is changing. It is also important to accommodate the changing interests of the academics involved in the course and the introduction of new academics. These activities help to keep the course up-to-date and relevant.

The model described in this article is a relatively simple one which can be adopted by departments within which a section of the academic staff have interests in, and a commitment to, Global Change Science. The introduction of such a course can broaden the appeal of the department significantly and lead to increased student numbers. It can also evolve to match the changing needs of the subject, the students, and the academic staff.

Acknowledgements

The author wishes to acknowledge John Garratt who had the original idea for the Chemistry, Resources and Environment course at York and did much to promote its development. The author also wishes to acknowledge members of the Department of Chemistry who have contributed to the successful delivery of the course over the years including Peter Hanson, Barry Thomas, John Vernon, David Goodall, Tom Halstead and others.

References

Schneider S H (1994). This book

Teaching Environmental Sciences to Undergraduate and Postgraduate Chemistry Students

M Filomena Camoes and Leonel A T Neves
CECUL, Department of Chemistry
University of Lisbon
1294 Lisbon Codex
Portugal

Introduction

In the Department of Chemistry at the University of Lisbon (FCUL) a one semester course on "Environmental Chemistry" was introduced ten years ago for undergraduate students studying for the Educational Chemistry Diploma. The course is taken in the 6th semester, and it is an opportunity to strengthen the scientific background of students, who until then have only taken one semester of analytical chemistry and who will soon be teaching in and working on areas related to a multidisciplinary subject.

A strong component of analytical chemistry is emphasized in the environmental chemistry course, since one of the main purposes is knowing the composition of real systems.

The adopted methodology combines regular lectures, showing the relevance of a theoretical approach, with laboratory sessions and study visits, and stresses the importance of good quality data upon which wise decisions can be made. Therefore, assessment is based on performance in the laboratory, the quality of reports and written tests, and a monograph that is elaborated on by oral presentation and discussion.

Ten years ago environmental issues were becoming more and more important. It was obvious that this message should be quickly and effectively passed on to as many people as possible, so it was thought that investment should be made in the education of young children. Accordingly, the Department of Chemistry decided to promote as top priority the improvement of the scientific and pedagogic standards of the undergraduates who would soon become school teachers.

The mass media are constantly reporting on environmental disasters and people are realizing that things are not what they used to be. Water appears to be

NATO ASI Series, Vol I 29
Global Environmental Change in Science Education and Training
Edited by David Waddington

polluted, dirt brings trouble to the nose and eyes, the rivers and lakes are dying, the air is smelly and harmful to the lungs, and there are many allergies.

Inevitably chemistry comes up as 'the bad guy', responsible for these problems. The school teacher must have the knowledge and the skill to transmit more accurate information than that obtained from the media, not telling 'stories' about events, but helping students to determine some truth from the facts, teaching how and why things are happening, how to evaluate, how to prevent pollution, and how to recover resources. Besides teaching children, it is of the utmost importance that the role played by the teacher is extended to help people in general to understand environmental problems. The community has to be educated in order to contribute to improving the quality of life.

Chemistry is everywhere in natural phenomena and in anthropogenic ones.

Nature itself is like it is in a big beaker: matter exists in a limited amount, and happening all the time are changes that lead towards a complex physicobiogeochemical chain of events. When equilibria are established in nature, the coexistence of the constituents of the different systems work together with harmony and a special kind of beauty, but if significant perturbations are introduced, pollution occurs.

The bigger the concentration of contaminants, the higher the rates of the reactions. Flying over a big, brightly lit city such as London at night, one has a good example of what might be involved in the management of big centres in terms of water supplies, sewage treatment, and air quality control.

What may have started as a healthy attitude - one family, one house, one farm - may soon result in an uncontrolled population with needs, requirements, and deleterious outputs. Neighbours react to a smoky fire, to a contaminated water flow, to a pit, to a barking dog. Extrapolating, we have what now exists: big cities, industrial complexes, and intensive agriculture, all of which consume and produce chemicals in the big beaker.

Intensive agriculture usually violates the cycles of nature, and the industrial complexes can be responsible for the destruction of plant and animal life in extensive areas. Both of these are important parameters of the increasing pollution of the earth.

It is urgent to help people to develop a state of mind, in which the selection of materials, consumables, and ways of disposing of residues are carefully considered.

Industry must use technologies that pollute less and contribute part of its profits to the development of cleaner technologies. For the preservation of humankind and quality of life, all human beings and human institutions must be good examples of civic and professional behaviour.

Reduction, Reutilization, Recycling

It is always necessary to remember that the producers - those working in industry as well as in agriculture - have to be assisted by experts capable of advising on the optimization of processes, while the consumers must be reassured, not only of the quality of goods, but also about the processes used and the environmental consequences of manufacturing.

Often it is too late when one finds out about the consequences. It took 20 years to realize that inhalation of unseen asbestos fibres produced lung cancer. Ozone depletion was only detected after many years of use of CFC's. How many years, if at all, will it take to overcome these and other problems? In many cases, the intensive remedy, by removing and cleaning, is possible and effective, although expensive. In some instances, the attempted solution of a problem brings another, sometimes worse problem.

Education and Training

To summarize what has been said: education and training are required at various levels in order to contribute to the same big issue.

Laboratory and field technicians play a very important role, since collection of samples, performance of measurements, and keeping equipment in good working condition are essential for good quality data upon which scientists can rely. Only education at an intermediate level that is followed by appropriate training will produce competent professionals needed for this work.

Intensive courses on global environmental change are recommended for people who, because of some sudden increase in professional demands, need extra training.

The scientist, who is expected to see further, needs higher education and a diversified curriculum. Degrees in physics, biology, geology, or chemistry, with a strong component of mathematics and modelling, appear to be useful in obtaining the recommended background.

Postgraduate studies in either field must ensure the production of experts, but experts capable of leading teams.

APPENDIX

Environmental Chemistry Course : 6th Semester
Diploma : Education in Chemistry
(3 hours of lectures and 3 hours of lab work per week)

Lectures

- Introduction
    - Chemistry of the environment
    - Distribution of the elements
    - Natural and anthropogenic pollution
    - Urban centres and industrial areas
    - Quality indicators
    - Biological indicators
    - Biogeochemical cycles
    - Energy, thermodynamics, and kinetics
    - Residence time
    - International organisations : standards and recommendations
    - Technical and scientific developments measuring and modelling
- Evolution of the Earth.
    - Origin and development of life
    - Major and minor components
    - Speciation
    - Essential and toxic elements accumulation factor and food chains
- The Atmosphere and Global Climate Change.
    - The Atmosphere composition and circulation
    - Air pollution
    - Aerosols
    - Atmospheric photochemistry
    - Fossil fuels and the Greenhouse effect
    - Acid rain
- Aquatic Systems and the Environment
    - Water quality and management
    - Waste water treatment
    - Aquaculture
    - Eutrophication
    - Chemistry of some elements
    - Historical introduction to oceanography
- The Earth's Crust
    - Soils
    - Agriculture and food supply
    - Mineral resources - metals and nonmetals
    - Sediments
    - Waste disposal
- Selected Topics.
    - Detergents
    - Pesticides
    - Polymers and plastics
    - Nuclear chemistry
    - Biodegradation

Laboratory:

- Water analysis bulletin (organoleptic factors phyisicochemical parameters toxic and microbiological parameters)
- DO , BOD , COD
- Salinity and chlorinity
- Sulphates in sea water
- Hardness of water
- Nitrates and nitrites in cold meats
- Pesticides
- Lead in human hair
- NOx
- Ozone

Study Visits:

- Not more than 3 out of the following
    - Papermill
    - Water treatment plant
    - National Meteorological Observatory
    - Oceanographic ship
    - Nuclear reactor
    - National Research Institute for Fisheries

The Science and Social Issues related to Global Environmental Change: an Undergraduate Honors Colloquium

Lynne M Carter
Graduate School of Oceanography
University of Rhode Island
Narragansett
Rhode Island 02882, U.S.A.

Introduction

The constant and sometimes dramatic changes in the climate of the earth have been recognized from time immemorial. What is new is the recognition that humankind has generated physical and chemical changes in the environment that can affect the climate of the natural earth system and the economy of the human earth system. Affluent societies emit carbon dioxide and chlorofluorocarbons as by-products of an increasing standard of living. Developing nations add to that burden yearly by burning forests the size of Tennessee. Population growth rates of some developing nations will push societies from living on this earth to merely surviving on it if they are sustained. We have put in motion an environmental experiment whose consequences we are only beginning to contemplate.

Assessing and understanding the causes and potential consequences of this environmental experiment are issues that need to be addressed by the physical sciences, the social sciences (political, social, human behavioral perspectives), and by innovative joint studies which include physical and social scientists. There are new physical science planning and coordinating groups, as well as major new scientific programs for understanding the causes and consequences of global environmental change. However, the complex science issues of global change, that the next generation must understand, are inexorably linked to social science issues: energy use and policies, by-products of technological development, population growth, and environmental impacts. The social aspects and concerns are far more difficult to define than the physical factors. The question "What actions can humankind implement to prevent, mitigate, or live with the scientifically projected environmental effects of global warming?" is much farther away from any real answers than, "What are the major causes of methane or carbon dioxide increases in the atmosphere?"

NATO ASI Series, Vol I 29
Global Environmental Change in Science Education and Training
Edited by David Waddington

Universities and other educational facilities are being called upon to define the potential human consequences and to address the response strategies of global environmental change. The traditional university role includes at least two functions: education and research. The educational role encompasses both student training and professional development. This short paper will not address the function of research or professional development in any detail. We must, however, remain critically aware that professional development is important for those who will not become professional scientists, but professional citizens whose daily choices and participation in environmental policy will direct society's course. As we can see, that course has important ramifications for environmental sustainability.

We have spent our educational time wisely only insofar as preparing individuals to deal with specific disciplinary realms and questions, but we are only now recognizing the need for educational programs that go beyond classical single discipline courses in bridging the gap between the physical and social aspects of this potentially all-encompassing issue. It is crucial that we now include both the single disciplinary educational opportunities and the interdisciplinary educational opportunities in higher education. One of the challenges to humankind today is to prepare the members of the next generation, by stretching their limits to be more capable of effectively dealing with the complex science/social issues that make up the scope of what we call "Global Change". This is the background to developing and offering the following course.

The Undergraduate Honors Colloquium: a Semester Long Course

The undergraduate Honors Colloquium on Global Environmental Change was one of a series of colloquia that are offered to honors students yearly. Normally students in their second year take a colloquium. However, this particular course had 40 students, ranging from freshmen to seniors, from a variety of disciplines. The expectation was that the students would leave the program better prepared to come to grips with national and international global change concerns. An important component of the course was the evening public lectures by invited experts. These lectures were given on 12 Wednesday evenings during the semester and were attended by the public and the university community, as this colloquium was a real effort in community-based learning.

The course was team-taught by Margaret Leinen (now Dean of the Graduate School of Oceanography) and myself. We organized the complex science and social issues into manageable units for study. While building each unit, we also developed and exposed students to the extent of interdisciplinary understanding, as well as to trade-offs and compromise that would be required to deal with these immense issues. Students left the colloquium with the following four skills related to global change:

- basic literacy in the physical scientific and social complexity of the problem
- recognition that scientific conclusions are based on an interpretation of data which can produce uncertainty
- skills needed to understand and evaluate consequences and policy options based on scientific projections
- the ability to ask the "right" questions of the scientific community, especially under uncertainty

The colloquium consisted of the public evening lectures by visiting experts plus two class meetings per week for the semester. One class meeting was a lecture; one a discussion. The topics were divided into three major sections and offered in order: Science Issues, Climate and Society, and Policy Options and Mechanisms (for specifics see lecture topics and speakers). The physical sciences were addressed by an overview of the entire topic and by discussions of specific problems, such as the greenhouse effect, ozone loss, and the role of the oceans. This scientific background set the stage for: 1) a definition of potential global environmental problems; 2) a discussion of the methods used to study it; and 3) an evaluation of the uncertainty in the scientific data and conclusions.

Climate and society issues were addressed next from two perspectives: anthropogenic activities and likely impacts, including how we learn about issues. The role of the press began this section. Other topics included an overview and discussion of the population growth, energy use, and technological development that have led us to the present environmental situation. An in-depth discussion of resource distribution issues included: food and fiber, energy, biodiversity, and tropical forests. A final consideration included the potential health effects of projected global environmental change.

The final section focused on policy options and mechanisms. Some of the overall questions addressed were:

> What are the economic issues and the mitigation options that are available?

Do we have the national and international ability to prevent, mitigate, or live with these potentially devastating problems?
How will developing and developed countries differ in their responses to predicted global environmental changes?
How have scientific findings influenced national and international policies?

The exploration of each of these topics included assigned readings, presentations by URI experts, public presentations by visiting speakers, in-class discussions with the visiting speakers, critiques of presentations, reviews of additional literature, and three written integration papers (one per major topic). At the time of this course there was no appropriate text, so the students read papers from the literature. Since then the bibliography for this course supported the development of a *Book of Readings in Global Change* (University of Maryland, International University Consortium). Since some of the information will change rather rapidly, the book will not be bound and will be subject to periodic updates.

In addition to a final essay exam and the other short discussion papers mentioned, the students were required to do a project. They were encouraged to do projects that would educate others in the community about global change issues and to develop interdisciplinary teams. Some of the projects undertaken included the following: a study of the university recycling program and identification of the problem areas; a special insert in the student newspaper on global change issues; a series of taped interviews with other students on campus to identify environmental attitudes held; development of an advertising campaign to alert community members to the problems of non-collection of CFC's during refrigerator and air conditioning recharge; a study of the implications of global environmental change predictions for Malaysia; and an energy audit of portions of the University campus.

The *process* that was undertaken in this course was just as important as the content in the final outcome. The students left this course with a better appreciation of not only the issues of global change, but also the importance of considering the issues from a variety of perspectives, such as those of future scientists, journalists, engineers, teachers, historians, industrialists, and citizens. They also became empowered by understanding the role that they can play as participants in environmental policy-making.

From the very beginning, it was crucial to require student interaction in discussions, as well as in a team project. While learning begins with awareness and knowledge, education in changing long-term behavior usually evolves through steps that can include comprehension, development of skills, application, analysis, synthesis, participation, and evaluation. Knowledge alone is isolated, comfortable, and often of a short-term duration. Students like to know they have the *right* answer. In addressing these complex issues there are no *right* answers; there are perspectives and trade-offs. Moving from the safe but isolated circle of knowledge to any of the other steps requires an output of energy and risk-taking. In order for students to be able to increase their level of risk, a safe, trusting environment had to be developed that encouraged interactions from them that involved something about what they really thought and not what they thought the *right* answer was. Their 'comfort level' with such interactions increased dramatically as the semester went on. It was important to keep emphasizing that the point of the course was to encourage what people should be thinking *about* and not what they should be thinking.

One cannot assume that people automatically move from knowledge to synthesis, or through any of the other steps. All too often the specifics of knowledge are lost after the final exam. Since the real point of education is to change human behavior, the process by which people are encouraged to take more risks by moving from a knowledge base to an ability to evaluate or synthesize is equally as important as the knowledge base itself. Without the forced interaction, the need to learn to communicate with one another in a meaningful way, and the non-judgmental class atmosphere, the educational process would not likely have been so successful. This course is still being cited at URI as a premier example of an important cross-disciplinary and interdisciplinary learning experience[1] .

---

1 A number of educational products and efforts have been generated by the course:

- *Video taped interviews* of half-hour duration each were made with the visiting experts. These video taped interviews are available for courses and educational purposes.
- The videotaped interviews are also the basis for a *distance-learning course* developed by the University of Maryland, International University Consortium (IUC).
- To support the distance-learning course, IUC has developed a *course guide* and a *Book of Readings* based on the colloquium bibliography.
- Two special issues of the Marine Technology Society Journal on *Global Change* have been developed to extend the concept of educating the community and the public.
- My office is responsible for developing a *graduate overview for teachers,* based on the colloquium. We hope this course will be offered during the school year '94-'95.
- I would be happy to share my course syllabus and bibliography.

Building an Undergraduate Earth Systems Science Education Program

Jeffrey J McDonnell †
College of Environmental Science and Forestry
State University of New York
Syracuse
New York
USA 13210

Introduction

This paper describes an undergraduate teaching program based on integrated Earth Systems Science (ESS). Although the view of the Earth as an integrated system is not new, teaching ESS as an independent interdisciplinary subject combining the essential interfaces between physical, chemical, and biologic processes is new, at least in the U.S. university scene. The motivation for developing the aforementioned program resulted from involvement in the National Aeronautic and Space Administration (NASA) and Universities Space Research Association (URSA) Earth System Science Education (ESSE) Program (Fisher, 1994; Johnson and Kalb, 1994). This program was initiated in 1991 as a five-year pilot effort to bring together university teachers and students, along with NASA scientists, to teach the fundamentals of global change science.

The program described herein is one that was implemented by the author and colleagues at Utah State University, one of 22 ESSE schools in the U.S.A. A strong motivating factor for initiating this program at Utah State University was the pre-exiting interdisciplinary centers and institutes already on campus (Table 1). The presence of these various centers meant that many of the institutional barriers to interdisciplinary teaching and research had already been overcome. At the teaching level, an existing and active degree-granting Watershed Science Unit was an ideal model for the facilitation of faculty teaching in departments other than their own.

The following is an overview of the ESSE program at Utah State University, as of June 1993:

---

† formerly, Watershed Science Unit, Utah State University, Logan, Utah, USA 84322

NATO ASI Series, Vol I 29
Global Environmental Change in Science Education and Training
Edited by David Waddington

Freshman Level Teaching

Within the scope of the ESSE program, the objectives for implementation of the freshmen level course were to (1) provide an overview of the earth as an integrated system, (2) motivate and stimulate interest in ESS issues, and (3) stress the interrelationships between scientific issues and their social, economic, and political contexts. In order to facilitate this broad and interdisciplinary form of teaching, faculty from Watershed Science, Forest Resources, Civil and Environmental Engineering, Plant Soil Science and Biometeorology, Geology and Geography, and Earth Resources were involved in course lecturing and material preparation. Weekly coordination meetings were arranged among six faculty members in order to coordinate and discuss interdisciplinary teaching strategies. In addition to standard lectures, several special teaching tools and techniques were used on an experimental basis. These included the use of 'toy models' (eg., SimEarth™), different video series (eg., Planet Earth) and role playing exercises.

CENTER	AFFILIATED COLLEGE
Center for Atmospheric and Space Sciences	Engineering
Space Dynamics Laboratory	Engineering
Ecology Center	Natural Resources
Utah Water Research Laboratory	Engineering, Natural Resources, Science
Institute for Land Rehabilitation	Natural Resources

Table 1 Interdisciplinary centers on campus that helped in ESSE implementation.

Table 2 lists the course outline used in the freshman-level course. Considerable debate on course content occurred among the faculty. This was based on departmental affiliations, interests, perspectives, and biases. The breakdown of course material into the different sections represented the best 'compromise' for getting across the necessary information in a fixed amount of time (i.e., one ten-week quarter). After an introduction to the concept of the earth as a system and an overview of the NASA interest in ESS educational issues, the basic processes surrounding the earth system were presented. The topics shown in Table 2 were not selected as representing an exhaustive list of key processes, but rather as a number of essential points that would be followed up later in the course during discussions of interdisciplinary linkages. After covering the basic material, linkages between climate and hydrology, biogeochemical cycling, and other coupled processes were discussed. This portion of the course was seen as the innovative aspect of the teaching during which the overlap between traditional disciplines was explored. Following a treatment of systems function and theory, the various tools and approaches to ESS research were examined. Remote sensing factored heavily into course materials in this section as a result of faculty interest and NASA involvement in the course. Finally, the human dimension of global change was woven into a previous discussion of basic physical and chemical processes by highlighting a number of global change issues, including acid deposition, population growth, decertification, and climate warming.

SECTION	LECTURE TOPIC
Introduction	
	Introduction to Earth System Sciences
	What is Global Change?
Solid Earth Processes	
	Solid Earth Basics
	Geologic Time and Rates of Processes
	Focus on the Cenozoic
	Carbon
	Radiation
	Atmospheric Circulation
	Climate Dynamics
	General Circulation Models
	Hydrologic Cycle
Linkages	
	Climate-Hydrologic Interactions
	Chaos
	Climate Modification
	Biosphere Components
	Systems Approaches
Tools, Approaches, and Summary	
	Remote Sensing and GIS
	NASA's Earth Observing Program
	Global Climate Modeling
	Highly Visible Geophysical Problems
	Social Economic and Political Concerns

Table 2 Survey-level course outline

Although successful overall, an number of problems were noted with this course. The team-teaching approach, while coordinated through weekly meetings, presented a problem to students more familiar with single instructor classes. Furthermore, individual team members had difficulty presenting difficult and technically-rich scientific concepts in a jargon-free and equation-limited format. Students within the class came from over 12 different disciplines, including many from the humanities, sciences, and natural and social sciences. Issues of breadth versus depth were cited on evaluations by students, but with no consistent argument for one treatment over the other.

Senior Level Teaching

A senior level course was developed around a more focused portion of the Earth System, namely Climate-Hydrologic Interactions. The student group comprising this course was much more homogeneous and included only students from Watershed Science, Forestry, and Civil and Environmental Engineering. As a result, the course was taught with a high level of scientific comprehension, including the use of differential equations and higher level mathematics. The objectives of the course were to (1) provide an in-depth view of climate-hydrologic interactions, (2) show how land-atmosphere moisture exchanges influence ecological and ecophysicological processes, and (3) examine hydrologic issues relating to global change, the role of measurements and modeling strategies.

The course was structured around a visiting seminar series (Table 3). Guest speakers were identified from colleagues at Utah State and at other ESSE schools and NASA centers. The invited lecture slots provided an overview of research topics in Climate-Hydrologic Interactions and was the basis for a lab series and a series of tutorials with individual students on topics covered by visiting speakers. While accommodating students in class, visiting speakers also provided a vehicle for exposing the larger university community to global change research and earth systems teaching philosophy.

WEEK	LECTURE TOPIC	LECTURER
1.	The USU-NASA Earth System Education Program	ESSE Faculty
2.	Evapotransipiration and the Importance of Scale	ESSE Faculty
3.	Hydrologic Water Balance Modeling	ESSE Faculty
4.	Regional Scale Surface Moisture	ESSE Faculty
5.	Paleo Flood Hydrology	Invited ESSE school
6.	Global Climate Modeling with CCMI	Invited NASA
7.	Sea Level Variations and Global Change	Invited Non-ESSE School
8	Climate-Hydrologic Interactions in the Geologic Record	USU Faculty
9.	Human Response to Local and Regional Climate Change	ESSE Faculty
10.	Water Resources and Global Change	Invited ESSE School

Table 3 **Senior course outline (weekly lecture topics)**

Several special hands-on tools were used in the senior class. Firstly, STELLA™ modeling software was the basis for the laboratory component of the course (see Fisher, 1994). Table 4 shows the lab syllabi which focused on model building and model analysis of climate-hydrological linkages. The STELLA™ approach enabled students to become involved in model building and sensitivity analysis in a highly focussed, hands-on way. STELLA™ exercises built upon students' knowledge of the software and technical details of climate-hydrologic processes, and culminated in an independently developed model/exercise. Overall, this portion of the course worked very well.

LAB NUMBER	LAB TITLE
1.	STELLA Basics and System Modeling Concepts
2.	Lake Water Balance in a Semi-Arid Closed Basin
3.	Principle of the Earth Energy System
4.	Seasonal Insolation, Surface Cover, and Surface Temperatures
5.	Seasonal Dynamics of Simplified Snow Covers
6.	Climate, Food Production, and Population Growth

Table 4 **STELLA lab course outline**

The invited visitors from other ESSE schools and NASA centers were an essential part of the senior level course and the Utah program. In addition to lecturing on topics related to their own research, visitors were used to lobby the administration on the benefits and importance of ESSE. Visitors filled in gaps in the local technical expertise for course teaching, and motivated and stimulated the interest of faculty and students in the program. As one of the ESSE schools, a NASA "mentor" was identified in order to help facilitate the flow of data and instructional material from NASA to the university for both the senior-level and freshman courses. This mentor visited the university and spoke to the senior class and participating faculty on NASA's perspective on ESSE.

Summary and Programmatic Issues

Overall, the ESSE program was, and continues to be a success. Nevertheless, several problems associated with organizing interdisciplinary teaching plagued the first two years of course implementation, most notably those associated with difficulties in large team-taught classes. As the program continues, several ongoing debates regarding course structure, content and implementation continue, including:

- What common scientific underpinning is needed to link broad interdisciplinary earth system education?
- How should the human dimensions of global change be covered, including economic, political, and social issues?
- How do faculty receive credit for teaching in departments outside of their home department?
- How does one insert enough flexibility into professional programs, like civil engineering, forestry, etc., so that students can enroll in ESSE classes?
- What level of mathematics should be expected of students at the undergraduate level, when the student body comprises both humanities and science?

Resolution of these issues at Utah State and other schools implementing this style of interdisciplinary teaching will greatly improve the chances of earth systems science becoming a viable and important undergraduate teaching endeavor.

References

Fisher G (1994) This volume

Johnson D and M Kalb (1994) The NASA-USRA university-based program in Earth System Science Education, Paper Presented at the 3rd Symposium on Education, American Meteorological Society Annual Meeting, Boston, Mass., pp. 38-41

Training atmospheric scientists : examples from Denmark

Henrik Skov and Jes Fenger
National Environmental Research Institute
Department of Emissions and Air Pollution
Frederiksborgvej 399
DK-4000 Roskilde
Denmark

Denmark is a small country with a population of 5 million and thus the amount of money spent on environmental issues is limited. This encourages universities and research institutions to work together and this paper illustrates one such collaboration. This is being done within the framework of The Environmental Research Program. Furthermore, a Ph.D. net-work between several research institutions and universities is being planned.

Denmark has 6 universities. Three of these are performing research in atmospheric chemistry in cooperation with two national institutions - Riso and the National Environmental Research Institute (NERI). Our work is based at Odense University and NERI.

The education of students up to Ph.D. level with special emphasis on global change will be presented in this paper. The purpose of the education is to obtain qualified atmospheric chemists with a high scientific standard and with experience in the field of global change.

Undergraduate students

First and second year students, at Odense, can choose a 30-hour course in Air Pollution. The aim is to give a qualitative introduction to important aspects of air pollution.

Although many books are written on air pollution none of them cover special Danish aspects at university level. Therefore, scientists at the National Environmental Research Institute, Copenhagen University and The Danish Technical University are collaborating to write a new textbook. The Danish title is translated to "Air Pollution". The contents of the book are:

NATO ASI Series, Vol I 29
Global Environmental Change in Science
Education and Training
Edited by David Waddington

Background

Chapter 1 Formulation of the problem
Chapter 2 Sources, emissions and levels of pollution

Processes in the atmosphere

Chapter 3 Physics
Chapter 4 Chemistry
Chapter 5 Deposition

Effects

Chapter 6 Ecology
Chapter 7 Health and comfort
Chapter 8 Materials

Sources

Chapter 9 Stationary sources
Chapter 10 Purification technology for stationary sources
Chapter 11 Mobile sources

Analytical techniques

Chapter 12 Emission
Chapter 13 Absorption and deposition
Chapter 14 Models of air pollution

Economics and Legislation

Chapter 15 Socio-economics
Chapter 16 Danish legislation
Chapter 17 International conventions

From the contents of the book it is seen that the course covers a wide range of subjects and students are introduced to many subjects and disciplines important for the understanding of atmospheric pollution.

There is also a course for third year students at Odense concerning Environmental Chemistry (the chemistry of soil, water and air). The aim of the course is to give the students a fundamental understanding of environmental problems, so for example students employed afterwards by industrial companies can take environmental issues into consideration in planning. Furthermore, this course is compulsory for students who want to specialize in atmospheric chemistry.

The course concentrates on Danish problems. However, to obtain a full understanding of air pollution, it is necessary to include global change issues like, for example, regional scale transport of pollutants from Central Europe to the Scandinavian countries and their effect here and the decreasing ozone concentrations in the stratosphere over the northern hemisphere. The textbook used is Environmental Chemistry (Manahan 1991) which covers most subjects in the course.

Bachelor projects

The Bachelor projects extend to a 250-hour study which includes a literature survey as well as laboratory work. It is assessed by a report. An example of a project is the construction of a map of a pollutant in the vicinity of Odense City. For example, the concentrations of nitrogen dioxide in the air has in several cases been studied by passive sampling (Palmes tubes). In order to give an adequate description of the importance of nitrogen dioxide in the atmosphere, the students have to understand and discuss for example the interrelationships of those gases such as the various oxides of nitrogen and ozone in the troposphere.

The Master thesis

The Master thesis is a 1 year project. The theses in atmospheric chemistry are mostly carried out at the Joint Research Centre, Ispra site, of the Commission of the European Union (JRC) or at NERI.

At the moment there are two MSc students at NERI. One student is measuring the concentrations of carbonyl species, and the other one ammonia, in the troposphere. At JRC, another student is studying the photochemical degradation of dimethyl sulphide (DMS) in the laboratory which is important for a possible global climate feet-back mechanism (Charlson et al 1987).

Among other subjects investigated, results are the photooxidation of HCFCs, the fate of formaldehyde, and the fate of alkenes and terpenes, all in the troposphere. To get a sufficient knowledge in atmospheric chemistry, all the students have studied Atmospheric Chemistry (Finlayson-Pitts and Pitts, 1986) parallel with the experimental work.

One point to be highlighted is that it is a very good idea to have students working outside the university before they study for the master degree. It has given them excellent opportunities to make the acquaintances of other research institutions and to work on subjects and with equipment which otherwise would have been impossible. Finally, students start on their Ph.D. study at an institution they already know and have a much faster start than usual, obtaining interesting results at a high scientific level.

Doctoral studies

Doctoral studies in Denmark have recently been reorganised. In the old system, the students confined their study to their project and did ad hoc courses. In the new system, students will take 300 hours of courses and have to choose from a wide range of courses. Thus, we hope that students will undertake a more multidisciplinary education and gain an understanding of topics other than their own.

Among Ph.D. projects studied so far the following examples are given.

- At JRC, the photooxidation of dimethyl sulphide and the nucleation of the photochemical products to form cloud condensation nuclei; the night time oxidation of isoprene initiated by NO_3.
- At NERI, the development of a three dimensional hemispheric air pollution model used for the Arctic with special attention towards the sulphur species of both natural and anthropogenical origin; the development of methods to measure ammonia concentrations and fluxes and hydrogen peroxide.

The projects cover disciplines like global climate change, changing photochemical activities in the troposphere and acidification together with autrofication.

Most projects in atmospheric chemistry performed so far have a multidisciplinary nature. For example the study of the role of DMS on the global climate on earth has involved physical chemistry; organic and inorganic synthesis; particle and cloud physics and analytical chemistry. This example illustrates well one of the key points of this Workshop; the multidisciplinary nature of global change science. It is, of course, impossible to obtain a high scientific standard in all these disciplines, and the aim of the education has been to reach a high scientific level in one of the disciplines and a basic knowledge in the others in order to be able to communicate across the professional barriers.

References

Charlson RJ, Lovelock JE, Andrea MO and Warren SG (1987) Oceanic phytoplankton, atmospheric sulphur, cloud albedo and climate *Nature* 326, 655-661

Fenger J and Tjell Chr. In preparation (1994) LuftForurening. Polyteknisk Forlag Denmark

Finlayson-Pitts BJ and Pitts JN Jr (1986) Atmospheric Chemistry. John Wiley & Sons. New York

Manahan SE (1991) Environmental Chemistry, 5 ed. Willard Grant Press, Boston

The Arctic Studies Program at the Arctic Centre, University of Lapland

John C Moore and Manfred A Lange
Arctic Centre
University of Lapland
Box 122
Fin 96101 Rovamemi
Finland

Introduction

The Arctic Centre (AC) is a relatively new institution at a young university. The recent trend of studying peripheral areas, such as the Arctic, actually in the region is a welcome and interesting development in education (see Lange 1994 for a discussion of the benefits of such developments). The AC was set up to provide several roles: research into the impacts of change in the Arctic, an information and data service for the scientific community and for decision makers in the political and private sectors, and a science centre exhibition for the public; and to provide an education in the many facets of the Arctic for students from all over the world via The Arctic Studies Program (ASP). The ASP was first introduced in the spring of 1992, with the first full year program run in the academic year 1993-4. Because the ASP is designed as an international course, all tuition is in English.

The ASP is an undergraduate course that aims to instil an interdisciplinary approach to the problems of the Arctic. The course content covers a wide range of disciplines from the humanities, social sciences, and natural sciences. Also, the students come from a wide range of undergraduate and postgraduate degree courses (Figure 1), and from an increasingly wide range of countries (Figure 2). Rather than equipping conventionally trained specialists (whether scientists, lawyers, or economists) with a global (or Arctic) multidisciplinary outlook, the ASP seeks to change the traditional mind-set of the specialist student into a much freer and more rounded one. The students themselves certainly appreciate this approach to the course (Frame, in press: results of student evaluation of the ASP), and the chance to broaden their horizons. In many ways the wide ranging and interrelated issues that they tackle in the course provide a much more realistic training for future careers and lives after leaving university.

The ASP is therefore in some ways an alternative to a masters course, especially for students who may not wish to follow an academic career - those students

NATO ASI Series, Vol I 29
Global Environmental Change in Science
Education and Training
Edited by David Waddington

who will play important roles as future law makers, industrialists, engineers, and decision makers. Thus in practical terms, courses like the ASP are of value in providing a broad multidisciplinary approach for students who will hopefully shape a more enlightened development of future society.

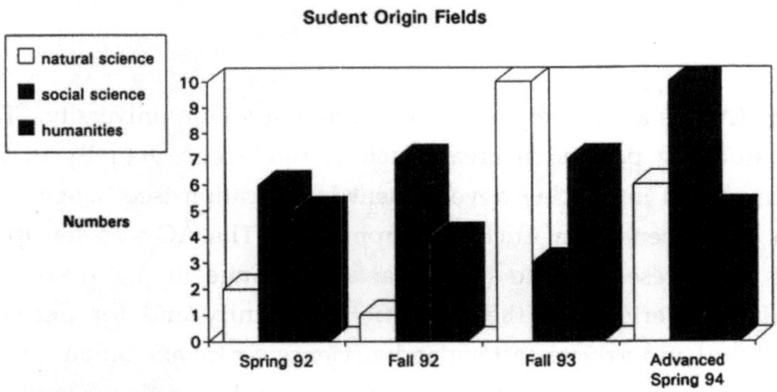

Figure 1. Breakdown of ASP students by field of their degrees in the first 3 ASP Introductory phases and the Spring 1994 Advanced Phase. The natural sciences include physics, biology, botany, environmental science, and geography; social science includes anthropology, economics, sociology; humanities include law, history, international relations, and education.

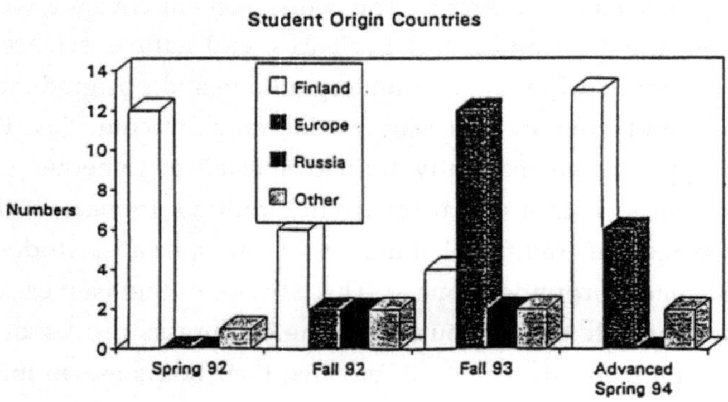

Figure 2. Breakdown of ASP students by their country of origin in the first 3 ASP Introductory phases and the Spring 1994 Advanced Phase. Europe includes EU and Eastern European countries; others include Canada, U.S.A., and Venezuela.

Course Structure and Integration in the Arctic

The ASP consists of two semesters: the Introductory Phase running from September to Christmas (Appendix A), and the Advanced Phase from February to June. The Introductory Phase is structured as a set of courses provided by lecturers in a wide range of topics. Many of the topics address some aspects of global change, whether it be the more directly obvious, such as the role of the Arctic in the global system or the ice sheets and paleoclimate, or less obvious topics (as in some courses in the social sciences and humanities) on aspects of development in the Arctic, such as the environmental and cultural issues involved in hydrocarbon exploitation in the Russian Arctic. The interrelatedness of the issues in the real world is a constant theme and is illustrated at the earliest stage of the course during a week long field trip to the Kilpisjarvi Biological Research Station in the far north of Lapland. There the students sample, often for the first time, such Finnish cultural experiences as the sauna, walk the fells, and meet Sami people in their homes; in addition to receiving lectures on Sami culture and ecology. Back in Rovaniemi, the balance of the course is approximately equally divided between natural science, social science, and humanities, with typically 5 hours of lectures and seminars per day. Students are also expected to study a language (Finnish, Sami, or Russian) or to take a literature assignment. Assessment is based on overall performance in the course, based on student input to individual lecture courses, a field report of the initial orientation exercise in Kilpisjarvi, and a final paper on any subject that is expected to highlight the need and benefits of interdisciplinary approaches to Arctic issues. Topics in the past have included sustainable development in the Arctic, lemmings, and future economic development on the Kola Peninsula.

The Advanced Phase has a rather different structure with a range of short focused courses, usually lasting a week, being offered to students who typically choose 5 to make up their study plan. The courses on offer in 1994 were "The Arctic and the Global System", "Arctic Ecology and Management", "The Baltic Sea Environment", "Winter Ecology and Physiology", Managing Biological Resources", "Military and Environment", "Legal Rights and Cultures of Minorities", and "Art, Culture and Design". Additionally an important part of the course is in the form of field trips such as to the Kola Science Centre and to the Yamal Peninsula in Siberia, and participation in international conference such as the Calotte Academy this May in Inari (Finland), Svanvik (Norway), and Nikel (Russia).

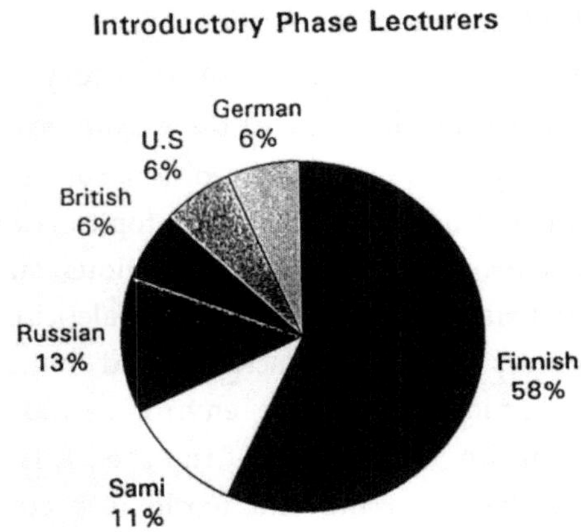

Figure 3. Breakdown of country of origin of ASP Introductory Phase teaching staff.

From the start, the national and international cooperation has been an important part of the ASP with teachers and experts coming from many countries (Figure 3): from institutes such as the University of Aberdeen, the Nordic Sami Institute (Kautokeino, Norway), and the Kola Science Centre; and from many in Finland (e.g., University of Oulu). (See Heininen 1994 for a fuller discussion of the mechanics and cooperation of the ASP.) A significant disadvantage of the ASP and probably of other undergraduate courses is the lack of opportunity to develop research projects that would exist as the basis for student thesis work. The Arctic Centre research section would like to address this problem by bringing in graduate students to do research in the AC, leading to either masters or doctoral degrees, or to a one year diploma. The Advanced Phase of the ASP requires students to complete a 15 page report that, while providing a challenge to undergraduates, is usually limited by time and other course commitments to fairly straightforward literature reviews and field work accounts. The future development of this research aspect of the ASP is aided by the AC proximity to, and good relations with, the Kola Science Centre (KSC) in the Murmansk Oblast of Russia. The KSC is the largest research institute in the Circumpolar Arctic and can provide numerous research opportunities through cooperation and student exchange. (Several KSC employees have participated in the ASP as students and lecturers.) The direct experience of the Kola Peninsula with its severe economic and environmental problems provides the ASP students with firsthand experience of the problems of Russia in general, and exemplify the problems of development in the Arctic.

Global Environmental Change Aspects of the ASP

For a number of reasons the polar regions attract ever increasing attention within the study of global change (see below). This fact should be reflected in the curriculum of a course dealing with the Arctic. We do this by presenting global changes in the Arctic as one of the overarching themes in the Introductory Phase, as well as in a focused short course of the Advanced Phase. In so doing, we stress the interconnectedness of the components of the Arctic system on one hand, and the linkages between the Arctic and the global system on the other. This theme also lends itself nicely to addressing the feedback between the natural processes and the social/human dimension issues in the Arctic. By addressing global change issues in the ASP, we also have the opportunity to deal with general issues of international research programs in the Arctic and their integration into other global change programs. Finally, we also cover questions relating to political decision making, as applied to policies that seek to avoid, or at least mitigate, global environmental changes in the Arctic. More specifically, we address the role of the Arctic in controlling the earth's climate, as a repository of climate information from the past that is available from ice cores, and the impact of industrial development in the region. The Introductory Phase, as may be expected, relies rather more on lectures and fairly direct teaching than does the Advanced Phase course, "The Role of the Arctic in the Global System", where students are expected to make much more active contributions during seminars and role playing exercises.

Dealing first with the paleoclimatic record, the personal experience of one of this paper's authors in the recent deep drilling program in central Greenland (GRIP) is used to present both the scientific findings regarding the pronounced variability of the climate in the past, and also the more personal impressions of being involved in a large scale international scientific collaboration. This combination is designed to try and instil the sense of excitement that can accompany real scientific research. One of the things that is not neglected in the course is the justification for the damage to the environment that necessarily accompanies such a project in the Arctic, and the political priorities that led to GRIP being funded, while other perhaps equally deserving science was not. Another aspect is the often lively debates that ensue during the course on the limits of current scientific knowledge and also on the limits of the imperfect scientific knowledge of the past and present, in modelling the future, and consequently the limits of science in political decision making.

The second role of the Arctic as a major player in the global climate system is covered with extensive reference to the current high profile of the greenhouse effect and the ozone hole. The IPCC reports on future climate change (Houghton et al., 1990, 1992) illustrate the very large changes in temperature that are expected to occur in the Arctic as a result of anthropogenic greenhouse gas emissions. The temperature rises predicted from GCM models are of the order of 3°C over the next 50 years. Such rapid changes exceed any changes in the historical record of climate, and almost certainly exceed any changes that have occurred since the end of the last ice age. The ice core records suggest that the last 8000 years of climate history have been remarkably stable compared with the previous 100,000 years at least. The reasons for such stability and the causes of its breakdown are under keen investigation at present, but it appears that changes in oceanic circulation play an important role. A majority of the sea ice in the Arctic Ocean spends the summer at temperatures within 2°C of the melting point, a rise in temperature would lead to a rapid decrease in sea ice cover in the Arctic Ocean as the positive feedback effect of lowering albedo by removing sea ice allows more heat to be absorbed causing more sea ice melting. Thus, it can be imagined that rather dramatic changes could take place within a very short period of time in one of the climate system's main parts.

The third aspect of the Arctic in environmental change concerns the human impact of industrial development in the Arctic. This is particularly obvious in the marine pollution of the Arctic Ocean with significant input of radiation around Murmansk and Novaya Zemlya, and large, very seasonal (due to freezing) inputs of PCBs and heavy metals from the main rivers of Russia into the Arctic Basin. The recent changes in the political situation in Russia have allowed investigation of the environmental damage around the large metal smelting plants in the Kola Peninsula to be internationally monitored. Damage to the ecosystems in such a fragile environment are likely to take generations to recover, if ever. Several research projects at the AC are studying the effect of air and water pollution on trees and sea ice organisms. It is, however, very striking that the attitudes towards environmental damage that are characteristic of most westerners are not shared by the Russian inhabitants of the Kola peninsula, most of whom are very happy to have well-paid employment in the mines, factories, and nuclear power plants, instead of doing the same job for half the pay in middle Russia (if they can get a job at all). While there is concern for the future state of the environment, the more everyday problems associated with providing themselves and their families with adequate food and other necessities mean that problems that may appear in 20 or 30 years time have low priority. This factor is a very important one to consider, and contrasts with the standard element of academic courses in which students are usually only presented with the results of the damaged environment. It is something that is best grasped during a visit to the region. The fragile nature of the Arctic environment is also a key element in studies of the impact of climate change in the region. Most of the Arctic plant life is limited not by climate, but by nutrient availability. The limitations of the often simplistic assumptions used by models of changes on plant species limits and agricultural practices resulting from climate changes, do not take into account other important factors such as the reduction in biodiversity that will arise as less cold hardy plants out-compete the traditional Arctic plant species.

One further aspect that we like to cover is the impact of change on the indigenous peoples of the Arctic. This again can be illustrated in Russia where there is currently interest in developing the Northern Sea Route (the North-East passage) to serve the Siberian Arctic Ocean ports and function as a trade route from Western Europe to the Far East. This could become much easier in future if climate warming reduces the Arctic Sea ice cover (a factor which is very sensitive to rises of a few degrees in temperature). Increased commercial traffic and pollution in the Arctic would clearly have impact on the traditional fishing and hunting activities of the indigenous people, and the possibility of much easier communication with the region would also bring changes to their way of life.

Assessment of the individual courses of the ASP depends on the requirements of the lecturers and is therefore quite flexible. For the Advanced Phase course on the Arctic in the global system, assessment was based on a 15 minute seminar given by each student on the last day of the course. (Topics included "Natural Climate Change", "Greenhouse Gases: Sources, Sinks and Trends", "Sea Level in a Warmed World", "The Role of Sea Ice in the Climate System", "Changes in Cereal Agriculture in a Greenhouse World".) Secondly, in the 2 weeks following the course, students produced an essay of 3-5 pages in length that emphasised the interdisciplinary nature of decision making in environmental issues and required the student to express a personal but well-founded view. This is illustrated by the following selection of essay topics:

"Explain the possible benefits of global warming in the polar regions and weigh them against the adverse effects of warming. What would be the balance between benefits and adverse effects in your view?"

"Suppose you were asked to make a choice between building a large nuclear power plant or to utilize an alternative form of energy production (hydropower, wind power) in the Arctic. What would be your choice, based on an assessment of relevant environmental factors, and how would you justify it ?"

"Consider that you are a high-ranking politician. You have to make a decision of a) initiating a profitable development project (e.g., exploitation of oil and gas) in an area that is occupied by indigenous people, which would considerably raise the standard of living of these people through shared benefits; or b) of 'leaving them alone', knowing that it will take them a long time before they will ever 'catch-up' with the 'modern world'. What would your decision be and why ?"

In summary then, the Arctic Studies program seeks to provide a rounded education for undergraduate students from across the range of the humanities and the natural and social sciences. Global changes and their relationships to the Arctic are a fundamental part of the course structure and provide one of the main thematic links between the individual courses that make up the ASP structure. Finally, by emphasising the links and feedback between the human and social worlds of the Arctic and the fragile, but still relatively undeveloped natural environment, we hope to leave a lasting influence on the students in whatever career they take up on leaving university.

References

Frame S (In Press) A season of discovery - reflections on the Arctic studies program and a winter in Lapland. Arctic Centre News.

Heininen L (1994) Arctic studies - for academic and practical purposes. In: *Teaching the North: the Challenge of Circumpolar Studies Programs,* L Heininen and L Muller-Wille (eds), Arctic Centre, University of Lapland Rovaniemi, 17 - 19.

Houghton JT, Jenkins GJ and Ephraums JJ (eds) (1990) Climate Change. The IPCC Scientific Assessment, Cambridge University Press, Cambridge.

Houghton JT Callender BA and Varney SK (eds) (1992) Climate Change 1992. The supplementary report to the IPCC Scientific Assessment, Cambridge.

Lange MA (1994) The Arctic in a changing world: new challenges for circumpolar universities. In: *The Changing Circumpolar North: Opportunities for Academic Development,* L Heininen (Ed) Arctic Centre Publ. 6., Rovaniemi, 151-165.

Appendix A

Arctic Studies Program
A Program for the Introduction to the Arctic

The Arctic Studies Program (ASP) starts with the Introduction to the Arctic (15 study weeks), which provides a broad introduction to the Arctic and its manifold aspects. This introduction aims to broaden students in arctic and northern affairs, to demonstrate the complexity of processes in the region, and to highlight the interrelationship between these processes and others at work around the globe. It also strives to illustrate the interdisciplinary nature of issues relevant to the Arctic or the Circumpolar North and the particular role played by the indigenous peoples living there.

The Introduction to the Arctic includes orientation comprising both field (an excursion to Kilpisjarvi in Finnish Lapland) and classroom work. The basic course consists of several thematic units, seminars, and, if possible, participation in an international seminar or conference. In addition, language courses (through the University Language Centre), literature, and offerings from other departments are available. Course requirements include the research and preparation of a 20 page study paper.

Structure of the program

1.	Orientation	(2 study weeks)
2.	The Arctic in the Global System	(3 study weeks)
3.	Indigenous People of the Circumpolar North	(3 study weeks)
4.	The Arctic as a Distinctive and Multiple-Use Region	(3 study weeks)
5.	Languages or Literature	(2 study weeks)
6.	Study Paper	(2 study weeks)

Academic Fields:

All themes and subthemes of the program are under the following three academic fields:

A. Natural Sciences, "Ecology"
B. Humanities, "Peoples"
C. Social Sciences

The teaching of the program runs either under the three academic fields or be interdisciplinary teaching. Interdisciplinary teaching will mostly concentrate on a relationship between ecology and social affairs.

A Weekly Class Schedule:

Teaching from each of the academic fields during the ten week study period will be done three days each week from Tuesday to Thursday. The teaching will include both lecturing and, once a week, an interdisciplinary seminar or discussion, which deals with a relationship between topics of two or three academic fields. On the 'rest days' of the week (Monday and Friday), there will be languages, individual studying of the literature, and a study paper.

Study Paper:

All students will write a study paper on a chosen topic, which should deal with the themes discussed during the program, emphasizing the interdisciplinary nature of issues and problems of the Circumpolar North. The paper should be based on written material, lectures, and discussions, and not exceed 20 pages.

The paper will be a university exercise, but it should be prepared like a scientific paper, e.g., with complete references. Student work will be supervised by one of the scientific staff members of the Arctic Centre or by one of the key lecturers of the ASP. The idea is that students prepare the paper during the course (by the beginning of December) and discuss it in a seminar in the middle of December.

Assessment System:

The studies of the ASP students will be assessed according to the elements of the course, e.g., a study paper, a report from the Kilpisjarvi excursion, languages, and each student's overall performance (participation in discussion, seminars, group work or other activities, and a seminar of study papers). Details of the assessment system are given to students in September at the beginning of the program.

Examples of some topics on which students are assessed are given in Appendix B.

ARCTIC STUDIES PROGRAM: FALL 1993

GLOBAL CHANGES AND THE ARCTIC: A TAKE-HOME EXERCISE

The following exercise is not so much meant to "test" your knowledge and whether you have been listening carefully to the lectures (I am assuming you have!), but they are rather intended to provide all of you with an opportunity to give me **your** view on issues that we have addressed during the last few days.

Please, do not take more than about ten minutes and do not write more than half a page to answer each question. The whole exercise should really not require more than one hour of your time. However, I would appreciate if you would write clearly, thus helping me to find out what you are saying.

1. What are some of the major processes that link the polar regions with the global system? Identify whether global warming would be enhanced or reduced if changes in these processes would take place in Antarctica or the Arctic.

2. Can you provide (at least) one example each for
   - disturbances on the Arctic system caused by human activities and for
   - adverse effects of global changes on Arctic societies?

3. How would you define the **three top** research questions that need to be addressed in order to better understand the role of the polar regions in global change?

4. Suppose you were a (very powerful) politician. What would be your three top policy actions (e.g., environmental laws) that you would introduce in order to minimise man-induced disturbances on the Arctic system?

5. Returning to our first carbon dioxide exercise (GC1), please provide a cost-benefit analysis for each of the four options in the action/problem matrix. Please note that "costs" in this context goes far beyond the usual monetary connotation and includes things such as "changes in lifestyle", time and labour, losses in personal choice, career options etc.

Please hand your answers in (you can collect them and have them sent to me) no later than November 26.

I am looking forward to your answers. Thanks!

Manfred.

THE ROLE OF THE POLAR REGIONS IN GLOBAL ENVIRONMENTAL CHANGES

TOPICS FOR ESSAYS

Preface

You are asked to pick one of the topics given below for a **three- to four-page** essay. The essay should not so much reflect any "learned knowledge" but rather your personal view. However, it is expected that the latter and what you express in your essays is well founded. So, simple "personal arguments" or any "hand-waving" won't do!

Your essay should be typewritten on A4 paper and should be handed in no later than **April 30** (12 o'clock a.m., sharp!!!). Please, contact either John or myself, if you have any questions.

We are looking forward to getting your essays!

Essay Topics

1. Suppose you were asked to make a choice between building a large nuclear power plant or to utilize an alternative form of energy production (hydropower, wind power) in the **Arctic**. What would be your choice, based on an assessment of relevant environmental factors and how would you justify it?

2. To what extent can/should research on global changes in the polar regions (i.e., Antarctica and the Arctic) be done with identical/similar methodologies and in what fields does it have to be done differently? Please explain and give concrete examples.

3. Explain possible benefits of global warming in the polar regions and weigh them against the adverse effects of warming. What would be the balance between benefits and adverse effects in your view?

4. Give a brief outline of a research project (including the rationale for choosing it) addressing the issue of global changes in the polar regions that you would consider to be of highest priority.

5. Consider to be a high-ranking politician. You have to make a decision of
a) initiating a profitable development project (for instance a project dealing with the exploitation of oil and gas) in an area that is occupied by indigenous people, which would considerably raise the standard of living of these people through shared benefits; or
b) of "leaving them alone", knowing that it will take a long time, before they will ever "catch-up" with the "modern world".
What would your decision be and why?

Education and Training in Environmental Science in Greece

Panayotis A Siskos
Laboratory of Analytical Chemistry
Department of Chemistry
University of Athens
Panepistimiopolis - Kouponia
157 71 Athens, Greece

The first Earth Summit held in Stockholm (1972), warned of the side effects of the growing economic development in industrial countries. At the same time, the European Communities first became involved in the environment protection. The new Treaty of European Unions, 1992, introduced as a principal objective the promotion of sustainable growth which respected the environment (Act 2) (Commission of the European Communities 1992).

The following characteristics of *sustainable development* were identified:

- it maintains the overall quality of life
- it maintains continuing access to nature resources
- it avoids lasting environmental damage.

Today there is growing world-wide concern at the continuing deterioration of the state of the environment and the serious degradation of global life support-systems. The ozone hole, the greenhouse effect, acid precipitation, air pollution, the loss of biodiversity are some of today's global problems of Earth science. The second Global Earth Summit, held in Rio in 1992, has defined the details of Agenda 21 for the protection of Environment (Keating 1993).

Greece, a member of European Union, is facing serious local environment problems and follows the European Environmental policy and legislation. It became clear that the implementation of legislation and the appropriate actions in protecting the environment need well informed people and politicians, trained activists and administrators and the training of scientists with a strong background in environmental issues.

In this paper, I will present the education and training programs in Greece related to environmental science and environment protection in schools and universities.

NATO ASI Series, Vol I 29
Global Environmental Change in Science
Education and Training
Edited by David Waddington

Secondary schools
Environmental aspects related to all branches of sciences are being taught in secondary schools in Greece. The book "50 Simple Things Kids Can Do to Save the Earth" (Javna 1990) translated in the Greek language, is given to pupils.

In 1990, the Ministry of Education introduced a law concerning *Environmental Education*. The teachers must emphasise the importance of environmental protection during the teaching of science. In addition, students are encouraged to study local environmental problems.

Undergraduate courses
At the university level, in all departments of sciences, courses (Core or optional) are provided in:

- Environmental chemistry
- Environmental physics
- Environmental biology
- Environmental geology
- Environmental technology
- Environmental education
- Ecology
- Sustainable agriculture

Curriculum requirements and teaching resources for these courses differ from university to university. More specifically, in the department of chemistry at the University of Athens, there are two courses on environmental chemistry. In *General Environmental Chemistry*, the following are studied:

- Basic principles of environmental science
- Water chemistry
- Soil chemistry
- Atmospheric chemistry
- Sources of pollutants
- Effects of pollutants

20% of the course time is devoted to the three global environmental problems (acid rain, ozone layer, greenhouse effect). In addition to the textbook given, a supplementary book, "The Changing Atmosphere: A global challenge" (Firor 1990) translated into Greek, is recommended as a reading resource.

The second course is entitled *Special Environmental Chemistry* and covers:

- Photochemical pollution
- The chemistry of burning fossil fuels
- Cleaner technologies
- Toxic wastes
- Solid wastes
- Liquid wastes
- Recovery and Recycling
- Environmental Analysis:

Sampling of solid, liquid and air. Determination of primary pollutants in air and water. Monitoring techniques.

Postgraduate studies

The coming academic year postgraduate studies will be started in almost all our universities. The traditional system of awarding Master and Ph.D. degrees is changing to the U.S. system of graduate studies. The following long courses, leading to formal qualifications, are given within the core of the program in our department at the University of Athens:

- Advanced Atmospheric Chemistry
- Soil Chemistry
- Monitoring and analytical methods
- Advanced Water and Marine Chemistry
- Health effect of air pollution
- Modern technologies for abatement of pollution

There is also field and laboratory work and research projects.

Short courses for scientists

The Greek Chemical Society has established short courses related to environmental problems, for chemists and other scientists are facing in their industrial work.

Such courses given so are are:

- Environmental microbiology
- Toxic industrial wastes
- Domestic wastes
- Treatment technologies of liquid wastes

Other organisations and scientific societies are also involved in training young scientists and technologists on different topics of the environmental science and technology.

There is a very interesting new development. The National Centre of Public Administration founded in 1983 an educational unit, the Institute of Continuing Training, in order to train on a continuing basis public employees and administrators in new concepts, technologies and legislation related to the practical environmental topics which constitute towards the implementation of the European Environmental Policy.

The following topics have been given the last year in different cities in Greece. The courses are intensive with a teaching period lasting two weeks (100 hours):

- Management of water resources
- Pollution and technologies for the protection of the environment
- Natural environment and natural risks
- Environmental management
- Assessment of environmental impacts.

In all above courses, formal or non-formal, it is not obvious at the first glance the global environmental dimension of the given courses. It is expected that the new training and education concepts, as presented in their workshop will affect the syllabus and the outline of new courses in Greece in the near future.

In addition, the new generation of scientists and activists with an global outlook will affect both the education and practice in facing environmental problems. Public awareness must be another target for educators. Politicians, local authorities and the public should be aware of the global dimension of environmental problems.

References

Birks JW, Calvet JG and Sievers RE, (eds) (1993) The Chemistry of the Atmosphere. Its impact on Global Change : Perspectives and Recommendations. American Chemical Society, Washington D.C.

Commission of the European Communities (1992) Towards Sustainability, Vol I, II and III, Luxemburg

Firor JW (1990) The Changing Atmosphere: A global challenge, Yale University. Translated in Greek by PA Siskos and E Ioannidou. Costrarakis, Athens, Greece 1992

Javna J (1990) 50 Simple things kids can do to save the earth. The Earth Work Group, Berkeley, CA, USA. Translated in Greek by R Branas. Pontiki AE, Athens, Greece 1990

Keating M (1993) The Earth Summit's Agenda for Change: A plain language version of Agenda 21 and the other Rio Agreements. Centre for our Common Future, U.S.A.

Manahan SE (1991) Environmental Chemistry, 5th ed Lewis Publishers Michigan U.S.A.

Siskos PA and Skoullos MJ (1990) Environmental Chemistry, Vol 1 and Vol 2, University of Athens, Athens

Climate Change Education and Training at Wageningen Agricultural University

Rudolf S de Groot
Center for Environment and Climate Studies
Wageningen Agricultural University
PO Box 9101
6700 HB Wageningen, The Netherlands

Introduction

In the Netherlands, Global Change research has a relatively long tradition, especially if we also consider research on acid precipitation and eutrophication as part of the "Global Change-issue". Research on the enhanced greenhouse effect and climate change increased strongly after the establishment of a special national funding structure for this issue in 1990: the National Research Programme on Global Air Pollution and Climate Change.

Initially, both research and education in Global Change at the Wageningen Agricultural University (WAU) focussed mainly on the direct physiological effects of air-(and water-) pollutants and greenhouse gases on plant productivity and cultivation. However, the number of departments involved in climate change research and education has grown rapidly since 1990 with many departments now looking into other aspects including the physical climate system, socioeconomic effects, and policy implications.

In January 1994 Wageningen University started a 5-year multi-disciplinary research programme on the "Greenhouse Effect and Climate Change", coordinated by the newly established Wageningen Institute for Environment and Climate Change Research (WIMEK)[1]. Fifteen departments from both the natural and social sciences are participating in the programme with over 40 projects divided among six main topics: 1) working and monitoring of the global climate system, 2) emission of greenhouse gases, 3) physiological effects on soils and plants, 4) effects on ecosystem stability and biodiversity, 5) consequences for land-use and water management, and 6) socio-economic aspects and solutions.

This research programme is the basis for the education and training provided by the University on the greenhouse effect and climate change. Climate change

---

1 WIMEK was established in 1993 in order to provide the institutional basis for coordinating and developing multi- and inter-disciplinary research on environmental issues at Wageningen University

NATO ASI Series, Vol I 29
Global Environmental Change in Science
Education and Training
Edited by David Waddington

education at Wageningen Agricultural University is given at three levels: a) an undergraduate, introductory course, b) an advanced/post graduate course, and c) occasional "masterclasses" and summer-schools. These three education programmes are described in the following sections.

Introductory/Under-graduate Course

Since 1990, the Agricultural University has provided a special course on the greenhouse effect and climate change, which is organised every two years by the Center for Environment and Climate Studies. It consists of a series of lectures (22 in total) which are given over a period of 3 months and divided over the 6 topics of the research programme mentioned earlier.

Advanced/Post-graduate Course

Work is now underway to develop a special course for post-graduate students on "global change". This training course will draw from the experience gained with the under-graduate course, and it is being developed by WIMEK in cooperation with the Institute for Environment Studies (Free University of Amsterdam) and the Open University which is working on an MSc course "Climate and Environment" (Jansen, 1993). The results of this NATO Workshop will be a welcome source of additional information for the development of the course, which is described in some more detail below.

Objectives

The complexity of the climate change issue calls for an integrated, multi-disciplinairy approach not only in research, but also in education and training efforts. Therefore, the intention is to concentrate on the interactions and feedback mechanisms between the various subsystems of the cause-effect-measure chains of global climate change. Incorporated in the course will be discussions of the basic principles of an integrated assessment of global climate change, as well as descriptions of uncertainties, risk analysis, the different perspectives, and policy implications.

The main learning objectives are to enable the student to:

- identify the main (feedback) mechanisms of the global climate system
- analyse some of the present major uncertainties in climate change
- evaluate the implications of the various climate change scenarios,
- develop preventive policies which reduce risks to tolerable levels

Provisional course-contents

The course will consist of four modules (A-D, see box 1). For each module several overview lectures will be given to introduce the main topics, after which the students will have the opportunity to break up in working groups to study the background material in more detail and to apply the obtained knowledge by writing a short essay and/or practicing with integrative computer models.

Box 1: **Provisional contents of post-graduate course on the greenhouse effect and climate change**

A) *Introduction: an integrated systems approach*

A description of the systems-approach and the many perspectives according to which one may describe and evaluate phenomena of global climate change, including scientific, social, and cultural perspectives.

B) *The climate change*

B1. Analysis of the climate system ("how does it work")
- radiation and energy balance, - climate change forcing mechanisms (e.g., solar radiation)
- biochemical feedbacks
- climate change from a geological perspective

B2. Global cycles:
- carbon, nitrogen, phosphorus, sulfur, water,
- linkages between the cycles/synergistic effects (e.g., ozone)

B3. Human impact on the climate system
- resource use (e.g., burning fossil fuels),
- land use (e.g., deforestation, urbanisation)
- agriculture (e.g., methane emission from rice fields),
- Transport

B4. Climate modeling and scenarios
- simple climate modeling (e.g., energy balance climate models)
- General Circulation Models (GCM's)
- model calibration and validation,
- climate predictions, scenarios and their limitations

C) *Impacts of climate change*

C1. Hydrological effects
- sealevel changes and its consequences
- changes in river discharge/runoff (floodings)
- changes in groundwater level and soil moisture
- changes in glaciers, snow accumulation, snow-melt
- changes in water quality

C2. Land degradation and soil erosion
- soil-erosion and salinisation
- changes in soil organic matter-desertification, coastal erosion

C3. Ecosystem stability
- ecosystem physiology (including direct effects of CO_2),
- ecosystem structure and complexity (effects on species and vegetation, including modeling of zonation and growth),
- landscape-ecological effects (integrated effects on major ecosystem complexes such as forests, magroves, coral reefs)

C4. Land and water use
- cultivation (agriculture, forestry, aquaculture/fishery, etc.)
- water management (including coastal defense)
- nature management and conservation
- recreation
- infrastructure (e.g., transport, urbanisation)
- energy use

C5. Socio economic impacts
- health (heat stress, diseases, water availability/quality),
- food availability (including models),
- extreme events and disasters (storm damage)
- economic costs

D) *Implications for planning and policy making*

D1. Uncertainties and risks
The integrated climate assessment framework used in this course should represent the reliability with which the various pieces of the climate puzzle are understood, and be able to propagate uncertainties through analysis and reflect them in the conclusions regarding the risks.

D2. Decision analysis
A comprehensive view of past and current global decision initiatives and of the available tools for decision-making is needed in order to evaluate the implications of various climate scenarios and the development of preventive policies that reduce risks.

D3. Integrated Assessment Models
Integrated climate assessment models are being developed (e.g., the TARGETS model by the RIVM), based on the systems approach as described above. This integrated modeling framework makes it possible to evaluate the consequences of several types of human influences simultaneously. It is thereby envisaged that synergystic effects may be brought into view. Much attention is paid to the issues of uncertainty, and how to deal with incomplete knowledge.

Course material

The integrated approach of this course demands a strong structuring of the course with regard to such educational strategies as motivation, creating an active attitude, steering, testing, and giving structural feedback. This complexity of approach is present in the level of integration, as well as in the transition from a global environmental problem to local causes, impacts, and actions. A text-workbook model is therefore proposed, whereby the textbook primarily deals with explanations of the background of the topics involved and the workbook primarily with the educational part of the course.

The **textbook** provides the understanding of the complex interactions and the basic concepts of the integrated assessment approach of the course. After an introduction on the systems-concept, the different aspects of the cause-impact-measure-chain are dealt with in more detail.

The **workbook** will present material that illustrates the contents of the textbook with actual issues and discussions. The main purpose of the workbook is to offer didactical support for students which is, among others, done by underpinning the theoretical concepts and ideas described in the textbook with a number of concrete and practical examples. The workbook also fulfils an important role in the translation from global to local action and impacts (together with the computer model PC Targets). Various integrated assessment studies related to the different scales of impacts of human-induced climate change are dealt with and will form the basis for student assignments. Some selected source material and extended references, that reflect actual problems and discussions related to a chapter or some parts of the textbook, will also be included. These cases or examples may be assisted by pre-produced modeling results.

A **computer model** (PC-TARGETS) forms an integral part of the course. The most important learning objective of the computer model is to apply knowledge of all aspects of climate change from an integrated perspective in a practical way. It will be an important aspect in the translation from global to local action and impacts ("think globally, act locally"). An active selection of socio-economic factors or emission scenarios will be possible on local and global scales. Also, the possible impacts of climate change are assessable on a regional scale.

Student **examinations** will consist of writing an essay based on literature and computer-model results. The learning objectives will be tested through the ability of the student to present and defend his or her scientific point of view.

Occasional "Masterclasses" and Summer-Schools

In December 1993, a so-called "Masterclass" was organised by the Center for Environment and Climate Studies for the first time (de Groot and Ooms, 1994). Similar to Masterclasses in the world of music, this event aimed to provide advanced students and PhD candidates the opportunity to obtain more in-depth information on this new research field and to exchange knowledge and ideas with well-known scientists. A review-committee selected 23 students and PhD candidates from 7 Dutch Universities, basing the selection on a short paper and propositions that the candidates had prepared around three themes: (1) how the climate system works and why it is changing, (2) possible consequences of the enhanced greenhouse effect and climate change for natural and cultivated landscapes, and (3) socioeconomic aspects and policy implications. In a three day period (1-3 December 1993), eleven "Masters" from The Netherlands, the UK, and Norway presented the latest information on their field of expertise in the morning sessions, while the afternoons were reserved for presentations of the propositions by the students and discussion with the masters and the audience. The Masterclass was a great success and may be repeated on a regular basis.

References

de Groot RS and Ooms B (1994) Masterclass challenges young scientists on greenhouse effect, in: "Change" June 1994:16-17, the Research and Policy Newsletter on global change from the Netherlands, Kluwer Ac. Publ.

Jansen D (1993) Course plan of Climate and Environment (24 pp). Open University of the Netherlands, Faculty of Natural Sciences. Heerlen.

Ooms B and de Groot RS (eds), (1994). Reader of the course "Greenhouse effect and climate change" (ca 400 pp, partly in Dutch). Center for Environment and Climate Studies, Wageningen Agricultural University .

Environmental Studies at the University of Colorado

James W C White
Director, Environmental Studies Program
University of Colorado at Boulder
Boulder
Colorado 80309, USA

Introduction

The Environmental Studies Program at the University of Colorado is an interdisciplinary degree program offering a Bachelor's degree (BA) with a major in environmental studies. The major consists of a small number of required courses, as well as a selection of advanced courses in two possible tracks: the Environmental Sciences Track, and the Society and Policy Track. The goal of the program and major is to provide students with a general education in environmental studies with reasonable rigor in either the physical sciences or the social sciences. The degree is structured such that a double major in a few established disciplines can be easily obtained, and that a minor in a discipline is almost assured. The courses are offered in the 17 participating departments. Over 80 courses are offered within the major, taught by over 60 faculty. There are over 500 majors in the program at this time. An internship program is also offered in order to provide students with the opportunity to gain experience in the field as well as to establish contacts in the job marketplace. The Program resides structurally within the Office of Interdisciplinary Studies, not within any host department (Appendix I).

The Environmental Studies Curriculum

The Environmental Studies (ES) curriculum consists of a common set of courses and a choice of two advanced tracks: the Society and Policy Track and the Environmental Sciences Track. Courses making up the common curriculum are introductory sequences in the physical sciences, as well as basic courses in the social sciences. The purpose of the common curriculum is to expose all majors to the physical sciences, and thus the physical workings of the Earth, as well as to the social sciences, and thus to the human dimensions of global change. Sequences in biology, chemistry, geology/geography, and economics are required, along with an environmental ethics course, either through the Philosophy Department or the Geography Department, and a course in public policy through the Department of Political Science. Also part of the common curriculum is one statistics course.

NATO ASI Series, Vol. I 29
Global Environmental Change in Science
Education and Training
Edited by David Waddington

The Society and Policy Track has three required courses in the areas of human ecology, environmental economics, and conservation practice, and then breaks into four possible specializations. These are Environment and Natural Resources, International Environment and Development, Decision-Making, Planning and Public Policy, and Environmental Analysis. Each specialization has a list of courses from which the student chooses. The primary departments contributing to this track are Geography, Economics, Sociology, and Political Science. Courses are also taught in the Departments of History, Physics, American Indian Studies, Psychology, and Environmental Design.

The Environmental Sciences Track consists of four required courses and three specializations. The required courses are one additional math (a calculus course), and courses in the principles of ecology and basics of hydrology. In addition, a set of two field courses is required. One is an introductory course on field observations and basic techniques; the other is an advanced field course in a choice of topics including, for example, hydrology and structural mapping. The field courses are taught in the Department of Geological Sciences. The three specializations are in water, biogeochemistry, and climate. Each specialization has a number of courses to choose from. The primary participating departments are Geological Sciences, Biology, and Geography. Courses are also taught in the Departments of Chemistry, Physics, and Atmospheric Sciences.

The University of Colorado is on the semester system, and a total of 120 credit hours are required for graduation. Lecture courses are typically 3 credit hours, so a normal load for a student who will graduate in four years is four to five courses per semester. The ES major has requirements totaling about 60 to about 75 credit hours. A typical ES major will take 20 to 25 courses in the major. Thus, it is important that the ES major begin the common curriculum as soon as possible in order to graduate on time. Advising, particularly at the incoming college level, is extremely important to the success of the student and the program, so the ES Program has a full time student advisor, as well as designated faculty advisors in several academic departments.

To complement the curriculum, the Environmental Studies Program offers its majors a number of internship opportunities, in which students earn academic credit while working in selected positions with private and public agencies and firms. The minimum time required is 10 hours per week for 16 weeks to earn three units of course credit. The primary goal of the internship program is to provide academically-supervised opportunities for majors to work in public and private organizations in order to relate theory to practice as an expansion of learning and understanding. Students also get valuable exposure to hiring opportunities. Internship opportunities are available with a wide variety of private and public agencies and firms.

The Environmental Studies Program is only one facet of global change teaching at the University of Colorado. Many of the departments which participate in the ES major offer degrees in areas of global change or offer their majors the option of concentrating in the area of the environment and global change. This situation helps to retain courses which are needed, as the ES major is not the only student served by the classes. In addition, the university has designated one of the freshman dormitories, Williams Village, as the "environmental dorm". Students who participate in the Williams Village Program can take classes in the dorm, some of which also make up the ES common curriculum. These classes have much smaller enrollments than the usual large lecture class and thus offer the student more chance for open discussion. Also offered in the Williams Village Program are discussion courses on aspects of global change, for example, climate change and atmospheric chemistry. These classes, also with small enrollments of 20 or less, provide an excellent discussion-based introduction into global change issues.

Changes to the ES major currently being planned include adding a seminar series which upper level students can take for credit. The seminars will be a mixture of science talks given by faculty working in the general area of environmental studies, and practical advice from persons currently employed in the field. The idea is expose the upper level students to the realities of working in this area. It is clear from interviews with students, that many do not have a good grasp of the job opportunities which await them, nor what they must do in order to qualify for those jobs. For example, while a student pursuing the Environmental Sciences Track many be required to take only one semester of calculus, it is clear that additional math is absolutely necessary for post-graduate work or for employment in many areas. Students are advised to take additional math, but may not see the need unless it is clearly presented to them by potential employers. While the ES major is not a job training program, we are very sensitive to the needs of the students in terms of providing them with employable skills. Another proposed addition to the curriculum are "magnet courses". These are courses which attempt to take advantage of the unique opportunities available at the University of Colorado. For example, the university has strong research programs in the modeling of climate change. A course in General Circulation Modeling (GCM) has been proposed, which will take advantage of that expertise and funnel it to the upper level undergraduate. Other proposed courses include remote sensing and urban air pollution. The University of Colorado has strong research and teaching programs in space studies and remote sensing, and nearby Denver is second only to Los Angeles in the severity of its air pollution problem.

As a final note, it should be pointed out that the Environmental Studies degree replaces an existing program in environmental studies, The Environmental Conservation Program. The changes are far more than in name only. The Environmental Conservation Degree, while very successful at attracting students (in excess of 500 declared majors), lacked scientific rigor and was faulted by the students for not providing a training leading to a job in the field. The Environmental Studies Program seeks to address both of these deficiencies. In particular, we have recently instituted a tracking system of graduates to assess how effective this new degree will be in training students to work in the general area of global change and environmental quality.

Discussion

Teaching in an interdisciplinary degree program, such as Environmental Studies, within a university system which is structured around established disciplines presents some unique challenges. Since the ES Program offers only a few classes itself, and rather relies on classes taught in traditional departments, one of the most important tasks of the Director is to ensure that classes useful to the ES major are being taught at the times needed and with the student spaces required. As rewards for faculty for teaching originate generally within departments, the ES Program has little leverage in accomplishing this task. Departments with large teaching loads and large numbers of majors, for example, see no benefit in adding ES majors to their already full classes. The ES Program benefits greatly from the fact that global change and environmental quality is currently the primary research theme at the University of Colorado, and that it is also a strong theme in many existing departments and hence classes within those departments. Our success in combining the physical and social sciences, for example, can be traced to the high level of enthusiasm at the faculty level to make this unlikely combination happen and work. For a university or college lacking this broad base of faculty support, managing a program like

Environmental Studies would require rearranging the reward structure for faculty, which in turn implies a committed and proactive administration.

The task of ensuring that the necessary courses are taught is easiest at the introductory level, in which the fundamentals of the functioning of the physical and human worlds are taught. The basics must be taught from first principles and understood thoroughly by the student, a concept which applies equally to the basics in the physical and social sciences. Teaching of the basics is already a part of established disciplines, and interdisciplinary programs can take advantage of these introductory courses. The common curriculum of the ES major uses this approach, although students are urged to take advantage of introductory courses which use environmental problems as examples and illustrations. Convincing departments that they should teach introductory courses of direct use to the ES major is therefore not an issue. The task is harder at the upper levels of college, however. ES majors need the information which is taught in upper division courses, but a course which is suitable for majors in a discipline may not be well suited for the ES major. Often, the examples used, the level and focus of detail, and the pace of the course are not appropriate for the ES major. One of the greatest challenges in supplying an interdisciplinary education in environmental studies is fitting all of the courses and concepts deemed necessary into the traditional four year college degree. The struggle of breadth versus depth is a constant one. As the goal is not to produce a shallow generalist, upper level courses become critical to providing the depth of knowledge needed, and the success of the ES program depends largely on these courses. Faculty teaching these courses must either be persuaded to use environmental examples as they teach to traditional majors, or else restructure the course to focus on the environment. The latter approach is becoming more common as students demand (and take) such courses and as the importance of understanding and coping with global change and environmental issues to the future success of society becomes clearer. It is also apparent that the long term stability of the ES program depends in large part on the development and teaching of these upper level courses.

An interdisciplinary degree in environmental studies should be a degree which is useful to the student in terms of gaining employment or in going on to postgraduate study. This sounds obvious, but in fact it is quite easy to lose sight of this goal. Our experience at the University of Colorado has been that we must monitor the rigor of the program, in order to ensure that the needs of the students are being met. The recent restructuring of the Environmental Conservation degree into the Environmental Studies degree is a reflection of this. One strategy which blends disciplinary and interdisciplinary educations is to encourage students to double major with a disciplinary degree. This approach recognizes that established departments and disciplines are dominant in both post-graduate education in the United States and also in many sectors of the job market. Examples of relatively easy double majors with ES include geological sciences, geography, and economics. Other double majors are possible within a four year time period, but require careful planning on the part of the student. Many departments offer a minor, which can also be a method by which students can pursue an interdisciplinary education, but also obtain disciplinary credentials.

Finally, the teaching of environmental studies poses challenges which go beyond the structure of the program and may conflict with traditional disciplinary programs. The central role of humans in global change presents one of the largest challenges, as it requires the merging of social and physical sciences. This is not an easy task. Yet the importance of ensuring success for this marriage is clear. For example, humans are an integral part of all biogeochemical cycles and the climate system. This is not inherently good or bad; we are long past the point at which such value judgments are useful. There is no sense in separating "natural" from "human caused" when looking at biogeochemical cycles and the climate system. Scientists may want to make this distinction in order to understand the systems better, but we are such a major player in the working of these earth systems that the distinction is not useful, and, in fact, gives people the impression that our actions can be cleaned up sufficiently that we would not perturb the "natural" cycles.

It is when we accept our real place in the workings of the earth that we realize that we must incorporate the dynamics of human behavior into the science of global change. The nature of human decision making, from the national and international levels involving politics and social norms, to the individual level involving risk assessment, must be considered, if we are to predict how we humans will act in the future (or for that matter, now). We can perfect our knowledge of the chemistry and physics of the Earth, but this will be of little use if one of the major forcing functions, human behavior, remains unpredictable.

This broad interdisciplinary nature of global change and environmental studies is one of the larger challenges in teaching global change. Classes that originate in science departments do not, in general, deal with the sociological, political, philosophical, or historical aspects of global change. These issues constantly come up in class discussions, but lacking a solid training in these areas, most physical scientists will simply acknowledge the importance of these fields in dealing with the questions at hand, and then focus the class back on the physical sciences. The same, with the roles of the physical and social sciences reversed, can probably be said of classes in the social sciences. In order for an interdisciplinary education in environmental studies to succeed, it must ultimately blend courses in the physical sciences and the social sciences. How well this can be done depends, in turn, on the ability of the faculty to be interdisciplinary. Given that today's faculty were trained in traditional disciplines, this is more of a challenge than one might think. Perhaps the real success of interdisciplinary programs such as Environmental Studies cannot be measured until a new generation of interdisciplinary thinkers has been produced and takes its place in college faculties. Until then, we must rely upon the vision and enthusiasm of a core group of faculty, and in particular, on the clear enthusiasm of the students, which ultimately drives programs such as Environmental Studies.

APPENDIX I

Environmental Studies Program, University of Colorado at Boulder
The College and Arts and Sciences at the University of Colorado at Boulder offers a Bachelor's degree with a major in environmental studies. Environmental Studies is a program, rather than a department, and utilizes existing course given in various Arts and Sciences Departments. The program is designed to provide a student with (i) a broad but rigorous education in environmental issues and problem solving, as opposed to a traditional, disciplinary-based training and (ii) the disciplinary rigor required to gain employment in the field, but taught in the context of an interdisciplinary education. The environmental studies major is administered through the Office of Interdisciplinary Studies, and draws from curriculum in the earth and natural sciences as well as the social sciences.

The program is composed of a required common curriculum which exposes all students to the basics of physical and social environmental sciences, and a choice between two tracks. The environmental science track has specializations in water, biogeochemistry, and climate, and the social science track has specializations in environment and natural resources, environmental analysis, and decision-making, planning, and policy.

The purpose of the major is to train students in the causes, scale, and remediation strategies of the major environmental problems of the United States and the world. Students will acquire an awareness of the complexity of factors relating to human interaction with the environment, especially the fact that environmental problems have both human and biophysical components, and knowledge of the general principles of human-environmental interaction, global habitability and environmental change, and sustainable human societies.

Students who wish to also pursue a traditional, discipline-based education are encouraged to double major or complete a minor in one of the participating departments. An Internship Program is offered to provide the upper level student with practical experience as well as valuable contacts in the field.

To complement its curriculum, the Environmental Studies program offers its majors internship opportunities, in which students earn academic credit while working in selected positions with private and public agencies and firms. Minimum time required is 10 hours per week for 16 weeks to earn three units of course credit. The primary goal is to provide academically-supervised opportunities for majors to work in public and private organizations on projects related to students' academic interests, in order to relate theory to practice as an expansion of learning and understanding.

All courses used for the major must have C- or better grades. No Independent Study hours apply to the major. No pass/fail hours apply to the major. A minimum grade point average of 2.0 in all C.U. courses is required.

The E.S. major has recently been redesigned to, in part, better equip students to work in the society/policy and/or sciences in the general area of the environment. The program actively tracks its majors after they graduate to access exactly the career opportunities. Students should consider that in some fields an advanced degree may be desirable or even necessary.

Based on past experience with the Environmental Conservation Program students have been employed in the following areas: water resources, waste management, environmental journalism, environmental law, cartography, public planning, computer mapping, market analysis, and handling hazardous materials.

Geographical Concept of Training in Environmental Management: Local, Regional and Global Aspects

Serge G Touchinski
Department of Environmental Management
Faculty of Geography
Moscow State University
119899 Moscow, Russia

One of the most important tasks in global change science is an appropriate description of the environment to enable forecasts to be made of future developments. Environmental management is a very complex and multidisciplinary field of enquiry and consequently the training of experts and scientists who can be the multidisciplinary team leaders in environmental management issues is very important issue.

A multidisciplinary approach in environmental management can have as its focus natural or social sciences. However, combining both foci could create a particularly successful route. One way to do this is to use a geographical approach, because by this way we would have the real combination between nature and society at local and regional levels. For global issues, we can focus our knowledge on earth sciences.

There are different approaches that can be applied in research and training activities in environmental management:
Technological concepts - based on the technological aspects of public production and its ecological basis, such as non-waste or low-waste technology and closed-cycle technology. This can be called industrial ecology.
Biological concepts - concerned with and focussed on ecological aspects such as biocenotic and biospheric interactions.
Socioeconomic concepts - based on the analysis of the economic impacts on the biosphere, including human society, economic structure and development.
Human-health concepts - constructed on health protection and preservation of the society from waste, industrial and other impacts.

The geographical concept of environmental management starts from assumptions of complex interactions between natural phenomena without or

NATO ASI Series, Vol I 29
Global Environmental Change in Science
Education and Training
Edited by David Waddington

with socioeconomic impacts. In contrast to concepts described above, the geographical approach based on earth sciences is concerned with natural variability under anthropogenic impacts, taking in account all aspects of environmental management.

A curriculum in Environmental Management with emphasis in global change problems is being developed here in two main directions. One is in training leaders of multidisciplinary teams to tackle global change problems. The second is in training the specialists for such teams.

We have included, for background knowledge, a strong education in the core of earth science disciplines with a concern for socioeconomic disciplines. So the main focus is on the natural sciences (mathematics, physics with elements of geophysics, chemistry with elements of geochemistry, biology, general ecology) and core earth sciences (geology, geomorphology, meteorology and climatology, hydrology, pedology and soil geography, etc.). All these courses are traditionally in courses on physical geography and in geophysics (for global problems) and some parts are in civil and environmental engineering courses (Appendix I).

This part of traditional education is supplemented by a very important and really core discipline, Environmental Basis of Earth Sciences - Developing and Changing Earth. This course is being constructed as selected units describing the most important processes that are studied in the earth sciences, time and space scales of influences of anthropogenic impacts based on the core projects that are included in the International Geosphere Biosphere Program (IGBP 1992).

An introductory Environmental Management course gives the current views of the problem and another course, the History of Environmental Management, gives a background of the interaction between civilization and the environment. The course in the History of Material Culture helps us to create the understanding of long-term structure in the global change processes. Core socioeconomic courses include environmental basis of production, environmental economics, environmental law, social psychology. At the final stage of training, there will be some seminars in theoretical and current problems of Environmental Management.

An extremely important part of an expert's work is the contact with human society to increase the knowledge in fields concerned with environmental problems, and we are planning to organize training in environmental information including elements of journalism, social psychology, environmental law and the new problems in environmental changes.

IGBP (1992) Climate Change 1992. Intergovernmental panel on climate change. Cambridge University Press

Appendix I

Curriculum in Environmental Management

Titles of Disciplines: courses and seminars	years of education 1	2	3	4	5
Nature Sciences	#				
Earth Sciences	#				
Environmental Management (EM):					
Introductory		#			
Developing and changing Earth			#		
Basic ecology			#		
History of EM		#			
Nature resources management				#	
Seminar on theoretical problems in EM					#
Environmental Basis of Public Production		#			
Environmental Law			#		
Environmental Economy				#	
Environmental Geochemistry		#			
Methods of Environmetal Survey:		#	#		
Environmental monitoring		#			
GIS		#			
Methods of field survey		#			
Analitical methods		#	#		
Data system analysis			#		
Environmental modelling			#	#	
Environmental expertize (Ecogeographical)				#	
Basis of Environmental Management					#
Special Topics and Seminars:					
Engineering geography				#	
Social geography and urbanization				#	
Nature preserves				#	
Human health and the environment					#
Environmentally sound technology				#	
Current problems of EM					#
Regional planning				#	
Environmental information					#

Environmental Monitoring: Special Course for Training Global Change Scientists

Serge G Touchinski
Department of Environmental Management
Faculty of Geography
Moscow State University
119899 Moscow, Russia

Measurements in environmental systems are complex, largely because of uncertainties in spatial and temporal predictions of environmental behavior. The environment is often discussed in terms of single samples or measurements taken in space or time but these results do not give a reliable view of environmental systems. Furthermore the existence of pollution sources may be responsible for nonlinearity in the dynamics of the environment on the background of the natural processes. It is very important to have a wide and adequate data-base of environmental measurements in different natural and man-made situations.

The aim of the Masters course (which will also be used for retraining) is to introduce the main concepts in the organization of monitoring networks.

The course includes 60 hours of lectures, seminars and laboratory work. There is also a 4-week period for fieldwork. Monitoring is the means for the observation, assessment and forecasting of the state of the environment, which can be applied to the management involved in sustainable development issues. The main principles are shown in fig. 1. Among the concepts in monitoring classification, the most convenient is that based on the scaling approach. Global monitoring deals with the background processes for large time and space scales. Ecological monitoring can be described as a matrix involving the scaling of processes or phenomena whether they are local, regional or global.

Scales of variability

One of the most difficult tasks is the optimization of observation networks, including spatial and temporal frequency and accuracy of measurements with actual variability of behavior of the ecosystem. However, there are different scales of fluctuations in environmental parameters, which are described

NATO ASI Series, Vol I 29
Global Environmental Change in Science
Education and Training
Edited by David Waddington

graphically in figure 2. In the figure, several types of fluctuation are shown which can be regarded to be on global, regional and local scales.

Experimental evidence is given in the lecture course with research findings in aquatic systems such as river, reservoirs, and lakes and in terrestrial ecosystems. The results help us to design systems of monitoring and observations with the necessary spatial and temporal scales, including a set of defined parameters, sampling sites and time selection, frequency and duration. It is possible to discuss the Complex Monitoring Program (COMMON) for predicting changes in ecosystems at different time and space scales.

Methods of observations

The main objectives for global background monitoring observations are given in the Table below. The specific features of background monitoring are the

- determination of very small concentrations of all parameters (tendency of trends) for a large territory with long time fluctuations;
- influences of pollution sources producing very strong anthropogenic noise in all of the measurements;
- periodicity and accuracy of measurements correlated with fluctuations and sensitivity of the methods applied.

Three conditions must be considered:

- the possible long distance from any pollution sources;
- the observation station must be in a typical biogeographical region.
- this territory must have a good history of research and observations.

Monitoring networks parameters

Parameter of network	Monitoring scales		
	local	regional	global
territory area, km^2	10 - 100	$20 - 2 \times 10^6$	up to $10^7 - 10^8$
discreteness of observation, km	0:01 - 10	10 - 500	up to 3000 - 5000
periodicity of processes	days-hours	month-years	up to centuries
frequency of observations	minutes-hours	months-decade	years
number of parameters	10 - 100	100 - 1000	more than 1000
accuracy (of limits)	units	% units	0.1 - 0:01% units

It is very important to consider the problem of choice for sampling in the field and accuracy of measurements (fig. 3).

In the background monitoring of global change, the most suitable methods for observations are bioindication for time series and remote sensing methods for spatial pattern of variability of the territory. In this part of the course, we have seminars concerned with the biomonitoring of large scale variations in the biosphere and in remote sensing.

Monitoring data analysis

Assessment models increase our understanding of the environment and to support decision making. However, there is a need to understand better the interplay between the capabilities of the model and the types of questions that the user or regulator needs to answer. The predictive nature of models is largely based on the choice of model structure, time and spatial scales, process description, and data requirements.

Statistics are an essential tool in all aspects of earth science: time series analysis is needed for the examination of variability or fluctuations structure including trend analysis, coherency and shift of phase in natural and in anthropogenic processes. Multivariate and other advanced statistical methods are often required to disentagnel effects in complex issues; and most environmental topics require that experimental results can be analyzed statistically:

- experimental design and analysis for field studies;
- multivariate methods and time-series analysis methods appropriate for monitoring;
- choosing the most informative parameters, describing the variability of ecosystems under anthropogenic impact;
- adaptation of optimum techniques and appropriate methods of measurements in ecosystems research;
- estimates of the state of the ecosystems for applied ecological studies:
- ecological state forecasting using methods of time series analysis for scenarios of possible natural and anthropogenic impacts;
- ecological standardization for geographical regions;
- ecological assessment;
- ecological zoning;
- ecological mapping.

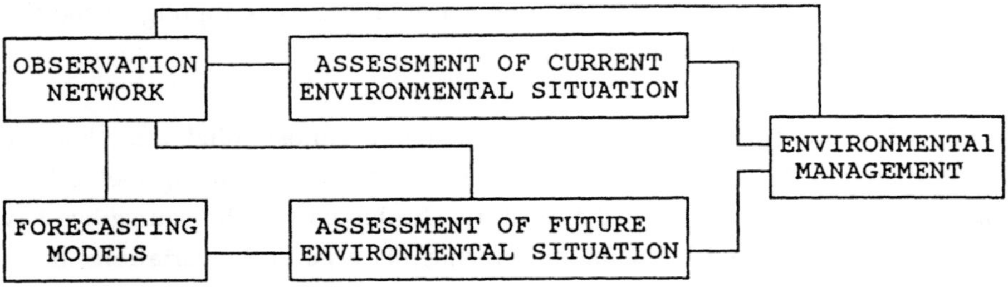

Fig.1 Conceptual scheme of environmental monitoring

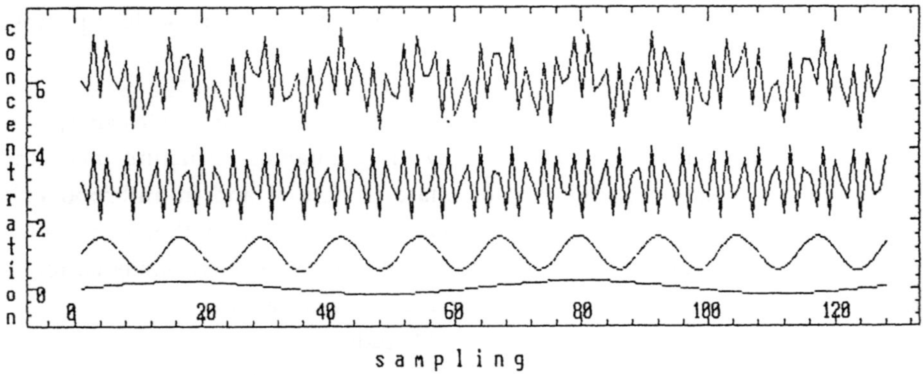

Fig.2 Examples of observations: local, regional and global background monitoring

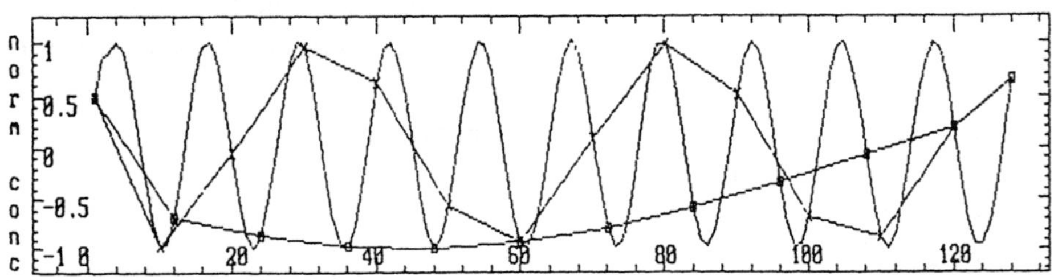

sampling

Fig. 3 One source of observation mistakes by using uncorrected sampling frequency

The development of a new masters level course: MSc Environmental Change and Management, University of Oxford, UK

John Boardman
Environmental Change Unit
University of Oxford
Mansfield Road
Oxford OX1 3TB, UK

Introduction

The Environmental Change Unit (ECU) at the University of Oxford is an interdisciplinary unit dedicated to collaborative research on the nature, causes, and impacts of environmental change, and to the development of management strategies for dealing with future changes. The Unit acts as a focus for cross-departmental environmental initiatives within the university. A primary example of this is the development of a new MSc degree.

The Unit was established in 1991 with a management committee drawn from the School of Geography, Department of Zoology and the Oxford Forestry Institute. Research within the Unit is structured within three programmes: long-term environmental change, energy and the environment, and climate impact on crops. It is planned to expand current work in two areas: health and the environment, and land degradation.

Planning the Course

The theme for the degree was relatively quickly arrived at based on the overall remit of the Unit, the expertise available in Oxford and the nature of existing degrees in the UK. It was agreed that "environmental change" would be broadly defined, that the course would concentrate on the present and the future, and that it would encompass both human and physical dimensions. The rejection of "past environmental change" as a theme was mainly because other degrees and training programmes cover this area. The concentration on the present would also be relevant to the management theme of the degree.

From the outset it was argued that if the degree were to succeed, it should aim to attract students who wished to be involved in the management of changing environments at a planning and policy level rather than at a technical level. The ideal student would be one with a strong disciplinary background in earth

NATO ASI Series, Vol I 29
Global Environmental Change in Science
Education and Training
Edited by David Waddington

sciences, economics, or ecology, for example, and who wished to broaden his or her experience. At the same time it was hoped that the degree would attract applicants with work experience in environmental management and who wished to update or gain a more theoretical background. The broadening experience was felt to be important: those with strong science backgrounds would be introduced to economic, legal, and policy issues.

The taught elements of the course (lectures, seminars, tutorials, laboratory and field sessions, and visits) occupy 10 to 12 hours per week over three eight-week terms. Courses are examined in the third term and the dissertation is completed in the period from June to September. All elements of the course are designed specifically for the MSc degree and are not part of undergraduate degrees. The initial planned intake is 15 students.

Course Content

The principle underlying the selection of course content was that the degree should contain broad-based core elements that would be undertaken by all students, plus sufficient optional and more specialist elements to satisfy students of very diverse background.

The MSc degree is a one-year course which will run from October to the following September when the dissertation is handed in.

The degree consists of a series of compulsory courses:

Course 1: Issues and Driving Forces

This is a lecture programme which considers the issues and driving forces underlying environmental change. After a general introduction considering central issues, such as sustainability, thresholds, population ecology, and energy, the course is sub-divided into terrestrial, atmospheric, and aquatic areas. These sections include issues such as desertification, salinization, greenhouse gases, (ozone and CFC's), and pollution.

Course 2: Managing the Environment

This is a lecture-based course which is divided into two sections: cross-cutting dimensions, which includes consideration of ethical, legal, and cultural aspects of environmental management; and the institutional framework. The latter deals with international institutions, special interest groups (NGO's, lobbies, and consultancies), and the framework of business and commerce. Finally, the course takes the use of energy as a case study in managing the environment.

Course 3: Methods and Techniques for Environmental Management
This is a lecture, laboratory, and field-based course which begins with general issues, such as experimental design, surveys and monitoring (both of the social and physical environments), and environmental and risk assessment. In the second and third terms, management techniques such as Geographic Information Systems (GIS), remote sensing, and modelling are covered.

The course includes two series of seminars, one by guest speakers which has been running since the founding of the ECU, and the other a student-led series.

A fortnightly series of visits to institutions working in the field of environmental change and management is also part of the course. These include Kew Gardens, National Rivers Authority, Bristol Energy Centre, and IUCN (Cambridge). A weekend field course at Slapton in Devon takes place at the beginning of the course and introduces the students to the management of coasts, nature reserves, and the threat of nitrate pollution to fresh water ecosytems.

Optional elements in the course are important as they allow students to follow their own interests at greater depths. All options develop from topics introduced in the main courses.

An option consists of four tutorials or seminars, depending on the number taking the course. For each option the student produces an essay which forms part of the degree assessment profile. Each student takes two options, preferably one in each of the first two terms. The options offered in 1994-95 are environmental law, conservation and diversity, transport and the environment, environmental education, energy and the environment, drainage basin management, further GIS, soil erosion and conservation, forests, food security and the environment, environmental economics, and cultural adaptation and the environment.

The final element of the course is a 15,000 word dissertation on an aspect of environmental change and management.

Discussion

The MSc Environmental Change and Management does not begin until October 1994, but some tentative conclusions can be drawn from the student selection process. In February-April 1994, 23 applicants were offered places from a total of about 85. Of these, seven are currently completing first degrees, but the majority have some form of work experience. Thirteen are non-British students, with the strongest representation from Australia (3) and the USA (3). Good applications were notably lacking from Africa, Latin America, and Asia, excepting Japan; largely, one suspects, reflecting the lack of educational opportunity and funding. Two scholarships for students from Thailand are available (Riche-Monde Scholarship), together with one rising to two in 1995, for students with an interest in Commerce/Industry and the Environment (Unipart International Scholarship). The bias toward first world students is unfortunate in a course dealing with environmental change, since many of the most urgent problems occur in the third world. The problem will only be addressed by obtaining sponsorship for high quality students from specific disadvantaged areas.

The application form asked applicants to give reasons why they wished to take the course. About 30 applicants were interviewed. It was revealed that there was a strong desire to undertake a broad-based course which also offered opportunities for specialisation. Breadth and interdisciplinarity were seen as positive reasons for applying. Many applicants commented favourably on the inclusion of such areas as economics and law; others welcomed the opportunity to become acquainted with techniques such as GIS and remote sensing.

The content of the course is strongly influenced by the requirement to spread the teaching and supervision load as widely as possible. This is because most of the teaching will be undertaken by staff with existing full-time university appointments. Research staff contribute a minor amount. The reason for this policy is to achieve staff continuity from year to year and to reduce the resource implications that would arise if new staff were to be appointed specifically to teach on the course.

From the course planning point-of-view, the breadth of expertise resulting from resource constraints is not unwelcome in a course of this nature and it also seems to be attractive to prospective students. However, the danger is one of fragmentation, with individual teaching sessions tied together, at best, by broad themes such as critical issues, managing the environment, and methods and techniques. Until students confront optional courses and their dissertation, there is a possibility that the educational experience will lack coherence.

Course planning produced several surprises. Within university departments a breadth of expertise concerning the environment was revealed. However, there were difficulties in filling some areas of the syllabus, such as environmental economics and law. What was also surprising was that there was far more expertise in the area of environmental change than in management issues. Apparently most academics find it easier to study change than to discuss how the environment can and should be managed!

Conclusion

The opportunity to develop a new masters course in Environmental Change and Management alongside the establishment of a new Unit at the University of Oxford is an exciting educational challenge. Not only does it allow the Unit to become a fully integrated teaching part of the university but requires it to integrate expertise in many areas of environmental teaching and research around a theme. The development process has not been without its problems, particularly in the area of producing a coherent course and obtaining sufficient resources to support the course. However, the admissions procedure has already shown the great demand for such a course and the high quality of potential students.

SECTION 3: RESOURCES FOR TEACHING GLOBAL CHANGE SCIENCE

Teaching Global Change Science to Non-Science Majors

George W Fisher
Morton K Blaustein Department of Earth and Planetary Sciences
The Johns Hopkins University
Baltimore MD 21218
USA

When the Apollo astronauts brought back the first photos from space more than twenty years ago, they gave us a new vision of Earth as a system. Through a gossamer-thin layer of air we saw a complex biogeochemical system of water, rock, and soil, inhabited by billions of fellow creatures, large and small, all dependent on one another. Those same astronauts showed us just how important that system is to us by bringing back images of the lunar landscape, barren and lifeless.

The Earth photos showed us that one of our most important jobs is to understand how the earth system works and how humans interact with that system. That task has two implications for the education community. We must, of course, train a new generation of scientists who are experts in the traditional earth sciences or in ecology, and who share a global perspective of how their discipline meshes with the rest of the earth system. But if we are to make effective use of the science these students are to produce, we will also need to involve the social sciences, the humanities, and even the arts as well. We need economists to help us understand how economic systems motivate behavior, and how we can use those systems to motivate people to behave responsibly. We need sociologists to help us understand different cultural attitudes toward nature. We need political scientists to teach us the political and diplomatic side of implementing well-intended measures. Artists, poets, and theologians can help people to see the world afresh and to understand the nurturing role that we now sense the need to adopt.

Consequently, we must teach some of the elements of earth system science to non-scientists, specialists who will help us deal with the practical issues of sustaining Earth, and a generation of citizens who must understand what we are about, if they are to pay for the task. We must convey our present understanding of global systems, but we must also impart an appreciation for the uncertainties in our understanding, the techniques by which we attempt to learn about earth

NATO ASI Series, Vol I 29
Global Environmental Change in Science
Education and Training
Edited by David Waddington

systems, and the degree of confidence that we have in our conclusions. In other words, we must communicate an accurate sense of the methods of science and a critical understanding of how effective those methods are, and where they can go wrong.

Too many introductory science courses focus on conveying current interpretations of a discipline as it is now understood. It is far harder, but essential, to convey a feeling for the culture of science, and for how science works.

The method of science - the interactive posing, testing, and refining of hypotheses or models - is easy enough to communicate in principle. But for students to really understand how the method works, and certainly for them to understand how models link to the natural world, they must themselves participate in building and refining models. In the sciences relevant to global change, working with models poses a problem because even the simplest models require levels of mathematical understanding that are typically well beyond those of most humanities and social science majors.

The new graphical differential equation solvers like Simulink[1] and Stella[2] provide an ideal vehicle for overcoming the mathematical difficulties involved in creating and refining models, and permit students to become personally involved in building and testing them. In the process, students can acquire a feel for how models (and nature) work which far exceeds that conveyed by more conventional methods. Few (1991) and other participants in the Earth System Science Education program developed by the Universities Space Research Association (Fisher, this volume) have made considerable use of these graphical equation solvers in laboratory exercises. But the point that I wish to make here is that these same programs are also excellent for communicating the methods of model development to students in large lecture courses that lack laboratory sessions. A much broader range of students can therefore be addressed. With the aid of equipment capable of projecting a computer display on a classroom screen, the instructor can show students the graphical representation of a model, explain the logic of the model, and generate graphical output intended to

1 Simulink is a trademark of The MathWorks, Inc., Cochituate Place, 24 Prime Park Way, Natick, MA 01760.

2 Stella is a trademark of High Performance Systems, Inc., 45 Lyme Road, Hanover NH 03755.

represent some natural process by running the model. If the results do not fit natural observations well, the instructor can then engage the students in a dialog to determine how to improve the model: by adding additional elements, by modifying parameter values, or both. By participating in that kind of a discussion, proposing changes in a model or in model parameters, then re-running the model to see precisely what the effects of those changes are, students can develop a real sense of how science is done and how the natural process being modeled actually works.

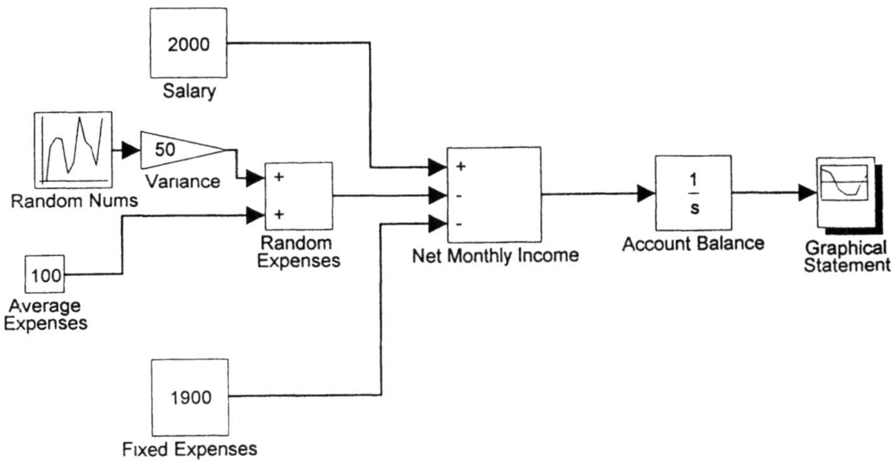

Figure 1: Simulink model of bank balance subject to random impulse buying, used to illustrate concepts of system modeling

To introduce the method to students, I generally use a simple model of a bank account that produces a graph of the account balance over time (Figure 1). The model has most of the elements needed to represent a biogeochemical cycle. There are sources (income), sinks (fixed expenses), parameters which control the size and variability of random expenses, a reservoir (an integrator which determines the current balance in the account), and a graph generator which shows the balance over time. Viewed in this context, the concept of integration seems natural to anyone who has ever balanced a checkbook, but to keep things friendly, I refer to the integrator as an 'accumulator', rather than an integrator. By tweaking the average value and variance of the random expenses and watching how the trend of the monthly balance changes, I can give students a feel for the dynamics of the system, and a sense of how to work with models of natural systems, all in a familiar and comfortable context.

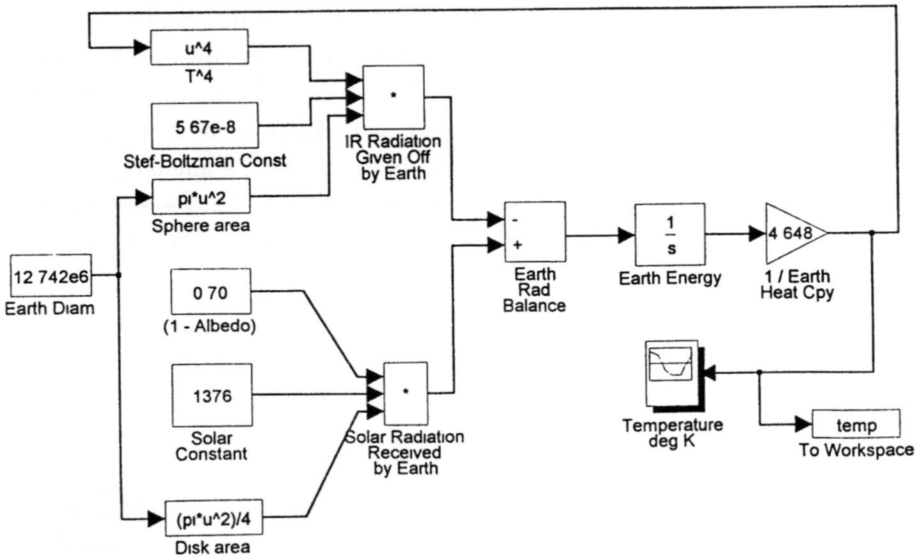

Figure 2: Simple radiation model for calculating Earth's surface temperature in the absence of an atmosphere

From that model, it's a short step to a simple model of the Earth's radiation balance without an atmosphere (Figure 2), and from there to a more sophisticated model that incorporates the effects of an atmosphere (Figure 3). After using the latter model to explain the role of the atmosphere in warming the Earth, we can go farther afield, changing the parameters to simulate conditions on Mars and Venus, and (by reducing the value of the solar constant) document the need for additional atmospheric warming in the early Precambrian.

I've developed Simulink models for processes as complex as ENSO (based on the model of Battisti and Hirst, 1989), and have found students willing to go a surprising distance into some of the more arcane realms of modeling when given a tool that instantly produces visual results to which they can relate. In the process, they seem to gain a real understanding of the modeling process, of how we can use models to refine our interpretation of complex natural processes, and of how the processes themselves work and respond to perturbations. I am convinced that this approach gives a better understanding of how Earth systems work and how we study Earth systems, than any other method with which I am familiar.

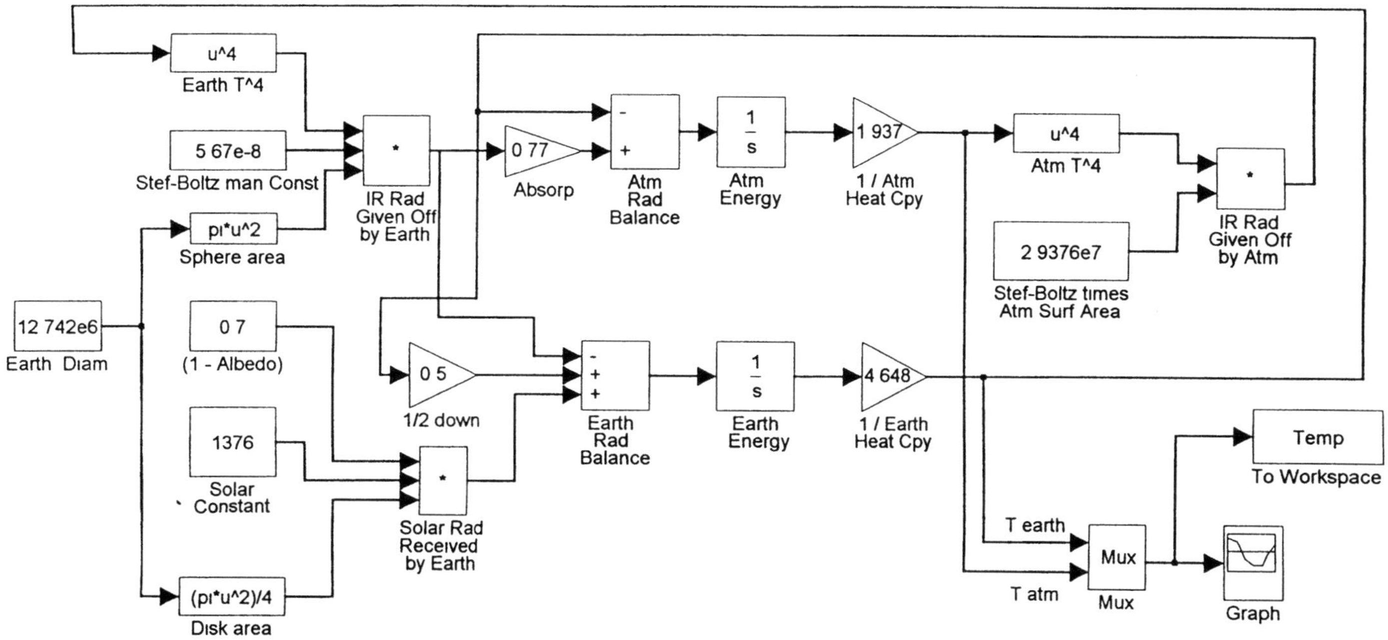

Figure 3: Radiation model for estimating Earth's surface temperature in the presence of an atmosphere

References

Battisti DS and Hirst AC (1989) Interannual variability in a tropical atmosphere-ocean model: influence of the basic state, ocean geometry and nonlinearity, Jour. of the Atmospheric Sciences 46: 1687-1712

Few AA (1991) System behavior and system modeling. Global Change Instruction Program, University Corporation for Atmospheric Research, Boulder CO

Teaching the Human Aspects of Earth System Science

Catherine Gautier, Diane Schweizer and William O'Hirok
Earth-Space Research Group
CRSEO - Ellison Hall
University of California
Santa Barbara, CA 93106, USA

One of the issues facing educators of Earth System Science is how to teach the human aspects of this discipline to students having mostly science backgrounds and interests. To address this issue, a "mini Rio Summit" is now an integral part of the Earth System Science curriculum at the University of California, Santa Barbara. The mini Rio Summit offers an avenue for students to investigate issues of development with limited environmental degradation and to confront some of the intricacies of international environmental decision making. The summit encourages dialogue among students, who represent nations participating in this summit, on topics of global change. It also broadens their awareness of political and socioeconomic considerations that individual countries must face when addressing global change issues. Students' evaluations of two summits (1993 and 1994) were positive and enthusiastic. Their response indicates that the summit is a successful method for bringing them more in touch with the complexities involved in dealing with both the physical and human aspects of global change.

Introduction

New Earth System Science (ESS) classes including Earth Sciences are emerging as part of the curriculum in many departments, and educators are faced with how to teach the human component of this discipline to students who have mostly science backgrounds and interests. To demonstrate the complexities associated with international decision making in the face of global environmental change, a "mini Rio Earth Summit" has been integrated into the ESS course recently developed in the Geography Department of the University of California, Santa Barbara (UCSB). Just as the Rio Earth Summit held in Rio de Janeiro, Brazil in June 1992, provided a forum for the nations of the world to voice and share their opinions on environmental protection and the move towards sustainable development, the UCSB mini summit provides a means for students to develop an awareness of concerns about global environmental change and to confront some of the intricacies of international decision making.

NATO ASI Series, Vol I 29
Global Environmental Change in Science
Education and Training
Edited by David Waddington

Earth System Science at UCSB is a recently developed (1992) course which provides the foundation to study the earth as a system and describes how it has evolved and continues to evolve under the constraints of nature and human activities. ESS presents the earth's atmospheric system as an integrated system of physical, biogeochemical, and human components, rather than as a group of independent subsystems. As with many such courses, the immediate question is how to incorporate human, social, and economic aspects of the evolving world in a curriculum which mostly concentrates on the physical basis, the biogeochemical aspects, and the linkages between components of the earth system. The mini Rio Summit concept has been developed in response to this need.

The Mini Rio Summit

a. Summit objectives and organization

The objective of the mini Rio Summit are: (1) to prepare students to be informed citizens in a global world, and (2) to enhance student awareness of difficulties associated with decision making on an international scale. The mini Rio Summit constitutes the final exam for ESS. Students are informed of the project on the first day of class. Throughout the quarter, class discussions are held concerning the content of the summit. One month prior to the initial meeting, students select the specific nations to be represented, with guidance from the instructor only to ensure social, demographic, and economic representation.

The mini summit is organized so that student groups represent particular nations or geographical regions which are economically, environmentally, and politically dissimilar. For example, Japan, China, Tanzania, Russia, Mexico, and New Zealand were the countries chosen by students to be represented at the 1994 mini summit. Students are required to become familiar with the environmental, economic, social, and political status of "their" selected country, as well as that of the other countries at the summit, so that they can debate an individual country's path towards sustainable development. *World Resources: A Guide to the Global Environment* (Hammond 1992 and 1994) and *Confronting Climate Change* (Miller 1992), are used as the main resources for preparation of the mini summit. These texts contain a collection of in-depth articles and environmental facts.

Students participate in the mini summit by putting themselves in the position of the nation they represent and by offering their opinions as to how topics of environmental concern could be handled, given the nation's current social, economic, and political circumstances. Each nation is required to think in terms of how it could move towards its own version of domestic sustainable development within the global community.

The mini summit stresses the importance of global interactions and communications. Students are informed of the positions of the other countries to better understand the paths of action of other countries and the consequences of those actions. As part of a global community, representatives are encouraged to call upon other nations for support and to offer their support to other nations, as well as to reproach them for unsound environmental practices.

There are two meetings of the representatives. The initial meeting is a "statement of position" in which students present an overview of each country's current environmental and economic circumstances. The second meeting takes the form of a panel discussion, which is designed to allow more in-depth discussion of selected topics and interaction among representatives. This second meeting is designed to be a precursor to a larger global environmental meeting.

b. Presentation of Environmental Position

During the first meeting, representatives deliver their country's position on topics which are of environmental significance to them. Topics include, but are not limited to, population, health, biodiversity, pollution, food, energy, and freshwater availability. The presentations include the political, economic, historical, and social circumstances facing the countries and are supported by recent statistics. The students are asked to offer specific reasons as to how the current situation has come about.

The position of each nation is presented to the representatives of the other countries as well as to additional members of an audience that consists of graduate students in the Geography Department. After the position of each nation is delivered, members of the audience are given the opportunity to initiate discussion and ask detailed questions concerning the contents of each presentation.

The presentation of the environmental positions increases student awareness of the complexities involved in developing environmental policy on an international scale. Students are able to identify particular areas of concern, present strategies for dealing with these concerns, and confront any corresponding policy problems. Successful steps that were decided upon and enacted in order to address the issues of concern are presented and discussed. Students become more familiar with the conditions of the other nations, which helps them formulate more encompassing opinions.

This first meeting of the mini summit provides a solid foundation for the second meeting, the panel discussion. The presentations enable individual countries to become familiar with many facets of the same issue. The panel discussion is held one or two days later, allowing students time to digest the information received and to prepare appropriate questions and concerns to be addressed.

c. Panel Discussion

The purpose of the panel discussion is to encourage communication among the individual nations, to outline the concerns of the global community, and to prepare for the necessary bargaining that typically takes place in establishing policies and treaties. During the presentation of environmental concerns, students usually develop a nationalistic attitude and become interested in exploring global environmental change from 'their' country's perspective, as well as from a global perspective.

The panel discussions are led by a moderator and include the representatives of each nation and an audience of geography graduate students. A week in advance, students are given a list of potential topics/questions to be addressed in the panel discussions. These questions are provided to help students prepare for the panel discussion and to give the discussion focus. The topics of the panel for the 1994 summit were:

1) cross-border pollution and water resource utilization
2) role of multi-national corporations and the environment
3) equality of energy resource use
4) saving the environment for future generations.

A couple of days prior to the discussions, several specific questions are provided to fine-tune the summit presentations.

The day of the panel discussion, the moderator initiates discussion by asking the panel one of the questions. Each 'nation' is given two minutes to respond. After all responses are heard, the floor is opened up for discussion and comments by, first, the representatives of other 'nations' and then the members of the audience.

The panel discussion delves into the intricacies of achieving sustainable development and leads to heated dialogue as to what are the optimal paths for each country to follow.

At the end of the 1993 panel discussions, each nation's representatives were asked to offer at least one plan of action they thought to be the most realistic for minimizing detrimental global environmental change. They were also asked to discuss the limitations of the plan and requirements for putting the plan into action. Some of these action plans were:

1) Fund research for alternate energy sources
2) Cease environmentally unsound fishing practices
3) Develop higher quality mass transit systems
4) Implement mandatory recycling programs
5) Encourage market mechanisms for pollution control
6) Switch from logging to harvesting forest products
7) Promote public education on achieving sustainable development

Overall, a panel discussion provides an excellent forum for students to express their opinions and concerns on global environmental change. If they do not agree with the policy of a particular nation, they are given an opportunity to voice their concern and to offer alternative suggestions. The interaction of many different viewpoints forces students to think carefully about their potential actions and how other countries are likely to respond to those actions.

Discussion

The response to both the 1993 and 1994 mini summit experiences was overwhelmingly positive and enthusiastic. Students felt the inclusion of the mini summit in a science course helped bring out the realization that the earth is both a physical and a human system, and that the two components are not as separate as they might first appear. Students indicated that they became more aware of the problems that can arise when trying to transcend national boundaries and work for the benefit of the international community. The mini summit works well because students generally develop a strong nationalistic attitude towards the nation they represent. Students are quick to learn how difficult it can be to put nationalistic attitudes aside in order to allow an international spirit to reach full bloom.

During the 1993 mini summit, all of the students' viewpoints were "green", indicating a tremendous desire on their part to limit environmental impact and strive for sustainable development. For the 1994 mini summit, a number of students were asked to introduce more controversy into the summit discussions by voluntarily expressing "non-green" attitudes. Ways to include economic and business students in future mini summits will be investigated.

Besides heightening student awareness of enacting public policy and educating them other nations' viewpoints, participation in the mini summit results in several fringe benefits. These benefits include sharper dialogue skills and an understanding of the importance of being well prepared for public debates. The summits were particularly successful as part of the upper division ESS course because the students were genuinely interested in the topics of discussion and devoted considerable time preparing their position papers, as well as their questions/concerns to discuss with other 'countries'. In general, the opinions expressed were all extremely encouraging and suggested that more courses should include this kind of exercise.

References

Hammond AL (ed) (1992) World Resources 1992-1993: A Guide to the Global Environment, Toward Sustainable Development. The World Resources Institute in collaboration with UNEP and UNDP. Oxford University Press, New York

Hammond AL (ed) (1994) World Resources 1994-1995: A Guide to the Global Environment, People and the Environment: Resource Consumption, Population Growth, Women. The World Resources Institute in collaboration with UNEP and UNDP. Oxford University Press, New York

Miller IM (ed) (1992) Confronting Climate Change: Risks, Implications and Responses. Cambridge University Press, Cambridge

Historical Geology: An Avenue for Teaching Global Change Issues

Nancy Healy-Williams
Marine Science Program*
University of South Carolina
Columbia, SC 29208 USA

Introduction

For several years a traditional historical geology course has been offered to non-science majors enrolled in the South Carolina Honors College at the University of South Carolina. During the spring 1992 academic semester that course was revised to reflect the growing trend of examining the earth as a system of interrelated components.

Global change is viewed by many scientists as representing a critical juncture in the history of the earth which presents numerous challenges to the world's populations. Historical Geology appeared to be a perfect avenue with which to explore earth system science in terms of past and future global change processes. The topics covered in the course by no means offer an in-depth examination of the issues of global change but do raise the level of understanding of the issues for non-science majors. The response to the course has been overwhelmingly positive.

Historical Geology was designed to reach several goals. First, students were to learn that earth has a history that is not just geological, but also multifaceted. The earth was to be examined as a system of interrelated components: biosphere, geosphere, hydrosphere, atmosphere, cryosphere, and the biogeochemical cycles. Second, it was hoped that students taking the course would attain a knowledge base from which they could make informed decisions regarding global change issues. Scientists have a duty to educate the general public to help them become more scientifically literate, particularly with the important issues of global change. This historical geology course represents one approach by which this goal can begin to be attained. What follows below is a brief discussion of the topics examined and the teaching methodologies employed.

* Present address: South Carolina Commission on Higher Education 1333 Main Street Suite 200 Columbia SC 29021 USA

NATO ASI Series, Vol I 29
Global Environmental Change in Science
Education and Training
Edited by David Waddington

Teaching Methodologies

The format consists of lectures, cooperative learning groups, outside reading assignments, class discussions, videos, and exercises drawn from "Education for Global Change". By utilizing several teaching strategies it was hoped that the students would have a keen level of interest in the subject matter. Listed in Table 1 is the topic schedule and reading assignments for a twice weekly class meeting. The textbook used is *Exploring Earth and Life Through Time* (Stanley, 1993).

The course begins by developing an understanding of the physical characteristics of the earth. For example, during the "Dynamic Nature of the Earth" topic students are introduced to the theory of plate tectonics and also a discussion of how altered continental configurations can affect climate patterns. Next we examine the materials of the earth and then delve more deeply into how these materials are used to decipher earth history in both qualitative and quantitative terms. Once a level of understanding is developed concerning the geosphere, these non-science majors (ranging from freshman to seniors) explore how the earth's systems are connected and how feedback mechanisms play an important role in the evolution of those systems. During this portion of the course, exercises are incorporated from "Education for Global Change", which allows a break from the lecture style delivery and also provides a format for class discussion (Stoltman and Waddington, 1994).

Table 1
Historical Geology

Introduction to the course

Lecture Topic Schedule*	Reading Assignments
Characteristics of the Earth	Chapter 1 (p. 1-3; 13-16)
Dynamic Nature of the Earth	Chapter 6, Chapter 7 (p. 157-167)
Materials of the Earth	Chapter 1 (p.3-6), Appendix 1
Sedimentary Environments	Chapter 3
Time and Geology	Chapter 1 (p.16-18), Chapter 4
Evolution and the Fossil Record	Chapter 1 (p9-12), Chapter 5
Introduction to Earth System Science	Outside reading
The Earth System: Global Atmospheric Processes	Outside reading
The Earth System: Global Ocean Processes	Outside reading
The Earth System: Global Hydrologic Cycle	Outside reading
The Earth System: Global Carbon Cycle	Outside reading
The Earth System: Paleoclimatic Change Processes	Outside reading
The Mesozoic Biosphere-Atmosphere System	Chapters 13 and 14
The Cenozoic Biosphere-Atmosphere System	Chapters 15 and 16
Final Exam	Modern Global Change Issues and Challenges

* Some topics are two class periods

The earth system science sequence is concluded by examining the tools and methodologies used by paleoclimatologists, paleoceanographers, and paleoecologists. Students are told that paleoenvironmental investigations are like solving a puzzle or mystery and to successfully do such investigations they must become great detectives. Consequently, students are frequently forced to rely on information they acquired during the physical geology sequence in order to answer questions posed by the instructor. A typical question asked is how scientists could possibly determine that there was a great interior seaway on North America during the Jurassic period. This type of question allows students to utilize and synthesize materials (presented earlier) into a cohesive framework.

Two classes are devoted to examining the evidence for present day global change. Topics included within this range from global warming to ozone depletion. Students view the video "A Hole in the Sky" (Nova Series) and enter into a discussion of the evidence presented. Unlike typical historical geology courses earth history is examined for the last 245 MA and approached in terms of the climate system. A teaching methodology utilized is cooperative learning exercises based on the Jigsaw Classroom (Aronson 1978) and adapted from a Geological Society of America workshop "Alternative Pedagogies in Geological Sciences". The Jigsaw Activity is described in Table 2 and represents the questions asked during the Mesozoic topic. The instructor interacts with the groups offering guidance when requested. One might ask if such exercises are successful and how well they are received by the students. The responses from my class were:

> "Do more jigsaw activities";
> "I learned more by doing the jigsaws than from listening to a lecture (no offence)";
> "The jigsaws made class enjoyable and the subject easier to learn".

Four such exercises are spread out over the lecture topics and from the student comments these activities are well worth the effort.

Table 2
Historical Geology Jigsaw Activity *

On the attached sheet are 4 questions about the Mesozoic to be completed as homework. Write a short answer (half page max.) to each question on a separate sheet of paper. Bring these answers to class on April 14. At that time we will perform the Jigsaw Activity outlined below.

1. At the beginning of the class period you will meet in your assigned home team. You and your teammates will each be assigned a different question to become an "expert" on.
2. Each expert will then join a group made up of "experts" on the same question. In your "expert" group discuss your answers to your assigned question and formulate a composite group answer. Plan ways to explain the answer to your home team.
3. Return to your home team. Explain the answer to your assigned question. Listen to your teammates' answers to their expert questions. Discuss the answers as a group.
4. Class discussion. I will respond to any questions from the class about the homework.
5. At the end of class, hand in the three questions on which you were not an "expert". You will receive credit of 10 points for every *genuine attempt* to answer the questions
6. Your "expert" question will be re-written as a new and improved answer, incorporating any ideas you received from your classmates.
7. Hand in the re-written answer to your "expert" question on Thursday, April 19. It will be graded for clarity, completeness, and accuracy on a 50-point scale.

QUESTIONS

1) In terms of animal life, what were the effects of the Late Triassic extinctions? Was this the result of climate change?
2) During the Triassic Period both dinosaurs and mammals (and mammal ancestors) lived together, yet this time period is considered the age of the dinosaurs. Why? What factors influenced this rise?
3) The Cretaceous period witnessed a change in floral patterns. What was this change and what reasons are given for its development?
4) What is some of the evidence for warm global climates during the Early Cretaceous?

* Adapted from Ann Bykerk-Kauffman, Cal. State Univ. Chico

The final exam is conducted during the last two days of the lecture schedule and consists of oral presentations presented by groups of students. The exam is adapted from another exercise in "Education for Global Change" (Stoltman and Waddington, 1994) and is shown in Table 3. Students are presented with the exam well in advance in order to prepare their arguments. Groups are assigned randomly and they are provided time in class to organize. Full preparation is done outside of class.

Table 3
Historical Geology
Final Exam
The Global Warming Threat: Is There a Problem?

This semester you have hopefully learned a great deal about the earth as a system consisting of numerous components. By now you should have an understanding that the earth is a dynamic place which has changed throughout its history. The question facing us now is whether or not the human population on earth is altering the system in ways which will adversely affect our lives and that of many other organisms. Is there a problem? If so what should we do about it, if anything?

Below is a list of four possibilities in answering the global warming question:

- We *take action;* the global warming *threat is real.*
- We *take action;* the global warming *threat is not real.*
- We *take no action;* the global warming *threat is real.*
- We *take no action;* the global warming *threat is not real.*

You will be assigned to a group which will be given the task of presenting the position for one of the four scenarios listed above. On Tuesday, April 26, each group will be allowed approximately 15 minutes to make their presentation regarding their scenario. On Thursday, April 28, the entire class will debate (by group) the four scenarios. It is important to remember that there is no pre-defined "correct" answer to a question of this importance and complexity. The four scenarios really represent the questions and opinions being expressed as the world's societies attempt to make a decision and develop a plan of action. The possible implications of each scenario have important ramifications depending on how real the threat of global warming is. One way to think about the scenarios is in terms of costs - possible costs involved both with the problem (global warming), and possible costs involved with actions taken to address the problem. As used here, costs extend well beyond the quantity of money involved. There can be costs associated with changes in life style and other aspects of the catch-all notion "quality of life". There can be costs involving time and labor, costs associated with losses in personal choice, career options, and so on. Costs do not necessarily have to be associated with taking action; there can be costs involved with non action. Think about how potentially serious the problem is and how likely is its occurrence. You can address your scenarios by examining how effective possible remedies or preventive actions are or will be. How willing do you think the world will be to bear the costs?

GRADING:

Each group will be assigned a letter grade for the effectiveness of their argument. This will include presenting a logical and documented argument for their "cause". Individuals within each group must participate in order to be included in the grade, i.e., the group may not rely on a single spokesperson for the group. A limited number may make the initial presentation and then the others may be involved more heavily in the debate.

IMPORTANT:

Anyone absent from class on either of the two days of this final will receive an automatic grade of F for the final.

Laboratory Exercises:

Historical Geology is a laboratory course which entails meeting weekly for two-hour lab sessions. The syllabus is provided in Table 4 and represents a combination of common historical geology exercises combined with global change exercises. The lab manual referred to in the table is by Frey, et al. (1992). The SimEarth labs are computer exercises designed around the software package of Maxis (SimEarth: The Living Planet). This software package allows students to manipulate earth systems as they evolve throughout geologic time. Working in pairs, students must perform specific tasks and explain the response of the system. This exercise proves very useful in demonstrating the feedback mechanisms of the earth's systems.

Table 4
Laboratory Topic Schedule

Stratigraphic Relationships	Chapter 6
Sedimentary Parameters and Sedimentary Rocks	Chapters 1 and 2
Stratigraphic Principles	Chapter 4
Time and Geology	Chapter 5
Introduction to Fossils	Chapter 3
Introduction to Earth System Science	Handout
The Changing Atmosphere	Handout
Paleoecology an Environmental Reconstruction	Chapter 12
SimEarth	(two lab sessions)
Mesozoic Life	Chapter 10
Cenozoic Life	Chapter 11

Conclusion

This course has been taught since 1992 and has been a great success with the students. Their feedback has indicated that they now have a better appreciation of the complexity of the earth system as well as the inter-relationships involved in the system. In terms of the primary goal of the course, it has raised the scientific literacy of the students involved. Many students took this class with some trepidation and the belief that geology would be stale and boring. Utilizing a variety of teaching methodologies and making the subject matter relevant to real world problems resulted in an enthusiastic response from the class. Some students actually indicated on their course evaluations that "they had no idea that geology could be interesting and fun". Taking the time and effort to alter class instruction can have its rewards.

References:

Aronson E (1978) The Jigsaw Classroom. Sage Pub Beverly Hills CA

Bykerk-Kaufman A, Savoy LE, Schneiderman JS (1993) Alternative Pedagogies in Geological Science: A Workshop. Geol Sci Amer Boulder CO

Frey Robert W, Pinet Paul R, Whitney James A (1992) Earth History An Introduction to the Methods of Historical Geology Hunter Textbooks Inc. Winston-Salem NC

Stanley SM (1993) Exploring Earth and Life Through Time. WH Freeman New York

Stoltman JP, Waddington DJ (1994) (eds) Science for understanding tomorrow's world. International Council of Scientific Unions

THE TRUTH ABOUT CATALYTIC CONVERTORS
A workshop activity in the area of Atmospheric Chemistry

C Anastasi
Department of Chemistry
University of York
Heslington
York YO1 5DD, UK

Introduction

A course titled *Chemistry, Resources and the Environment*, which includes many issues within Global Change Science, has been described elsewhere in this book. There are many interesting and distinctive features associated with this course, not least the various methods of teaching used. There are the traditional methods of lectures, seminars and practical work; small group teaching via tutorials are also very effective teaching methods, while site visits and 'expert' guest speakers provide valuable broadening activities. Workshop activities arguably provide the most effective methods of teaching in this course. Here, students participating in a particular course unit come together for an exercise that typically lasts about 4 hours. Small groups (typically 3 or 4 students) are formed so that students can share knowledge and experiences in working through an exercise that addresses some aspect of the course.

The informal nature of the environment and the supportive action of working in a team, with the help of tutors as required, is a very powerful learning vehicle. Students are able to develop the ability to interact effectively in a team working towards a common goal, and to learn valuable communication skills: verbal through presentation of group work; written through formal reports for assessment by the tutors.

What is described in this paper is a workshop which services one of the optional units, *Chemistry in the Atmosphere,* which contributes to the *Chemistry, Resources and Environment* course. The workshop covers three different aspects of Global Change Science: acid rain, photochemical air pollution and the greenhouse effect and, in particular, the impact of catalytic convertors on each phenomenon. It is given in some detail to illustrate the depth of understanding required and a particular strategy, which is helpful in teaching and learning 'open-ended' areas of science such as those associated with Global Change.

NATO ASI Series, Vol I 29
Global Environmental Change in Science
Education and Training
Edited by David Waddington

Background to catalytic convertors

Exhaust emissions from vehicles contain gases harmful to the environment including unburnt hydrocarbons (HC), carbon monoxide (CO) and oxides of nitrogen (NO and NO_2, known as NO_X). Catalytic convertors were introduced into the exhaust stream to reduce these emissions and are now a standard item in many new cars. The most common is a three way catalyst with a mixture of platinum metals as the active component. The catalysts causes a reaction between oxygen and NO_X, and between CO and the HCs. Oxides of nitrogen are reduced to harmless nitrogen, and the carbon monoxide and hydrocarbons are oxidised to carbon dioxide and water. When working optimally, catalytic convertors are able to remove 90% of the unburnt HC, CO and NO_X. Lead in petrol 'poisons' the catalyst material and to accommodate these devices has meant the introduction of lead free petrol, thus removing the environmental impact of lead at the same time.

Catalytic convertors have been presented as potentially important devices in alleviating the acid rain and photochemical air pollution, where the gases of interest are NO_X; and HC, CO and NO_X, respectively; there has also been a suggestion that these devices enhance the greenhouse effect because they convert HC and CO to CO_2, the latter an important greenhouse agent. This workshop addresses the impact of catalytic convertors in each of these important environmental issues.

The Impact of Catalytic Convertors on Acid Rain

The acid rain debate is one which has gone on for over two decades. The term (acid rain) is commonly used to describe the deposition of atmospheric pollutants, whose acidity causes damage to forests, soils and water, buildings and materials, and agriculture crops. It is an unsatisfactory term. Some pollutants may be acidic when deposited, others become so only after deposition. Moreover, the pollutants are not necessarily carried to the ground with rain, snow or fog; they may also be active in a dry form as gases or aerosols.

Workshop activity

Sulphur containing compounds and oxides of nitrogen are known to contribute to the acid load in the atmosphere and a ratio of $H_2SO_4/HNO_3 \approx 2$ for has been measured in the UK. Using this value and the information given in Table 1 calculate the percentage reduction in acid loading in the atmosphere by installing catalytic converters on all petrol driven cars in the UK. State the assumptions inherent in the calculation.

Table 1: NO_x emissions by source in the UK in 1990

Emission source	Thousand tonnes	Percentage of total
Domestic	68	3
Commercial/public service	56	2
Power stations	777	28
Refineries	36	1
Other industry	231	8
Railways	30	1
Road transport (petrol)	802	29
Road transport (diesel)	579	21
Civil aircraft	14	1
Shipping	131	5
TOTAL	2729	100

The Impact of Catalytic Convertors on Acid Rain - Tutors notes

The H_2SO_4/HNO_3 ratio has been measured to be 2, i.e. 33% of acid is NO_x-derived. The table shows petrol driven vehicles contribute 29% of the total NO_x.
If catalytic convertors are fitted to all petrol driven vehicles, the *reduction* in 'acid' loading of the environment is 8.6 %.

Assumptions and other comments:

- HNO_3 and H_2SO_4 are equally effective with respect to environmental impact.
- All NO_2 ends up as HNO_3.
- The convertor operates at optimum conditions at all times.
- There are no cars already fitted with catalytic convertors.
- The H_2SO_4/HNO_3 ratio is only an average and there can be significant variations.

The Impact of Catalytic Convertors on Photochemical Air Pollution

Photochemical Air Pollution (PAP) or Photochemical Smog is a phenomenon resulting from the action of strong sunlight on NO_X (NO and NO_2) and hydrocarbon polluted air, leading to reduced visibility and the formation of ozone (a key smog indicator), and other oxidising agents. The problem was highlighted several decades ago in the Los Angeles basin in the USA. However, it soon became apparent that enhanced ozone concentrations were also prevalent in other North American cities. It is now recognised to be a widespread phenomenon, the symptoms appearing when meteorological conditions (e.g., temperature inversions and anticyclonic weather) or geography, or both, are favourable.

Typical atmospheric measurements of the key components HC, NO_X and O_3 are shown in Figure 1. The primary pollutants HC and NO are high in the morning due to an increase in human activity. As the photochemistry progresses, NO is converted to NO_2, and the latter is photolysed leading to O_3 rising in the afternoon. The O_3 level observed in the afternoon is directly related to the HC and NO_X values in the 6-9 am period.

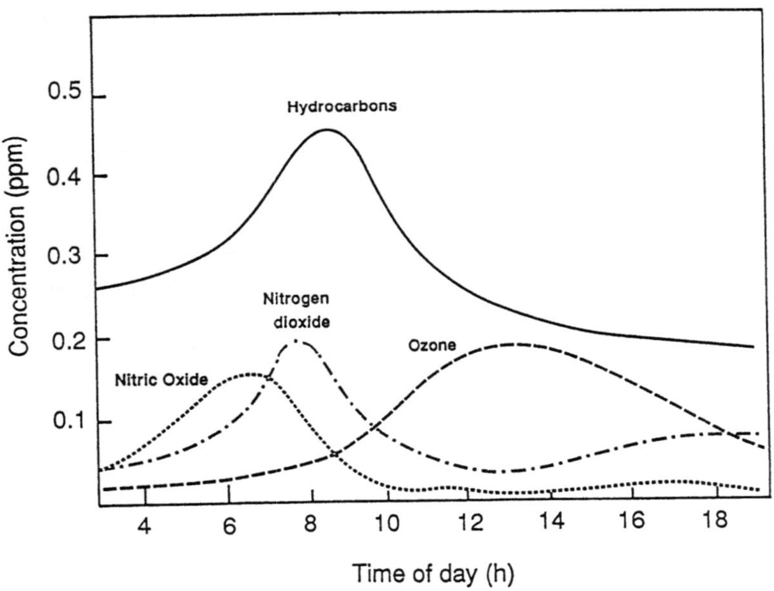

Fig. 1 Ambient measurements of key primary (hydro-carbons, NO and to a lesser extent NO_2) and secondary (O and NO_2) pollutants in an urban environment.

Workshop activity

Figure 2 shows ozone isopleths generated using an atmospheric model. A straight line may be drawn through the apex points of the isopleths dividing the figure into predominantly hydrocarbon-rich and NO_X-rich regions. State what control strategies would lead to a reduction in O_3 levels for these two regions.

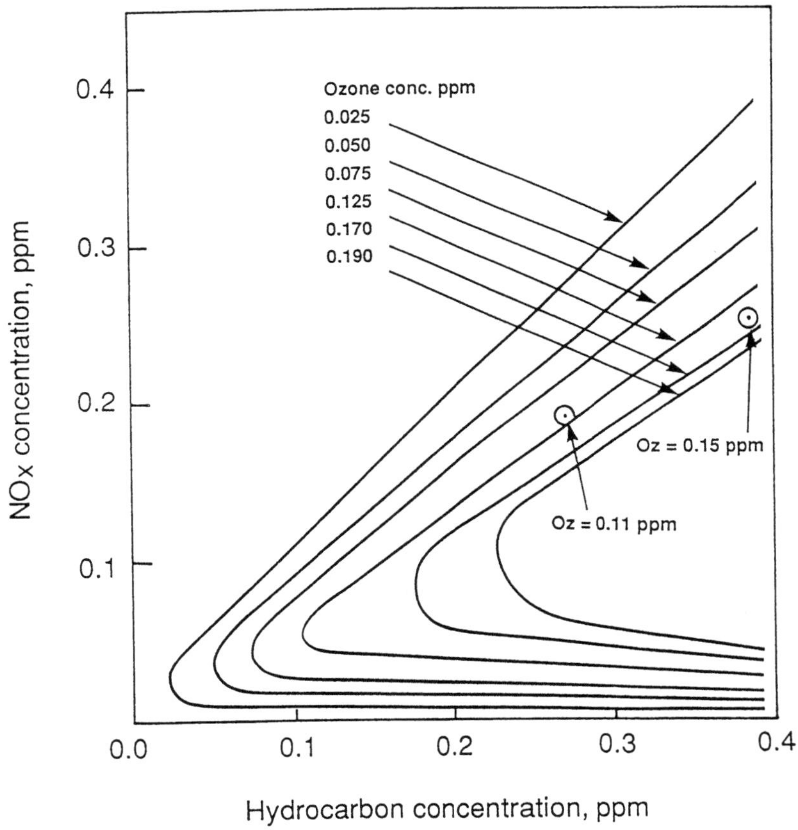

Fig. 2 Ozone isopleth generated using the mechanism in Table 1 under the conditions outlined in the text.

Control strategies for photochemical air pollution - Tutors notes

To the left of the apical axis a reduction of hydrocarbon levels results in the greatest decrease in ozone yields; to the right of this line, a reduction of NO_X levels will lead to the greatest decrease in ozone formation.

As indicated earlier, Figure 1 shows actual HC and NO_x measurements for a typical day in a European city. Calculate the average 6-9 am HC and NO_x values and plot them on the O_3 isopleth, taking note of the O_3 value under these conditions.

Table 2 below summarises the HC and NO_x emissions by source in the UK for 1990. Use this data to establish the reduced values for HC and NO_x emissions if all petrol driven vehicles are fitted with catalytic convertors. Plot these on the isopleth and comment on the 0_3 predicted under this scenario.

Table 2: HC and NO_x emissions by source in the UK in 1990

Emission source	**HC/percentage of total**	**NO_x/percentage of total**
Domestic	2	3
Commercial/public service	-	2
Power stations	-	28
Other industry	2	8
Refineries	-	1
Processes and solvents	50	-
Gas leakage	1	-
Forests	3	-
Railways	-	1
Road transport (petrol)	33	29
Road transport (diesel)	8	21
Civil aircraft	-	1
Shipping	1	5
TOTAL	100	100

The Impact of Catalytic Convertors on Photochemical Air Pollution - Tutors notes

From Figure 1, the mean NO_x ($NO + NO_2$) over the period 6-9 am is Ø.25 ppm and the mean HC over the period 6-9 am is O.39 ppm. From Figure 2, an HC value of O.39 ppm and an NO_x value of O.25 ppm yields an ozone value of ~O.15 ppm (in fair agreement with that observed in the afternoon peak of ~O.18ppm shown in Figure 1).

Using the HC and NO_x values given in Table 2, use of catalytic convertors can reduce the HC from petrol driven vehicles by 0.33 x 0.9 or 30%; HC remaining in the atmosphere is O.27 ppm. Similarly, use of catalytic convertors can reduce NO_x from petrol driven vehicles by 29 x O.9 or 26%; NO_x remaining in the atmosphere is O.19 ppm. From Figure 2, an HC value of O.27 ppm and NO_x value of O.19 ppm yields an ozone value of ~O 11 ppm. Fitting of catalytic convertors on all vehicles would, in this case, reduce the ozone levels to the maximum 1 hour exposure level suggested by the World Health Organisation.

Assumptions and other comments:

- The isopleths given were constructed for Midsummer's Day in an urban environment; the model used detailed chemistry with three representative HC's.
- Different models will yield slightly different isopleth diagrams.
- As the catalytic convertors age they become less efficient and the reduction of O_3 is less.
- In this case, a reduction in HC, with NO_x left alone, would lead to a greater reduction in ozone.

The Impact of Catalytic Convertors on the Greenhouse Effect

The greenhouse effect is strongly linked to the concentration of CO_2 in the atmosphere and, in particular, its strong absorption of long wave radiation emitted by the earth. If sufficient CO_2 is released into the atmosphere by human activities, the radiation balance of the earth could be disturbed, leading to higher global temperatures. The observation then, of a steady increase in atmospheric CO_2 from an estimated 290 ppm in 1860 to a current value of 335 ppm, has prompted many studies on the possible effects of a continued rise in ambient levels. The estimated increase in CO_2 is the subject of much speculation, but some models predict a possible doubling of its concentration in the next century. The corresponding rise in temperature is also the subject of much speculation, and once again modelling exercises suggest between 1 and 12°C increase in the surface temperature of the earth.

Workshop Activity

Human activities release 20.1 Gt CO_2 yr^{-1} into the atmosphere. There are 600 million cars in the world, of which approximately two thirds are petrol driven. The cars are assumed to have an average mileage of 20,000 km/yr, at 10km per litre (1 litre weighs 0.0007 tonnes). 1.2% of the fuel by weight is emitted in the exhaust as unburnt HC (assumed to be octane). The amount of CO in the exhaust is equivalent to 10.8% of the fuel (and the remainder is CO_2 and H_2O). Calculate the *additional* carbon dioxide loading that is emitted into the atmosphere as a result of fitting catalytic convertors to all petrol driven cars. State the assumptions inherent in your calculations.

The Impact of Catalytic Convertors on the Greenhouse Effect - Tutors notes

Petrol used per year is 8×10^{11} litres, which weighs 5.6×10^8 tonnes.
1.2% of fuel is unburnt and emitted in exhaust; this weighs 6.7×10^6 tonnes.
90% of unburnt fuel (assumed to be octane) is converted to CO_2, which yields 1.9×10^7 tonnes CO_2.
10.8 % of fuel is emitted as CO in exhaust; this weighs 6.05×10^7 tonnes. 90% of CO_2 is converted to CO_2 which yields 8.6×10^7 tonnes CO_2.

Using a catalytic convertor on the unburnt HC and CO in the exhaust yields 1.05×10^8 tonnes CO_2, which *constitutes an increase of ~ 0.5 % in the 20.1 Gt CO_2 emitted by human activities*

Assumptions and other comments:

- Catalytic convertors can use up to 10% more fuel, and thus the amount carbon released would be higher.
- The assumption that petrol is octane is an over-simplification, although impact of using a true petrol content description would be small.
- CO, and to a lesser extent unburnt HC, are converted to CO_2 in the atmosphere, but the temporal and spatial distribution would be different, i.e., the overall impact of catalytic convertors would be even less than the simple calculation suggests.

Capacity Building in Marine Remote Sensing Through Computer-based Modules Serving a Global Network

Dirk Troost
UNESCO
1 rue Miollis
75732 Paris, France

Ian S Robinson
Department of Oceonography
University of Southampton
Southampton SO9 5NH, U.K.

David Blackburn
Bureau International
des Poids et Mesures
Sevres, France

The UNESCO-BILKO project for training in the analysis and application of marine image data has surprised its originators by the worldwide response which it has generated. Using a basic image processing package (BILKO) operating on personal computers, the project provides computer-based lessons to marine educational establishments around the world. The need and requirements for training in marine remote sensing are examined as the context in which this initiative arose, and some of the obstacles to effective training are identified. The evolution of a network of users and producers of lessons is described, the wide range of lesson topics is reviewed, and the international penetration and usage of the programme are analysed. The mechanics of handling the network are also considered. New developments in the programme are described, including a new Windows version of BILKO. It is considered that the most significant contribution of the project lies in the capacity of the growing network to spread knowledge and hands-on training in ocean remote sensing to all parts of the world.

Introduction

In 1989 UNESCO distributed to marine science laboratories and educational establishments around the world, a booklet containing eight self-contained lessons on aspects of satellite oceanography, accompanied by a set of magnetic disks containing a simple image display and analysis software called BILKO, and a number of digital image datasets relating to the lessons (Callison et al., 1989). Responses were requested, including an invitation to suggest ways of utilising the software and of developing its application as an educational tool. As an experiment of computer-based education, distributed unsolicited to a diverse list of recipients, there was no way of anticipating what, if any, response would be obtained. Five years later the project has grown and evolved to a far greater extent than the expectations of its originators.

NATO ASI Series, Vol I 29
Global Environmental Change in Science
Education and Training
Edited by David Waddington

Not only was the BILKO software well received as a learning aid for ocean remote sensing, particularly by those whose computing power was limited to basic PC's, but also teachers of marine science responded to the example lessons by producing their own lessons which made use of the BILKO toolkit. As a steady flow of image data and associated lesson material has been offered to UNESCO, so it has been possible, after editing for quality control, to publish these in three further modules. More are in preparation. The resulting body of lessons cover phenomena from all parts of the ocean, using image data from a wide range of sensors, interpreted by local experts from every continent. In this way a network of lesson users and producers has grown almost spontaneously. Whilst the team responsible for the original UNESCO initiative have provided the guidance for this development - the driving energy for it - has come from the community of users itself. International training courses are now making use of the lesson modules. Some of these courses have led to the generation of further lessons, and from one summer school a complete module of eight lessons is in preparation. What started as a small experimental project in computer-based learning is turning into a global network of ocean remote sensing teachers, with a life of its own.

Compared with some other international training initiatives in marine science, the significant impact and influence of this programme has far outweighed the modest resources which initiated it. This paper aims to describe this training programme in more detail and to document its worldwide penetration into marine science educational training establishments. It places the activity in the context of the global training needs for marine remote sensing. However, the underlying objective is to understand why the initiative has been effective, and to identify whether it has wider lessons to teach in general about achieving successful outcomes from international training endeavours.

Training for Satellite Oceanography

(i) Worldwide training requirements in ocean remote sensing

Remote sensing is no longer a specialist technology pursued by a few, but is an observational tool used from time to time by most oceanographers. Consequently all marine scientists need to be trained in the methods of remote sensing. Most need just to learn how to use satellite data, and to be aware of the capabilities and limitations of satellite sensors, although some will need specialist training if they are to contribute to satellite oceanography programmes.

For example, imaging sensors provide a sampling capability which is significantly different from conventional *in situ* ocean measurement methods. They make it possible to acquire "snapshots" of the instantaneous spatial distribution of constantly changing ocean parameters, at high resolution over wide areas. Users of such data need to be taught not only about the measurement methodology and the underlying scientific processes, but they also need to learn how to interpret ocean images. Another important characteristic of satellite-borne sensors is their capability for measuring global distributions of climatically important parameters in a regular, consistent way over many years. Present international marine research programmes to study global ocean productivity and the carbon budget of the ocean, to measure world ocean circulation and the transfer and storage of heat by the ocean, and to examine basin-wide interactions between the ocean and atmosphere, all depend on satellites to provide the spatial context and overview for local and regional experiments. It is evident that effective training must incorporate hands-on experience with image datasets. Pictures in textbooks will not suffice.

The requirement for satellite oceanography training arises in an international context because the use of satellites tends to bring together scientists from many countries into global programmes. Indeed, remote sensing is a particularly useful tool for scientists in countries lacking large resources of conventional oceanographic equipment, because it can provide ready access to large quantities of satellite observations of local seas at relatively low cost. Moreover, it enables isolated ocean scientists to make their own significant contributions to international programmes, for example by obtaining and contributing vital "ground truth" for calibration of satellite datasets. For this reason there is considerable merit in providing training programmes which are international in scope (see e.g., UNESCO, 1988).

Finally, there is also a requirement that education in ocean remote sensing should be offered at a variety of levels. As well as the obvious need to teach ocean scientists at graduate and postgraduate levels, there is also much to be gained from providing the rudiments of satellite data analysis to non-specialists: senior high school students, engineers concerned with environmental issues, planners, politicians, and environmentally aware members of the public. Satellite images have an immediate visual impact and offer an extremely effective way of communicating information about global ocean processes.

(ii) How is training to be provided?

If the need for training in ocean remote sensing is so evident, what obstacles are preventing its provision both locally and internationally, and what can be done to promote it? For a number of reasons, computer-based distance learning appears to be a sensible way to tackle the issue. Firstly there is a need for teachers. Although the importance of satellite data is recognised by all oceanographers, there are relatively few experts with the confidence to offer anything but the most basic of training. Therefore, training materials must be designed for use in a distance learning mode without dependence on local experts. It may be necessary for teachers of marine science to educate themselves before they can teach others, and computer-based distance learning methods enable teachers to learn along with their students.

Since the basis of remote sensing lies in physical science, there is a danger that marine earth scientists and biologists may not be motivated to study it. This would be a great pity since satellite measurements can contribute to all branches of ocean science. Such reluctance can be overcome by offering students a direct hands-on approach to learning, instead of, or in addition to, a standard textbook approach based around the theory of satellite oceanography which does not always inspire scientists to use the technique subsequently.

Unfortunately, the material required to provide hands-on experience of satellite data is not readily available in many educational establishments. Whilst most teachers of oceanography will have access to some satellite images, they are less likely to have easy access to a wide range of digital datasets. But just providing satellite pictures is not enough for students. Pictures tend to lead to qualitative analysis, whereas the goal of the ocean scientist is to make quantitative deductions about the ocean. Training programmes should therefore provide opportunity for the analysis of digital image data as part of the learning process.

The UNESCO Tredmar Computer-based Learning Module

(i) The BILKO image processing toolkit

Given the international requirement for training in remote sensing, constrained by the obstacles identified above, there is evidently a place for distance-learning educational material which can teach students of oceanography and practising marine scientists how to make use of satellite data. With this in mind, UNESCO commissioned the creation of a basic image analysis system which could form the heart of computer-based educational material. The result was BILKO, an image display toolkit software, written by R.D. Callison and M. Dobson of Pharos Scientific, Dundee, UK. It runs on personal computers operating under MS-DOS. To enable it to run on the majority of PCs throughout the world, and hence be widely accessible, it was deliberately kept simple, initially requiring just 640 kB of memory and an EGA display, although a later version makes use of the VGA display.

Another assumption underlying the toolkit was that it would be used by students who had no previous computing experience, so it needed to be very user friendly if its users were to "learn by doing". Therefore, all operations are reached through a simple screen menu, and a contextual help facility is fully built in, aspects which were not as common in PC software six years ago when BILKO version 1 was written. Students should be able to work on their own without supervision, and as far as possible the software was designed to eliminate situations where the machine might fail to respond, and an isolated user would be forced to abandon the work with consequent discouragement.

A fuller description of the software will be found in Callison et al., 1989. Usage of the software has been reported from around the world, and those lacking any access to more powerful image analysis machines have used it regularly as a research tool as well as for its intended educational application. Although its specification now seems dated in relation to recent rapid advances in PC technology, it continues to be widely used. Undoubtedly it was the essential foundation for all the educational activity which has developed from it. However, by itself it was not enough to generate the worldwide response mentioned at the start of this paper.

(ii) Lessons based on Bilko

Lessons must be provided which enable students to use BILKO to learn both the manipulative skills for displaying and enhancing satellite image datasets, and to extract quantitative information leading to scientific interpretations of their oceanographic significance. The demonstration lessons in the first module distributed by UNESCO set a pattern for those subsequently generated by others within the evolving network. The lessons are in printed form, accompanied by the image datasets in digital form on diskettes. An introductory tutorial, several lessons, and disks containing the software and the data make up a complete module. The lessons are structured to be self-contained and accessible by students with little or no local teaching backup. The user is assumed to be isolated and must find all the necessary resources within the module itself. This view of the target recipient underpins the editorial monitoring and review of all lesson material.

Each lesson commences with a statement of its overall aim and specific objectives so that students working in isolation can judge the extent to which they have learned what is intended from the lesson. The lesson text provides background information about the study area and details of the location of the image and how to convert the pixel value on the screen (a number between O and 255) into an ocean measurement. The heart of the lesson consists of a series of steps instructing a student on which functions of BILKO to use and which image files to load. It contains questions to draw out an interpretation and oceanographic analysis of the information contained in the images. Although learning is intended to flow from exploration of the digital image by interacting with the computer, some answers are provided so that students are not left without clear guidance.

(iii) A network of lesson users and producers

The first module was issued with no firm expectations of what might follow. Early exposure of the material at marine science educational workshops soon generated the realisation that lessons based on the software toolkit could just as well be written by recipients as by those organising the project centrally. When the initial mailing was distributed in mid-1989 to UNESCO marine science contacts throughout the world, recipients were encouraged to copy the material and share it with others. They were also requested to feed back comments, and to consider writing their own lessons. There was sufficient positive response of lessons offered for a second module to be planned, and by the time that was distributed in early 1991 (Robinson et al., 1991), it was evident that several more modules would follow, such was the steady flow of new lessons (Robinson et al., 1992).

In this way the network emerged naturally. Because the project operates on a low budget, depending primarily on the goodwill of those directing it under UNESCO's auspices, there was never any vested interest in forcing the programme in a particular direction. Had there been a limited response to the first module, it would have been the last, and the project would have terminated then. It continues to be managed with this philosophy. The user feedback provides the basis for new ideas.

No charges have been made for the modules. It was considered that the benefits to be derived from sales income would not justify the additional administration involved. Also the first module was sent unsolicited, and it is unlikely that the network would have grown so rapidly if modules had had to be purchased. The copyright is owned by UNESCO, and onward sale by recipients is forbidden, but this has not seemed to hinder the sharing and copying of material in a spirit of goodwill. Furthermore, if the project had been set up on a commercial basis, then authors might legitimately look for payment for their contributions which normally represent a considerable investment of time and effort. Instead, authors recognise the benefits to themselves and others of contributing on a sharing basis.

The individual effort in writing a single lesson can be multiplied many times over by receiving the lessons of others, extending the range of training material to cover satellite sensors, oceanic processes, and geographical regions, which would otherwise be beyond the individual's knowledge.

Global coverage achieved by the Project

(i) Distribution of the lesson modules

From the beginning, feedback was considered essential, not only to fuel the development discussed above, but to monitor the extent of use, and thus account for the investment of money, effort, and time by the project team. Without sales figures, the return of a response form by each recipient is important, including those who receive a copy second-or third-hand. To prevent the waste associated with distribution to uninterested recipients, each new module is distributed only to those who have responded from a previous module. Enquiries from names and organisations not on the distribution list are referred, if it is appropriate, to a nearby centre with an established commitment to the network. This has sometimes had the added benefit of encouraging local collaboration and networking between scientists and teachers.

In this way, some regional groups have developed their own momentum. For example, a Spanish-Latin American network has led to the translation into Spanish of modules 1 to 4, and the generation of a new module to be published first in Spanish. In the United States, users are experimenting with the transfer of lessons via electronic mail and data transmission networks. Both of these activities are feeding back into central development of the project.

Based on responses returned to UNESCO, the penetration of Module 3 (the most recent for which the feedback process is complete, is shown in Table 2.

Table 2: Penetration of Module 3

No. of countries receiving Module 3	71
No. of institutions using Module 3	502

It can be assumed that the number of individual users runs in the thousands. It is also pleasing to see that the material is being used in at least 71 countries. Table 3 shows the ten countries which make the greatest use of the lessons. Although the network tends to be most developed in those nations which already have a strong marine science infrastructure, users in the other 61 countries make the network a truly global one, reaching places where satellite oceanography is just beginning to develop.

Table 3: Countries with highest numbers of users

USA, UK, Canada, France, Netherlands, Mexico, Spain, Brazil, Germany, Italy.

(ii) Uses of the lesson material

The reported use of the BILKO image processing toolkit and the associated lessons covers a wide range of situations. Many copies are used individually by scientists, teachers, and students, learning more about satellite oceanography in the distance learning mode envisaged by the authors of the project. Some of the scientists are research scientists in situations with no access to powerful image analysis computers, and use BILKO for simple research applications as well as educationally. The material is also in use by government planning officers and consultants.

Another way in which it is frequently used is by groups of students in classwork situations. The software is either loaded individually onto each of a cluster of personal computers, or else it is placed on a network fileserver. Typically, the groups being taught in this way are undergraduate or postgraduate students of oceanography or coastal science taking a specialist unit in remote sensing. However, some teachers have developed simplified lessons which they offer to first year undergraduates, or even to high school students.

Regional group training activities and summer schools in satellite oceanography and marine remote sensing have also started to make regular use of the material as a simple way of providing hands-on training to support the more conventional lectures. Courses known to have made use of the material include those held in Nairobi, Kenya; La Reunion; Caracas, Venezuela; Enschede, Netherlands; Lepe, Spain; and Rio Grande, Brazil. When teachers running such courses request copies of the material, they are invited to use the opportunity of the course to produce some further lesson material to feed back to the network. This has resulted in a number of valuable lessons produced by groups of teachers and students. In one case, a course at Lepe, Spain, has led to the preparation of a whole module, which will soon be published.

(iii) Lesson Contributors

The variety of lessons published so far is evident from the titles listed in Table 4. They are based on a range of oceanographic phenomena and coastal features

Table 4: Title of lessons

MODULE 1 (July 1989) Images of open ocean sea-surface temperature Thermal images of ocean fronts Coastal Zone Color Scanner data off Iceland Comparison of coincident AVHRR and CZCS data U.K. Shelf seas thermal structure Airborne thematic mapper data - Southampton Water.	**MODULE 2 (February 1991)** Analysis of a SAR image of the Straits of Dover Seasonal plankton distribution - Arabian Sea Coastal discharge - Brazilian Coast Wind-driven coastal upwelling, South Australia Thermal structure of the southern Coral Sea Display and analysis of ADCP (velocity) data
MODULE 3 (February 1992) Extent and content differentiation of coastal plumes (Malaysia) Using Landsat for coastal management - mapping and determining major substrate categories Using Landsat for coastal management - vegetation mapping Detection of potential areas for coastal aquaculture using the Indian sensing satellite - IRS 1A Study of Indian lagoonal features using IRS-1A LISS data Temporal and spatial variations in sea-ice concentration in the Southern Ocean Global waves and winds from Geosat Ocean eddy dynamics generated by a numerical model	**MODULE 4 (September 1993)** Imaging *in situ* bio-optical and physical parameters Spatio-temporal evolution of coastal upwelling in the Senegalese and Mauritanian littoral Variation of AVHRR-derived sea surface temperature with satellite zenith angles Seasonal and yearly variation of surface-temperature distributions in the North Sea Analysis of the turbid plume of the Gironde Estuary (France) using SPOT data Detection of sea-surface life with an Airborne Synthetic Aperture Radar Bathymetry prediction from satellite altimeter data

illustrated with examples from around the world, and several different remote sensing methods are covered. There are also some lessons which do not use remotely sensed data, but employ two-dimensional datasets from other sources, since the BILKO software is suitable for lessons based on any image-type data. The authors of the lessons are also from around the world, as illustrated in Table 5.

Table 5: Country of origin of lessons

Lessons published by marine scientists from: Australia, Brazil, Canada, France, Germany, India, Netherlands, Singapore, United Kingdom, USA.
Lessons in press from: Argentina, Chile, Mexico, Peru, Spain, Venezuela

Future Developments

As long as additional lesson material continues to be contributed and the network remains active, it will be necessary to provide editorial coordination of the project. However, personal computer technology is rapidly developing, and hardware is now widely available which is far more powerful than that for which BILKO was originally designed. If the network of lesson users and producers is to be effectively served, it will be necessary to keep abreast of these developments. It has therefore been decided to provide a new version of BILKO which will operate within the environment of Windows 3.1, and Microsoft user interface software, which is now widely available around the world and which is effectively setting standards for personal computer software.

Another exciting possibility is that the lesson material itself can be supplied through the computer, in another window, instead of on the printed page. From this concept flows a number of additional benefits. Hypertext functions can be used to make the lessons more interactive, responding to the needs of the individual user. Linkage between the text window and the BILKO window will lead the student naturally through the lesson, without the need to consciously switch between the two. Because the whole module will be computer-based, the material will be optimised for direct electronic transmission through computer and educational media networks. There will also be scope in future to include multimedia capability into the lessons (e.g., sound and video), and to access data sources other than satellite images (e.g., oceanographic databases on CD-ROM). In this way the network might evolve beyond remote sensing to provide lesson material on oceanography in general. The software development required to make this possible is now in progress.

Discussion and Conclusion

As a worldwide educational initiative, the provision of the BILKO toolkit as the basis of distance learning material in satellite oceanography can be judged a success. It demonstrates the effectiveness, even in the context of advanced technical training, of matching the teaching materials to readily available hardware, rather than attempting to reproduce the sophistication of advanced research tools in the lessons. Beyond the success of BILKO, however, it is the emergence of the user-producer network which has taken the authors by surprise. It may be useful to consider why this aspect of the project seems to have developed a life of its own.

Firstly, by opting for the lowest common denominator in terms of equipment, few potential users have been excluded, and contributions to the developing network have been elicited from all parts of the world without requiring a strong dependence on those nations which have major space and remote sensing programmes. Training in satellite oceanography is to be encouraged in all maritime nations because remote sensing should provide a means for marine scientists from all nations to play their own unique part in the task of understanding earth system science and global change, and in managing the marine environment. It is therefore of significance that lesson material should derive from genuine training contexts in all parts of the world and not be delivered "top down" from outside the network. The BILKO network has grown because it allows local teachers to participate actively as producers, and not just passively as users. It provides opportunities for isolated teachers and scientists to make an internationally valued contribution.

Secondly, the framework provided by UNESCO to administer the network has not sought to dominate its growth, but has tried to provide underpinning support. This has consisted of the original software development and its modification in response to suggestions from the network; the free provision of the modules to committed users, some of whom have acted as local distributors; the provision of a central contact point through the UNESCO TREDMAR office; the establishing of editorial standards for the educational material; and the provision of tutorial material to help lesson creators. The organisational structure has not been established for its own sake, and will remain in being only as long as there is a substantial need to be met in the form of continued demand for educational material and the provision from within the network of new lessons to be edited.

Thirdly, the number of marine science educators in individual countries is usually small, and the concept of sharing educational resources through a network enables them to keep abreast of the rapid pace of technological development in remote sensing. More generally, the network approach offers an effective way to create an international resource base of computer-based learning material in marine science.

However, if the project is to remain effective as the scope of the network grows, it is becoming apparent that it will need to be properly supported. So far it has relied on a small UNESCO budget and the goodwill of the editorial and software development team. Lack of dedicated personnel has caused delays in editing and compiling the modules, and this is frustrating for authors. Provision of additional resources for coordination are therefore necessary to keep pace with the network growth. The core software development also needs to be commissioned at a price appropriate to the standard expected by the network, even though it is hoped that within a windows environment, additional software functionality can be donated by network members. Finally, new lessons require suitable image data, which will not be readily and cheaply available, if space agencies extend pricing structures which are prohibitive for many users of the network. Since the network could be a very effective way to provide skills training for the anlysis of data from new sensors, there is evidently scope for collaboration between the space agencies and the network, including the provision of free data for educational purposes.

References

Callison RD, Robinson IS, Blackburn D, Cracknell AP & Cummings D (1989) Some marine applications of satellite and airborne remote sensing - A computer-based learning module, MARINF/70, UNESCO (Mar. Sci. Div.), Paris, 90pp + disks.

Robinson IS, Callison R, Blackburn D, Boxall S, Garcia C, Monaghan J, Nunes Vaz, Sathyendranath S & Troost S (1991) Applications of marine image data - second computer-based learning module, MARINF/81, UNESCO (TREDMAR Progr), Paris, 85pp + disks.

Robinson IS, Callison R, Carter D, Nichol J, Sheppard C, Sudarshana R, Vifehoff T, Wells N, Blackburn D & Troost D (1992). Applications of marine and coastal image data from satellite, airborne and in-situ sensors - Third computer-based learning module, MARINF/83, UNESCO (TREDMAR Progr), Paris, 101pp + disks 22.

UNESCO 1988, Year 2000 challenges for marine science training and education worldwide, UNESCO reports in marine science, 52, UNESCO, Paris, 129p (available in 6 languages).

Teaching Global Change Topics by Linking Classroom and International Field Research Experiences

Douglas F Williams
Marine Science Program
University of South Carolina
Columbia, SC 29208, USA

Introduction

Under the auspices of the nationally recognized strengths of the Marine Science Program and the Baruch Institute at the University of South Carolina, a special seminar-style course (MSCI 399) titled "Comparative Functioning of Biogeochemical Cycles in Freshwater and Marine Systems" was taught to a team of undergraduates called the Baikal Undergraduate Research Group (BURG). Our hypothesis was that with the proper motivation and mentoring, outstanding undergraduate students could seize exciting opportunities to conduct cutting-edge research, undertake collaboration on an international scale, and develop leadership and organizational skills usually not thought possible at this early stage in their careers.

Goals of MSCI 399

The formal course work of the global change topics course involved readings coupled with weekly written assignments and oral questioning in discussion format to provide undergraduate science majors with foundation knowledge about such topics as global greenhouse processes, global climate change, the biogeochemical cycle of carbon, phytoplankton dynamics and nutrient chemistry, the preservation of sedimentary organic matter, and the physical limnology of large lakes, especially as related to Lake Baikal. In consideration of the students' backgrounds, the program was designed to familiarize students with the challenges of integrating information from a variety of sources and disciplines to attain an higher understanding of complex scientific problems. The extended goal was to link this knowledge with a field research experience in order to gain hands-on exposure to a global change problem. In addition to the formal course work, we also discussed ways for students to achieve critical thinking, establish research priorities, seek funding for research, and experience the process of proposal writing.

NATO ASI Series, Vol I 29
Global Environmental Change in Science
Education and Training
Edited by David Waddington

In anticipation that the program would be an on-going, multi-year training effort, the class rank of the BURG students purposefully ranged from first to third year students whereby "senior", more experienced students would be encouraged to assume leadership roles as "peer mentors" as they gained in knowledge, experience, and maturity. In addition to the seminars, we also discussed the nature of collaborative research, methods of seeking research funding, and the importance of disseminating research results through publications and scientific meetings. Special consideration was also given to the exciting cross-cultural aspects of working with Russian limnologists, biologists, and geochemists before, during, and after the expedition. Using a nearby freshwater system (Lake Murray), the BURG students also received pre-expedition hands-on training in the methods of sampling to be employed on the expedition. The classroom and laboratory experiences outlined above were enhanced through a team of faculty mentors both inside and outside USC in the areas of geochemistry, paleolimnology, organic geochemistry, phytoplankton dynamics/sediment trapping, sedimentology/palynology, ecological modeling, and so on. The course outline is given in Table 1.

Table 1. Course Outline for the Marine Science 399 Spring 1994 "Comparative Functioning of Biogeochemical Cycles in Freshwater and Marine Systems"

Topic 1	The Global Climate System
Topic 2	The Carbon Cycle and Global Greenhouse Processes
Topic 3	Introduction to Physical Limnology of Large Lakes
Topic 4	Nutrient Chemistry in Freshwater Systems
Topic 5	Phytoplankton Dynamics and Nutrient Cycles
Topic 6	Limnology and Phytoplankton Productivity of Lake Baikal
Topic 7	The Biodiversity of Lake Baikal
Topic 8	The Origin of Lake Baikal and Rift Basin Sedimentation
Topic 9	Principles of Ecological Modeling
Topic 10	Introduction to Organic Geochemistry of Lacustrine Sediment
Topic 11	Evidence for Paleo-pCO_2 from Marine, Lake and Ice Core Records

The Field Research Experience

To link classroom discussions of Global Change with the field research aspect of the program, we scheduled 3 weeks of ship time on Lake Baikal in May-June 1994 to coincide as closely as possible to when the frozen surface of Lake Baikal is melting (Sitnikova et al., 1984; Bolgrien, 1993). As the surface waters are exposed to atmospheric CO_2 after 4 months of isolation and as the microscopic phytoplankton flourish, we wanted to determine how the carbon reservoir of Baikal was affected and what geochemical signals in both the organic and inorganic carbon reservoirs of the surface waters are recorded and preserved in the underlying sediments. Previous work with our Russian colleagues has shown that silica and organic carbon from diatoms are dominant components of modern Lake Baikal sediments. Previous work on the carbon cycle in a small Swiss lake by Hollander and McKenzie (1991) has shown that the organic carbon from phytoplankton faithfully records the seasonal changes in the concentration of CO_2. With these results, the authors proposed that the organic record of lake sediments must act as a monitor of past changes in atmospheric CO_2 (paleo-pCO_2). By verifying the response of Baikal's surface waters to changes in atmospheric CO_2, we wanted to define some of the parameters for determining whether a paleo-pCO_2 record is represented in the sediments of Lake Baikal.

The BURG students with their American and Russian faculty mentors conducted the sampling at 14 stations (Fig. 1; Table 2). The exact number of stations and the depth of sampling were determined by the depth of maximum

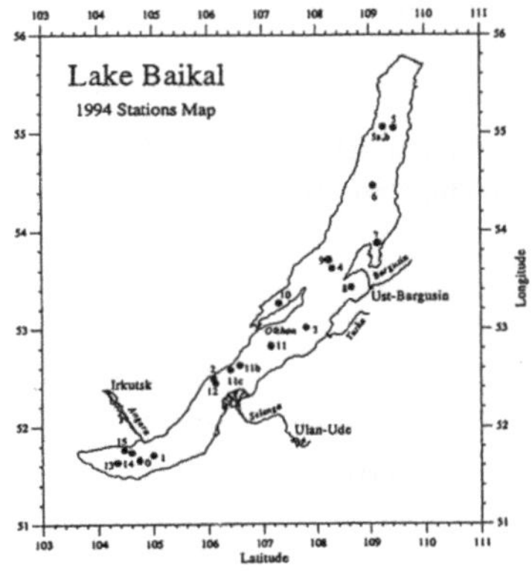

phytoplankton. In addition we attempted to document the nature of the organic materials falling through the water column (sediment traps) and arriving at the lake bottom (box coring). Aboard ship, one or two students were responsible for a particular sampling task and subsequent analytical work. All students were familiar with our reasons for conducting each sampling task and understood that all data would be pooled, discussed, and interpreted as a group in consultation with associate faculty of the project. It was anticipated that all of the students would continue working on the samples as part-time assistants for the summer and fall 1994-spring 1995 semesters. Prior to the expedition, a detailed analytical plan that took into consideration existing facilities and expertise was decided jointly by the students and our Russian colleagues.

At least two BURG students are using data from the expedition for their senior theses in the South Carolina Honors College. Two students remained in Irkutsk for the remainder of June-August (ten weeks) to participate in additional expeditions and to work with their Russian mentors at the Limnological Institute and Institute of Geochemistry. We are now working to utilize existing satellite imagery and remote sensing to track certain variables of Lake Baikal related to the *in situ* stations. Our eventual goal is to obtain an annual time series of the changes in phytoplankton flora (biomass, diversity, etc.), nutrient chemistry, and geochemistry of the organic and inorganic carbon reservoirs

Table 2. Sampling Protocol for May-June 1994 Baikal Expedition

Type of Sampling	Analytical Determinations
1. plankton towing using nets arrayed to sample the maximum phytoplankton abundance of the upper 250 m	special identification biomass measurements stable isotope analyses
2. water column sampling using Niskin bottles	total dissolved CO_2 alkalinity, pH nutrient chemistry (nitrate/phosphate)
3. water column sampling of particulate organic carbon (POC) by filtering pumped water	stable isotope analyses
4. sediment trap deployment	total organic carbon/nitrogen content stable isotope analyses biogenic silica content inorganic particulate content
5. box coring of uppermost surface sediment	total organic carbon/nitrogen content biomarkers (organic geochemistry) stable isotope analyses biogenic silica content
6. CTD deployment	pore water geochemistry profiles of temperature, pressure, light transmission, dissolved oxygen

Acknowledgments

I personally want to thank the many Russian and American individuals and faculty mentors who helped make possible the formation of the Baikal Undergraduate Research Group: Sarah Fowell, Oleg Timoshkin, Nina Guselnikova, Nina Bordeenko, Evgeny Karabanov, Valentina Sidelova, Mikhail Grachev, Mitchell Colgan, Dwayne Porter, Steve Davis, Robert Thunell, Eric Tappa, Tom Chandler, Bill Johnson, Willard Moore, Tim Shaw, David Hollander, Jim Morris, Hank McKeller, Jim Collister and Dave Wavrek. Of course, none of the expedition's success would have been possible without the effort of Captain Oleg Kalinin and the crew of R/V *Vereschagin*. On behalf of BURG, I also gratefully acknowledge financial assistance from the Marine Science Research Fund of the Provost's Office, Dr. William Murray and Samuel Freeman Charitable Trust, the Explorer's Club Research Fund, the REU program of the National Science Foundation, the South Carolina Honors College, the Belle W. Baruch Institute for Marine Biology and Coastal Research, and the USC Marine Science Program. In addition I thank the Limnological Institute (M. Grachev), the Institute of Geochemistry (M. Kuzmin, T. Bounaeva, and E. Almukhamedov), the Baikal International Center for Ecological Research, the United Institute of Geochemistry, Geophysics and Mineralogy of Novosibirsk (V. Efremov and N. Dobretsov), and all of the Siberian Branch of the Russian Academy of Science for many forms of assistance.[1]

References

Bolgrien DW 1993 Delineation of the Hydrodynamics of Lake Michigan and Lake Baikal Using Satellite-Derived Surface Temperatures. PhD Dissertation, Biological Sciences, Univ. Wisconsin, Milwaukee

Hollander DJ and McKenzie JA 1991 CO_2 control on carbon-isotope fractionation during aqueous photosynthesis: a paleo-pCO_2 barometer. Geology v 19, p. 929-932

Sitnikova GV, Furman MS Yanter NN 1984. Remote and ground-based methods for researching the dynamics of ice processes in Lake Baikal. In: Remote and Ground-based Studies of the Dynamics of Natural Processes in Siberia. Inst. of Geography, USSR Academy of Sciences, Irkutsk, p. 72-81

---

1 USC is a rounding member of the Baikal International Center for Ecological Research, along with the Siberian Branch of the Russian Academy of Sciences, the Royal Society of London, the Belgium Academy of Natural Science, the Japan Baikal Research Association, and the Swiss Federal Technological Institute.

Global Change: A Context for Teaching Science

Joseph P Stoltman
Department of Geography
Western Michigan University
Kalamazoo, MI 49009, USA

David J Waddington
Department of Chemistry
University of York
York, YO1 5DD, U.K.

Introduction

This workshop is concentrating principally on the training of chemists who will be concerned with global change, either in research or management. Such scientists usually come from single-subject honour degrees in physics, chemistry, earth science, and biological sciences, yet they are expected to understand and appreciate the role all sciences contribute to the understanding of the key issues in the environment, which are known under the umbrella title 'global change'.

The problem is exacerbated by school and junior university courses in which the sciences are also taught in terms of specialisms using boundaries defined a century ago. Further, the content is often delivered as a series of received truths, in terms of simple models. Environmental concerns have therefore very little impact on many school and university science courses. This leads, in turn, to many students seeing science as irrelevant to their needs and being attracted to other subjects in the curricula which *appear* more interesting and suitable for their careers.

Education in Global Change Project (EGC)

The Education in Global Change (Stoltman and Waddington, 1994) is a new project sponsored by the International Council of Scientific Unions[1] for students in senior secondary schools and colleges (ages 16-20). EGC was created by a group of science educators drawn from France, India, the United Kingdom, and the United States, with considerable help from scientists involved in the International Geosphere-Biosphere Programme : A Study in Global Change, to ensure that the materials were accurate and fair.

---

1 *International Council of Scientific Unions* has a membership of scientific academic and research councils and international scientific unions (for example, the International Union of Pure and Applied Chemistry, IUPAC).

NATO ASI Series, Vol I 29
Global Environmental Change in Science
Education and Training
Edited by David Waddington

These materials are put together in six units:

The Changing Atmosphere
Clues from the Past: Glimpses of our Future
The Global Cycle
Population and Land Use
Oceans
Remote Sensing: Windows on Global Change

Each unit has a series of **student activities**, which can be used to introduce important scientific principles, principles such as met in courses at this level in thermodynamics (energetics and equilibria), in reaction kinetics (catalysis and mechanisms), free radical chemistry, chemical analysis, the electromagnetic spectrum.

The principles are set in the context of key contemporary environmental issues, which, in turn, help students towards the understanding of issues such as the protection of the ozone layer, the enhancement of the greenhouse effect and the destruction of the rain forests.

We want, at the same time, to show that science is a force for good and that scientists have an important role in enhancing the quality of our lives. To show this, students must be able to appreciate what advice scientists can give to decision makers, be able to judge the quality of that advice and appreciate what effect the advice, if accepted will have on everyone. It is important that students are able to make up their own minds about these and issues yet unfocussed. Thus, many of the activities have been made as open as possible to allow students make up their own minds about the conclusions it is possible to make from the evidence.

To summarise, the activities are designed to:

- cover key scientific concepts;
- illustrate science in the context of important environmental issues;
- encourage student participation;
- encourage skills such as data interpretation, decision making and communication; and
- illustrate the positive role played by scientists in society.

The units can be used in different ways:

- as a whole unit, occupying about 10-12 hours
- in existing courses, as single activities to illustrate a specific scientific principle or environmental issue by reassembling
- as items from more than one unit to produce a unit, specific to school or college.

Each unit begins with an **introduction** for the teacher which includes a synopsis describing the student activities. These are brief descriptions so that teachers can readily select those activities they wish to use for their courses. We do not expect the activities to be used exactly as printed. Many teachers will wish to tailor them for their own purposes. The **activities** include laboratory experiments, data collection, data interpretation, literature surveys, short talks, and writing. There is a minimum of teacher-led work, with most time devoted to the teacher assisting the students in discussing procedures and judging the significance of their findings. EGC is *not* extra work. Rather the activities provide *alternative* means for the teacher to introduce or illustrate principles which are in their courses in a new and exciting global change context. Since some teachers are themselves likely to come across the global change context for the first time, great care is taken in the units to ensure they are given adequate background information in the **Teachers' notes.** Here are also found the answers to questions and discussion suggestions, hints about setting up experiments and suggestions for teaching strategies.

Examples of Activities

(i) Key scientific concepts

In one activity in *The Changing Atmosphere*, the greenhouse effect is used as a way of introducing the properties of the electromagnetic spectrum. The students design and carry out an experiment to illustrate blackbody radiation in which they use powdered graphite, some oil and two thermometers. In the same unit, they collect and then interpret absorption data for a series of gases, including oxygen, nitrogen, methane, carbon dioxide and CFCs. The activity introduces them to the properties of molecules which lead to different absorption patterns and on to the idea of the atmosphere functioning as a greenhouse window. The properties of the electromagnetic spectrum are reinforced when its role is examined in another activity, in *Remote Sensing: Window on Global Change*.

In *Global Carbon Cycle,* students carry out a series of basic experiments using a bottle of soda water. The closed bottle is treated as a system at equilibrium and when the cap is removed, the dynamic equilibrium is disturbed. A series of questions and experiments unfold. How much carbon dioxide will water hold? What happens when the system gets warmer? Thus the solubility of gases is introduced related to an atmospheric context. The reaction chemistry of carbon dioxide is also introduced in this way.

(ii) Student participation

We have designed the activities so that students participate in their design and interpretation, illustrating many of the dilemmas facing scientists. *Land Use and Population* examines the human dimension of global change. It begins with an analysis of population change and the demands populations place on the global environment both for resources and disposal. The contribution of land use to the production of several global change elements, such as carbon dioxide and methane, are analysed as case studies. Care is taken to examine cause and effect relationships, and to develop and test hypotheses that require the use of statistics and information from sciences other than geography.

In *Clues from the Past: Glimpses of the Future,* students use actual data to determine long term changes in the atmosphere, and evaluate their importance in projecting what might happen in the future. The data include evidence such as the size of tree-rings, pollen and, perhaps most dramatically, samples of ice drilled deep in Antarctica. The activities include a laboratory experiment and game designed to give practice in interpreting data.

Remote Sensing: Window on Global Change relies upon satellite and remote imaging technology to analyze the changes that have occurred with land cover in several regions of the world. The students are given the opportunity to examine the images, make measurement and use the image as a source of data, process the data and test hypotheses related to both the temporal and spatial implications of global change. The use of temporal data permits the students to make predictions, and to check those predictions against the uncertainty that accompanies scientific investigation. A case study is presented of Rondonia State, Brazil. This real life laboratory is revealed *via* remotely sensed images that are used to measure the increasingly greater amounts of deforested land between during three time periods. The students observe and analyze the changes, relate the changes to other regions of the world, and begin to measure the cumulative effects of human action on land cover and the relationships between people and environment.

In *Global Carbon Cycle,* students interpret data, and construct, both qualitatively and quantitatively, a carbon cycle that includes considering the limits of data and the types of errors scientists must consider.

In *Changing Atmosphere,* students learn about the reactions of nitric oxide and ozone. Then they are asked to predict what effect a new fleet of supersonic aircraft may have on the stratosphere since both nitric oxide and ozone will be present. They have to decide what data are required and what assumptions they need to make. They are faced with a real problem, such as scientists face when they predict the effects of scientific and technological changes on the environment. For example, this activity, based on the work of Harold Johnston, Douglas Kinnison and Donald Wuebbles (1989) gives students a salutary lesson in the difficulties of looking far into the future to explore and predict compounding effects. Finally, the students are asked to write a short account for a newspaper read by a general clientele. There is nothing to compare with the difficulty of trying to explain science to a non-scientific audience, and the student soon realizes that to do so, you have to have a real mastery of the science yourself!

The *Education in Global Change Project* has brought scientists and science teachers together to collaborate on teaching resources for schools and colleges which illustrate key scientific principles and, at the same time, show the essential role of scientists in society. This, combined with the international structure of the project, demonstrates that science and the global environment have no national boundaries.

References

Johnston HS, Kinnison DE and Wuebbles DJ (1989) Nitrogen oxides from high-altitude aircraft: An update of potential effects on ozone. J Geographical Research 94: 16351 - 16362

Stoltman JP and Waddington DJ (eds) (1994) Science for understanding tomorrow's world: Global Change. International Council of Scientific Unions.

Table 1
Education in Global Change Units

The Global Carbon Cycle
At the heart of the problem of global change lies the role of carbon dioxide and its fate in the atmosphere. This unit includes authoritative data on the various reservoirs and fluxes in the global carbon cycle. Activities require students to interpret the data, to carry out simple experiments and to discuss some of the global change consequences of the cycle.

Remote Sensing: Window Global Change
Remote sensing plays an important role in detecting changes in land cover, types of vegetation, and ocean surface conditions. This unit includes exercises in the use of remote sensing techniques ranging from aerial photographs to satellite imagery to monitor changes occurring on the earth's surface.

The Changing Atmosphere
The unit gives students a firm understanding of the effect of trace gases produced naturally and by human activity, on the stratosphere. It includes the greenhouse effect and the role of the ozone, methane and chlorinated hydrocarbons in the troposphere and stratosphere using a series of data interpretation exercises.

Clues from the past: Glimpses of our future
The unit introduces students to a range of natural archives which can give scientists clues about climate change in the past. Data analysis and practical exercises show students how scientists interpret the results of studying tree rings, ice cores and pollen grains. The units end with activities to underline the significance of these studies both for scientists interested in climate modelling and for policy makers in society.

Population and land use
Population and the activities of people are two significant variables in the investigation of global change. Global population has increased dramatically during the 20th century and growth projections suggest continued increases into the next century. Changes in population have an important effect on land use in many parts of the world, with more land being used for agriculture in particular. The effects of increasing population and the contributions of changing land use to global change are investigated.

Oceans

The importance of the oceans when considering global change are illustrated in two sections. One, entitled "Blooming Algae", gives an overview of the way in which algae in the oceans interact with carbon dioxide in the atmosphere, during which ideas on positive and negative feedbacks are introduced. In the second, "Salty Sinks and Sources for Carbon Dioxide", students examine data giving possible carbon dioxide production and absorption in the oceans.

SECTION 4: WORKSHOP DISCUSSIONS

Education and Training in Global Change Science

A review of the working groups

John Boardman
Environmental Change Unit
University of Oxford
Mansfield Road
Oxford, OX1 3TB, UK

Lynne Carter
Graduate School of Oceanography
University of Rhode Island
Narragansett, RI, USA 02882

Rudolf de Groot
Centre for Environmental and Climate Studies
Wageningen Agricultural University
PO Box 9101
6700 HB Wageningen, The Netherlands

David Waddington
Department of Chemistry
University of York
York, YO1 5DD, UK

There were three working groups, one concerned principally with early undergraduate courses, another with senior undergraduate courses, and the third with postgraduate courses. Each group was asked to structure its discussion along the following lines:

What are the significant objectives of courses in global change science at the level of education being discussed by the group?
Where are we now?
Where should we be going?
How do we get there?

The groups worked independently, coming together for plenary sessions to report on progress and to give advice to one another on possible discussion points. Each will be described and, at the end, there is a synthesis.

WORKING GROUP 1: SENIOR SECONDARY AND FRESHMAN UNIVERSITY COURSES

Working Group 1 addressed the role of global change science in senior secondary and early college courses. The focus was on introductory level courses concerned more with breadth than with depth. The courses are intended for all students, not only those planning a career in science, but also those planning non-scientific professional careers.

NATO ASI Series, Vol I 29
Global Environmental Change in Science
Education and Training
Edited by David Waddington

The purposes of these introductory courses are to familiarize all students with the earth system concept and the role of humans in that system and to contribute to the development of responsible citizenship by

- improving their scientific understanding (content) related to global change issues,
- teaching them how to use scientific evidence in addressing issues in global change,
- enabling the application of knowledge about the earth system to the evaluation of public environmental information, and
- encouraging them to appreciate personal behavior and increased participation in public policy related to global change issues are necessary.

Where are we now?

A growing proportion of the population is interested in understanding a wide range of global environmental issues. Interests range from a future as an environmental scientist to that of a concerned citizen wanting to participate in the increasingly technical public policy debates. The educational needs students in senior high school and the first year in college or university are being met, at least in the US, through offering an introductory or survey course of one or two semesters that addresses the wide range of interactions between humans and the natural environment. There are fewer examples from other countries. However, even in the US, there are a number of problems that face the introduction and maintenance of these introductory courses. Issues and questions include: interdepartmental 'turf' and support; granting science credit to students when the course also discusses human dimensions; who should be required or allowed to take such courses; developing an introductory course with available expertise; departmental responsibility for cross-disciplinary courses; and course focus for a mixed audience (e.g., science and non-science majors).

Many of these potential difficulties must be handled course by course, and university by university. The mixed audience issues, however, were generic enough that some comments were offered by the participants which is described in the next section.

Where should we be going?

The working group felt that all educational institutions should play a role in increasing staff and student awareness of the interactions between humans with the environment. The information to achieve this should be available at all levels of education. While the focus of this report is on formal education, the group considered that informal education related to global environmental issues should be offered to all citizens at all levels through a variety of outreach efforts that hold particular implications for teachers in primary and secondary education.

The following *Overview of an Introductory Course in Global Change* represents the sort of course that could become a regular offering from many educational institutions. The working group recognized that it is impossible to address all of the issues that could fall under a heading of global change in any one course, but that any survey course aimed at the senior high school or early undergraduate student should include four important perspectives. The four perspectives are earth system processes, interactions, the role of humans in the global system, and tools and approaches. The themes and order of perspectives chosen will depend on the expertise available at the educational institution as well as the structure of the course and its function.

Overview of an Introductory Course in Global Change

Specific aspects of the four perspectives important to any introductory course on global change are:

1. ***Earth system processes:*** *links between and processes in the atmosphere, hydrosphere/cryosphere, biosphere, and lithosphere*

   Depending on the global change/human interaction issues emphasized in the course, the basics of different aspects of the earth system involved must be understood. For example, if ozone depletion is the issue being addressed, then atmospheric circulation must be understood; or if global warming is the issue, then oceanic circulation, biospheric interactions and the greenhouse effect must be prominent.

2. ***Interactions:*** *natural and anthropogenic influences on element cycling in the earth system*

   The important focuses of this course are the interactions between the human system and the physical sea/land/air system. The emphasis in this category should be towards an interdisciplinary (not a multidisciplinary) approach to learning. Using global warming as an example, natural and anthropogenic contributions to greenhouse gas build-up (i.e., volcanic activity, plants, such as rice growing which gives off methane, and fossil fuel use) and the feedbacks in the system need to be understood and discussed.

3. ***Influence of humans in the global system:*** *anthropogenic influences, impacts, and responses*

Specific issues that contribute to and are impacted by global change can include population, energy, resources (depletion, distribution, recycling), acid rain, desertification, global warming, ozone loss, land cover changes, biodiversity, sea level rise, water distribution, socio-economic issues, equity concerns, and response options including individual and collective, economic, national and international, and technical.

The issues that are chosen as a theme or a focus for any specific introductory course could be any combination of the above. Again using the example of global warming, one might include a discussion of the economics of fossil fuel use versus alternative energies, and all of the social and environmental implications of our energy choices.

The definition of what issues are considered global environmental change are variable, thus the list of choices. The issues chosen will determine which earth processes are included, which interactions are focused on, and therefore the teaching strategies, such as laboratory or field work and role playing, to be utilized in the course.

4. ***Tools and approaches***

In teaching this interdisciplinary course to the chosen audience, it is important to utilize a number of approaches, one of which addresses the understanding of tools that scientists use to collect data. Familiarity with appropriate scientific tools is an important perspective in understanding uncertainty of scientific data and the resulting information. It was deemed crucial that participants have an historic perspective, as well as an introduction to such scientific techniques and viewpoints as

- scientific method and experimental techniques
- remote sensing of the environment
- Geographic Information Systems (GIS)

- computer models and simulation of environmental change
- chaos theory and limits to predictability
- earth history, palaeoclimatology, and past environmental change
- analysis of uncertainty and perception of risk

The choice of what technologies to discuss, again relates to the topic chosen. In global warming, how carbon dioxide is measured in the atmosphere, past and present, would be important to understand.

The wide range of potential audience for this sort of an introductory course implies a disparity of backgrounds, which in turn suggests a difficulty in choosing appropriate teaching focuses and strategies. Some of the options suggested for dealing with different groups in the same audience include:

- additional work or sections for advanced students in their own speciality and interest, and
- use of creative teaching methods, such as role playing exercises, interactive computer technology, team or individual projects, and other methods which allow active participation by students in their education. Some examples are given in the papers.

How do we get there?

The approach to getting a course / unit like this offered in an institution will be specific to each institution. However, there are a number of models that have been employed with various levels of success.

Top down: A senior faculty member initiates or is willing to support strongly junior colleagues to mount a new course.

Bottom-up: Strengthen or extend existing courses. This can lead to the evolution into a whole new course. One discipline may first show by example; then others copy it by extending their courses. Eventually there should be a groundswell of support for a fresh new coordinated course in global change science.

Staple gun[1]: Use existing courses or pieces of courses. This is often used for a unit or grouping of courses and is easier if the university has a modular system, such as that utilized in the North American model. If a university has a tight cellular structure as in many European universities, there are considerable staffing problems for new courses.

---

1 Staple gun - a vivid metaphor much used in our discussion, implying the collation of a new course by stapling together modules from pre-existing courses. (Schneider 1994)

The working group also wanted to comment on the broader aspects of the goal for all educational institutions to be actively pursuing programmes for all levels, aimed at increasing awareness of our relationship with the environment. A variety of strategies could be employed to achieve that goal. One strategy could be to develop a series of focused offerings, for example, a specific course for school teachers, businesses or the public. Another strategy could be employing distance-learning methodologies (i.e., electronic, TV, video) for those unable to attend. Finally, an important group to reach are those scientists that have the information but have no experience in offering other than technical programmes.

A university group could arrange special sessions at professional meetings to provide available resources to encourage outreach efforts by its faculty and staff.

WORKING GROUP 2: SENIOR UNDERGRADUATE COURSES (3RD AND 4TH YEARS) IN GLOBAL CHANGE SCIENCE

Working Group 2 addressed the role of global change science in advanced undergraduate courses (for 3rd and 4th years). The courses considered were those aimed specifically at science students with a good, though discipline-oriented, knowledge of science at the freshman level. The objectives were not dissimilar to those from Working Group 1: namely that students in global change science should

- improve their scientific understanding in contexts related to global change to enable them to be prepared for studies at postgraduate level and for professional careers
- become aware of key issues in global change
- be trained to use systems analysis and models, and risk analysis
- be given as wide a view as possible of the impacts of global change on society through evaluation of implications of global change scenarios and therefore development of preventive policies to reduce risks.

Where are we now?

The development of programmes that cater for students in advanced undergraduate courses are at very different stages, not only in different countries but also between institutions. It is perhaps the stance taken by the institution towards innovation in this field, which has the greatest impact on development of such programmes.

Different approaches are being tried to develop interdisciplinary courses, and most of the current work is in producing courses within single disciplines, which have a heavy bias towards global change examples. The Lisbon (Camoes and Neves 1994) and York, UK (Anastasi 1994), models are of this character. The U.S. system, in which faculties and schools hold relatively more influence than individual departments, allows for a multidisciplinary approach, drawing on existing courses in different faculties and departments to provide the content. Examples in this category include courses at the University of Colorado (White 1994) and at Stanford University (Schneider 1994). The drawback to this system is that there is little or no control on content and inevitably this leads to a lack of coherence. Nevertheless, it is a system that introduces students to a subject of global change science and enables the faculty to build up experience and expertise in this sector of science education.

Where should we be going?

There is considerable evidence that interdisciplinarity in science education is seen by many to cut across what are regarded as deeply held, established tenets. One such tenet is that it is important at this stage of education to give students a good grounding in discrete scientific disciplines and that there is only time to do this in one, or at the most two, disciplines. Secondly, interdisciplinary courses cut across established institutional funding arrangements to the perceived detriment of existing departments. Other difficulties are related to associated research where full credit is not given to achievements that transcend boundaries.

The group did not feel that any difficulties provided insurmountable problems. While agreeing that it is important for future global change scientists to have a thorough grasp of important concepts of the different sciences, it is nevertheless possible, given goodwill, to ensure that concepts, which are irrelevant to global change science training are eliminated. Secondly, by using examples and global change science while teaching fundamental scientific principles, time is saved and courses become more lively, relevant and accessible.

The 'staple-gun approach' has many limitations and the creation of specific courses is important if measurable progress is to be made. This requires a dedicated organisational structure including tenured faculty within an existing department or institution, but whatever structure is chosen, it must be flexible in its approach. The group did not develop a general outline for a 'global change course', believing that at this level it is dependent, at least in part, on the experiences and expertise of the faculty. However, judging by the courses described in presentations, there is a common-core which can be abstracted.

How do we get there?

Again, this question was mainly discussed from the 'institutional' point of view in order to provide a stable base for the development of truly interdisciplinary global change courses.

To achieve institutional stability for interdisciplinary global change courses, a so-called 'successional model' was suggested (Schneider 1994) with the various steps, starting with multi-disciplinary seminars and the instalment of a 'programme committee' to draft staple-gun courses leading to the development of new 'core' courses. Simultaneously, new discipline tracks and interdisciplinary senior seminars should be developed, and the possibilities for co-term MSc degrees widened and encouraged. Ideally, all this could lead to the development of new departments/schools or centers/institutes. It was noticed that this model is not truly successional and that there may be 'jumps' along the road.

The group split their discussion into two parts
a) the identification of the main obstacles, followed by
b) an attempt to describe some possible solutions.

(a) Some problems

Some problems concerned with structural administration of the institution may be

- a lack of 'political muscle' within a single department or in a group to overcome institutional inertia
- financial constraints
- a lack of credit given to the individual for teaching in interdisciplinary courses organised by another department.

(b) Solutions

- Depending on circumstances, the original ideas may come either from junior faculty or from faculty leaders. Regardless of which, in order to get courses started, course providers should discuss complex issues. This requires an experienced and well-respected leader and a group of analytic thinkers. Although not ideal, staple-gun courses are a realistic, and probably in practice, an essential first step.

- Marketing and networking: because we are advocating a new approach, the idea has to be sold to existing institutions, both to develop the course content which will mean trespassing on existing disciplinary and institutional borders, and to obtain structural funding. Demonstration projects and field-excursions were suggested as possible activities for demonstrating marketing and networking strategies.
- The link between research and teaching is as essential in global change science as in other science courses, if not more so. Credit for interdisciplinarity in research has to be put on the same footing as that for conventional research.
- Again, as in other courses, it is important to be sensitive to the future needs of students, whether for direct employment after the course or in pursuing their education through higher degrees.
- There is need for regular evaluation and feedback on the course, the teaching, and on student progress.
- Summer courses/schools and exchange programmes are important at this educational level to provide off-campus ('real world') experience in global change science. They could be regionally oriented, for example, short courses at the Arctic Center, University of Lapland (Moore and Lange 1994), or more thematic or problem-driven, such as excursions or short courses to sites where the relation between climate change and desertification is studied, in Spain (Medalus) or to the permanent research site in a boreal forest in Norway (Climex-experiment). The work at Wanginengen University also deserves special attention (de Groot 1994), as does the course on Climate and Environment, which is planned by the Open University of the Netherlands in collaboration with the Open University in the UK and the Wanginengen University.

WORKING GROUP 3: POSTGRADUATE PROGRAMMES IN GLOBAL CHANGE SCIENCE

Working Group 3 addressed the issues involved in designing postgraduate programmes, particularly courses leading to masters degrees, whose major component is course work and which allow for specialisation in global change science.

The purpose of such courses is to provide the necessary subject content and practice to develop skills which will enable students to contribute to the teams of experts who are attempting to understand problems in global change science.

Heavy emphasis was laid on the point about teams as it is inconceivable that global issues will be dealt with in organizations by solitary individuals. Although global change science may concentrate on human-induced perturbations of the global system, inclusion of background trends (natural variations) is necessary.

The programme will accept students from different academic backgrounds. They may be from single subjects (e.g. biology, chemistry, physics), from interdisciplinary courses, or from non-science programmes (e.g. economics, anthropology).
The end product wanted and the motivation of students will vary. Some participants will wish to pursue research; others will be professionals, training or re-training in global environmental science; others will be motivated by curiosity. This, in turn, should lead to a variety of different programmes and, within the courses, to a flexible approach to programme design.

Where are we now?
As in the case of the other working groups, interesting differences between national systems were noted. Particularly relevant in the context of postgraduate programmes is the greater reliance in US systems on course textbooks, in comparison with systems in Europe where they play a lesser role. Taught elements play a larger role in US courses, even in doctoral programmes. However, there is a small number of masters courses which cater for global change scientists.

A programme's very novelty will lead to a large burden on the staff promoting it. There is little experience to draw on. Each activity is likely to have to be thought through afresh, teaching and learning methods questioned, new resources produced, and new assessment procedures designed. The normal problems facing faculty in setting up courses are thus magnified. Although it is very likely that programmes will be put together, in the first instance, by the staple-gun approach, there will be constraints, in some cases severe, on what can be achieved by staff who are often discipline-based. Because the courses that are being designed are new, the constraints, observed by Working Party 2, are likely to be exacerbated.

Two examples are to be found in the UK. The Oxford MSc degree course (Boardman 1994) is being created from existing sets of courses, and is based in a research unit. At York (UK) the MSc degree course in Environmental Economics and Environmental Management, while the brainchild of the Department of Biology and Economics, has as its core a new Department of Environmental Economics and Environmental Management with students also taking relevant courses in Biology and Economics. As has been stated for other courses, this one is designed with two goals in mind: (i) to provide a professional qualification for those intending to work or already working in the area of environmental management, and (ii) to prepare those who wish to enrol in a research degree in environmental economics and environmental management. The course combines insights from two key disciplines: economics and biology. Drawing largely from the sub-fields of environmental economics and ecology, it offers a set of skills not at present available in other courses. From the field of ecology, it offers insights into the functioning of ecological systems, and the nature of the ecological feedback effects of economic utilization of those systems. From the field of environmental economics, it offers insights into the optimal utilization of scarce environmental resources, with and without markets. The combination of the two enables students to learn how to incorporate environmental feedbacks into the economic decision-making process in a way that satisfies the goals of both resource managers and resource economists. To be eligible for direct entry into the MSc programme, applicants must normally have studied economics and biology to an advanced level in their first degree. Students who do not satisfy these requirements will normally be required to do a foundation year - comprising a set of upper level undergraduate courses tailored to the individual needs of the student (University of York, 1995).

Where should we be going?

A masters course which is principally concerned with course work is probably the most appropriate model. It will be an advantage if the course is constructed so that it can be taken either full- or part-time over a larger time span. The latter mode will be an advantage to those students in employment.

A main issue in course design is the provision of elements of both breadth and depth. A suggested balance which gives equal weighting in time and assessment is outlined below.

Section 1 *Breadth* : students take all sections of this part of the course. Details are eschewed, as it will depend on the precise aims of the course and the resources at the institution. However, in general terms it should cover aspects of the following broad areas: earth as a system; causes of change; effects of change; socio-economic consequences; response; human behaviour.

Section 2 *Depth* : students choose options in discipline areas of special interest to them such as environmental law, environmental economics, and advanced scientific-based courses.

Section 3 A research *project* leading to a dissertation

How do we get there?

There are problems and constraints to consider as the course is being designed and during its early years. The problems are similar to those described by Working Group 2. The development of the course may be constrained by

- the expertise of staff already in post
- the background knowledge of students
- lack of funding
- the lack of developed and well-tested teaching resources, e.g., software and datasets books
- the necessity to develop appropriate student assessment techniques

There are the institutional problems which include

- academic inertia - a particular problem for interdisciplinary programmes and the staff involved in these programmes
- structural inertia (institutional barriers between departments contributing to interdisciplinary programmes)

Once again the solutions are largely similar to those discussed by other groups. The following were considered essential.

- Enlist international support for an exciting new programme, and especially obtain financial support for students from less-developed areas

For this it is helpful to be able to show that the programme already attracts a large number of students

- Develop new teaching resources, such as software, books and multimedia resources, and, where possible, collaborate with other institutions by sharing materials developed in-house
- Limit the number of options at the start and expand the number as the course becomes established
- Strive for flexibility in the design of programmes, so that Section 1 of the core can also be used as an *outreach* course.

A synthesis

It is unsurprising that the discussions overlapped and several salient points are common to all levels of education:

1. The introduction of global change science courses competes with existing deeply-entrenched courses in departments with long-established disciplines. Such departments have relatively stable infrastructures of funding and staff.

2. There is an unresolved (and perhaps unresolvable) debate, within the global change science area, on the relationship between the depth and the breadth of understanding to be sought in courses.
   There was, at times, an assumption in discussions that breadth could be maintained at secondary and freshman level and depth would only become a serious issue late in the undergraduate course. This *may* be true for some countries where choice of speciality is delayed through a student's school and university career and a general education is encouraged. It is not true for many European countries and for those less-developed countries which use a European education model.
   It was recognized that it was essential for courses to allow career choice and to permit progress to further education, and it was appreciated that employers would continue to look for specialists, in this instance specialists who have an understanding of global change science.

3. There is much evidence that global change science courses are popular both with students and with staff. The courses are highly motivating, and even

students who have an in-built fear or dislike of quantitative scientific concepts find, because their interest is held, that they are able to understand the principles. Those university departments that advocate single subject courses should take note. Probably they will discover that collaboration to produce multidisciplinary courses won't be at a high cost to themselves. They may even get more students to teach!

4. All courses need new resources for teachers, often in the form of student activities, so that the students can take a positive role in their own learning. Teachers need help with organising teaching strategies to deal with this. Some examples of good practice were given in the papers.

5. Pervading the discussions in every group was the importance of what were termed outreach courses. Particularly important are those in which school teachers are given help with teaching global change science, thus, in turn, helping the next generation to be able to participate more securely in the inevitable debate on global change issues. This will become increasingly important in the next decades, when less developed countries such as China and India, which also have significantly larger populations, will be approaching a standard of living comparable to that enjoyed now by more developed countries.

References

Anastasi C (1994) This book
Boardman J (1994) This book
Camoes MF; Neves LAT (1994) This book
de Groot RS (1994) This book
Moore JC and Lange MA (1994) This book
Schneider S (1994) This book
University of York (1995) Graduate Prospectus, 1995 Entry
White J (1994) This book

NATO ADVANCED RESEARCH WORKSHOP: JUNE 1994

PARTICIPANTS

DR CHRIS ANASTASI
Group Planning, Shell International Petroleum Company, Shell Centre, London SE1 7NA, U.K.

MAX BERAN
TIGER Programme, Institute of Hydrology, Wallingford, Oxfordshire, OX10 8BB, U.K.

DR JOHN BOARDMAN
Environmental Change Unit, University of Oxford, Oxford, OX1 3TB, U.K.

PROFESSOR FILOMENA CAMOES
Department of Chemistry, University of Lisbon, 1700 Lisbon, Portugal.

LYNNE M CARTER
Graduate School of Oceonography, University of Rhode Island, Narragansett, RI 02882, U.S.A.

DR RUDOLF S DE GROOT
Wageningen Center for Environment and Climate Studies,
Wageningen Agricultural University, PO Box 9101, 6700 HB Wageningen, Netherlands.

DR MONTSERRAT FILELLA
Department of Inorganic, Analytical & Applied Chemistry, University of Geneva,
Quai Ernest Ansermet 30, CH-1211 Geneva 4, Switzerland

PROFESSOR GEORGE W FISHER
Department of Earth and Planetary Sciences, The John Hopkins University, Baltimore, MD 21218, U.S.A.

PROFESSOR WILLIAM S FYFE, FRS
Department of Earth Sciences, University of Western Ontario, London, Ontario, N6A 3K7, Canada.

PROFESSOR CATHERINE GAUTIER
Earth - Space Research Group, Computer Systems Lab, Ellison Hall, University of California, Santa Barbara, CA 93106, U.S.A

KATHIE GERSTER
Oakridge S.S., 1040 Oxford Street, London, Ontario N6H 1V4, Canada

DR HENRIETTA LIDIARD
Environmental Geochemistry Research Unit, Department of Environmental Sciences, University of Bradford, Bradford, West Yorkshire BD7 1DP, U.K.

DR JEFFREY J MCDONNELL
College of Environmental Science and Forestry, State University of New York, Syracuse, NY 13210, U.S.A.

DR JOHN C MOORE
Arctic Centre, University of Lapland, FIN - 96101, Rofaniemi, Finland.

DR MICHAEL POWELL
Department of Earth Sciences, University of Western Ontario, London, Ontario N6A 3K7, Canada

DR STEPHEN SCHNEIDER
Department of Biological Sciences, Stanford University, Stanford, CA 94305-5020, U.S.A.

DR SERENA SCHWARTZ
Departments of Bio & Atmospherics, Oceanic & Space Sciences (AOSS), University of Michigan, Ann Arbor, MI, U.S.A.

DR PANAYOTIS A SISKOS
Department of Chemistry, University of Athens, Panepistimiopolis 15771, Athens, Greece

DR HENRIK SKOV
National Environmental Research Institute, DK 4000 Roskilde, Denmark

PROFESSOR RICHARD SOMERVILLE
Climate Research Division, Scripps Institution of Oceonography, University of California, San Diego, La Jolla, CA 92093 - 0224, U.S.A.

DR JOSEPH P STOLTMAN
Department of Geography, Western Michigan University, Kalamazoo, MI 49008, U.S.A.

DR DIRK G TROOST
Coastal and Marine Sciences, UNESCO, 1 Rue Miollis, 75732 Paris Cedex 15, France.

PROFESSOR SERGE G TOUCHINSKI
Department of Environmental Management, Faculty of Geography, Moscow State University, 119899 Moscow, Russia.

PROFESSOR DAVID J WADDINGTON
Department of Chemistry, University of York, York, YO1 5DD, U.K.

DR JAMES W C WHITE
Department of Arctic and Alpine Research, University of Colorado, Boulder CO 80309, U.S.A.

The following wrote papers for the meeting but were unable to attend:

DR NANCY HEALY-WILLIAMS
Department of Geological Sciences, University of South Carolina, Columbia, SC 29208, U.S.A.

PROFESSOR WALTER LEAL FILHO
European Research Centre on Environmental Education and Development, Department of Environmental Sciences, University of Bradford, Bradford, BD7 1DP, U.K.

DR DOUGLAS WILLIAMS
Department of Geological Sciences, University of South Carolina, Columbia, SC 29208, U.S.A.

WORKSHOP COMMITTEE

MAX BERAN

PROFESSOR WILLIAM FYFE

PROFESSOR STEPHEN SCHNEIDER

PROFESSOR DAVID WADDINGTON

The ASI Series Books Published as a Result of Activities of the Special Programme on Global Environmental Change

This book contains the proceedings of a NATO Advanced Research Workshop held within the activities of the NATO Special Programme on Global Environmental Change, which started in 1991 under the auspices of the NATO Science Committee.

The volumes published as a result of the activities of the Special Programme are:

Vol. 1: **Global Environmental Change.** Edited by R. W. Corell and P. A. Anderson. 1991.
Vol. 2: **The Last Deglaciation: Absolute and Radiocarbon Chronologies.** Edited by E. Bard and W. S. Broecker. 1992.
Vol. 3: **Start of a Glacial.** Edited by G. J. Kukla and E. Went. 1992.
Vol. 4: **Interactions of C, N, P and S Biogeochemical Cycles and Global Change.** Edited by R. Wollast, F. T. Mackenzie and L. Chou. 1993.
Vol. 5: **Energy and Water Cycles in the Climate System.** Edited by E. Raschke and D. Jacob. 1993.
Vol 6 **Prediction of Interannual Climate Variations.** Edited by J. Shukla. 1993.
Vol. 7: **The Tropospheric Chemistry of Ozone in the Polar Regions.** Edited by H. Niki and K. H. Becker. 1993.
Vol. 8: **The Role of the Stratosphere in Global Change.** Edited by M.-L. Chanin 1993.
Vol. 9: **High Spectral Resolution Infrared Remote Sensing for Earth's Weather and Climate Studies.** Edited by A Chedin, M.T. Chahine and N.A. Scott. 1993
Vol 10: **Towards a Model of Ocean Biogeochemical Processes.** Edited by G. T. Evans and M.J.R. Fasham. 1993
Vol. 11: **Modelling Oceanic Climate Interactions.** Edited by J. Willebrand and D.L.T. Anderson. 1993.
Vol. 12: **Ice in the Climate System.** Edited by W. Richard Peltier. 1993.
Vol. 13: **Atmospheric Methane: Sources, Sinks, and Role in Global Change.** Edited by M. A. K. Khalil. 1993.
Vol. 14: **The Role of Regional Organizations in the Context of Climate Change.** Edited by M H. Glantz 1993
Vol. 15 **The Global Carbon Cycle.** Edited by M. Heimann. 1993.
Vol. 16: **Interacting Stresses on Plants in a Changing Climate.** Edited by M B. Jackson and C. R. Black. 1993.
Vol 17: **Carbon Cycling in the Glacial Ocean: Constraints on the Ocean's Role in Global Change.** Edited by R. Zahn, T. F. Pedersen, M. A. Kaminski and L. Labeyrie. 1994.
Vol. 18: **Stratospheric Ozone Depletion/UV-B Radiation in the Biosphere.** Edited by R. H. Biggs and M. E. B. Joyner. 1994.
Vol 19 **Data Assimilation: Tools for Modelling the Ocean in a Global Change Perspective.** Edited by P O Brasseur and J. Nihoul. 1994

Vol. 20: **Biodiversity, Temperate Ecosystems, and Global Change.**
Edited by T. J. B. Boyle and C. E. B. Boyle. 1994.
Vol. 21: **Low-Temperature Chemistry of the Atmosphere.**
Edited by G. K. Moortgat, A. J. Barnes, G. Le Bras and J. R. Sodeau. 1994.
Vol. 22: **Long-Term Climatic Variations – Data and Modelling.**
Edited by J.-C. Duplessy and M.-T. Spyridakis. 1994.
Vol. 23: **Soil Responses to Climate Change.**
Edited by M. D. A. Rounsevell and P. J. Loveland. 1994.
Vol. 24: **Remote Sensing and Global Climate Change.**
Edited by R. A. Vaughan and A. P. Cracknell. 1994.
Vol. 25: **The Solar Engine and Its Influence on Terrestrial Atmosphere and Climate.**
Edited by E. Nesme-Ribes. 1994.
Vol. 26: **Global Precipitations and Climate Change.**
Edited by M. Desbois and F. Désalmand. 1994.
Vol. 27: **Cenozoic Plants and Climates of the Arctic.**
Edited by M. C. Boulter and H. C. Fisher. 1994.
Vol. 28: **Evaluating and Monitoring the Health of Large-Scale Ecosystems.**
Edited by D. J. Rapport, C. L. Gaudet and P. Calow. 1995.
Vol. 29: **Global Environmental Change Science: Education and Training.**
Edited by D. J. Waddington. 1995.
Vol. 30: **Ice Core Studies of Global Biogeochemical Cycles.**
Edited by R. J. Delmas. 1995.

Springer-Verlag and the Environment

We at Springer-Verlag firmly believe that an international science publisher has a special obligation to the environment, and our corporate policies consistently reflect this conviction.

We also expect our business partners – paper mills, printers, packaging manufacturers, etc. – to commit themselves to using environmentally friendly materials and production processes.

The paper in this book is made from low- or no-chlorine pulp and is acid free, in conformance with international standards for paper permanency.

PROBLEMS OF THE HEGELIAN DIALECTIC

VOLUME 225

MENAHEM ROSEN
Centre National de la Recherche Scientifique, Villejuif, France

PROBLEMS OF THE HEGELIAN DIALECTIC

Dialectic Reconstructed as a Logic of Human Reality

SPRINGER-SCIENCE+BUSINESS MEDIA, B.V.

Library of Congress Cataloging-in-Publication Data

ISBN 978-94-010-4792-0 ISBN 978-94-011-1799-9 (eBook)
DOI 10.1007/978-94-011-1799-9

Printed on acid-free paper

TABLE OF CONTENTS

Acknowledgements

First of all, my thanks to Professor Jaakko Hintikka, without whom this book would perhaps never been published—I wish to emphasize his open-mindedness and his willingness to give a chance to alternative ways of thinking and writing. My thanks too to Annie Kuipers, Acquisition Editor of the KLUWER Academic Publishers, who had the patience to bear a young author's anguish, while presiding over every step of the editing task.

I have worked on this study for a long time and in various places. I gratefully thank my teachers, colleagues, students and friends who have contributed to my endeavours. I wish to extend special thanks to a number of them. Professor Georges Labica allowed me to sit in on his stimulating seminars and to make use of the Wordprocessor available in his C.N.R.S.' Department. Mireille Delbraccio, his devoted secretary, was of a precious help to me, either by her precious friendship, giving judicious advice when needed, or by her effective support concerning the use of the Wordprocessor. H. Khalfon, psychoanalyst, showed kindness and perseverance in commenting on a very rough draft of this study written at that time, K. Teepe helped me a lot in the translation of my work, *originally written in French*, while M-A. Prigent fulfilled an important practical function—I am grateful to each of these parties for their faithful support. None of these people is responsible for my failure to make better use of their advice.

Also of a special help were the clever participants of my two private seminars, who were a real source of stimulation and insight. I mean the seminar on Dialectic for Psychoanalysts, with the participation of: H.Khalfon, E.Aubert, Cl.Fourcade, M. Lathourakis, D.Levaguerèse, M-C.Meplan and K.Teepe; and the seminar on Dialectic for Marxists and the 'Friends of Hegelian Dialectic', with the participation of P. and O. Alvarez, L. and K. Alvarez, R.Gutkind, J-F.Lauginie, B. Menasce and A. Pichon.

Naturally, my thanks too to my teachers and colleagues of the Hebrew University of Jerusalem, where I was born as a thinker and a philosopher. I mean especially Professors Y.Yovel, Z.Bar-On and N.Rotenstreich, from whom I learnt so much.

I thank Mr. Rudolf A. Makkreel, Editor of the *Journal of the History of Philosophy*, for permission to reprint material which appeared in vol.xxiii of that journal, N°4 Oct.85, under the title 'Identité, différence et contradiction dialectiques selon Hegel' (here as Part One: 'Dialectical Identity, Difference and Contradiction'). I thank also Mrs. Anne Williams, Executive Secretary of the *Canadian Journal of Philosophy* for permission to reprint material which appeared in vol.19 of that journal, N°4 Dec.89, under the title 'Le problème du commencement dans la philosophie de Hegel' (here as Part Two, 'The problem of the beginning').

Note on technical terms and notations

1. Use of Capitals and definite articles:

Generally, the Hegelian concept is written 'the Concept';
the Hegelian system is written 'the System'.

I speak also of 'the Individual' and 'the Idea'.

When it may help, Hegelian categories are capitalized:
Being, Nothing, Becoming, Essence, Identity, etc.

'Understanding', 'Reason' denote the corresponding human faculties and, as such, differ from 'understanding' (comprehension) and 'reason' (ground).

2. 'A/B' means that A and B are opposites, such as the pairs:
Being/Nothing or Identity/Difference.

3. The terms 'Ambiguity', 'Ambivalence', 'Double meaning', are not given special definition, but are understood as usually employed in common everyday or scientific language: dialectic being 'plastic' and 'fluid', any strict definition would immediately find itself on the move...

The same is true of 'Paradox'.

As for the term 'Antinomy', I can briefly add that by 'Antinomy of experience' (Part Four) and 'Antinomy of language' (Part Five and Hegel's *Logic*, vol.1 p.76 (I,103)) , I mean a contradictory relation so that the 'external' is a necessary constitutive moment, but also a disturbing one that causes the Hegelian plan to fail in part. As such this 'external' subsists as a 'residue', never really transcended—whence the 'tense' inner opposition.

PREFACE

In this book, I deal with some fundamental **problems** of the Hegelian dialectic. For this purpose, I take a middle course between total scepticism, which considers dialectic as a devastator sophistry with no respect even for the non-contradiction principle, and authoritarian dogmatism, which claims to solve any question with the magic wand of the Hegelian **Aufhebung**. That is, I decide to be **critical**, defining concepts anew, bringing out sources, determining conditions of possibility and fields of validity, accepting or rejecting when necessary. Following G.R.G. Mure's thinking, from an **inner** point of view I examine whether, in carrying out his work, Hegel remains faithful to the different principles he proclaims, and I find substantial deviations. And, following W. Becker's thinking, from an **external** point of view, that is, from a formal, empirical or existential contemporary angle, I try to determine the extent to which we may legitimately talk about the fruitfulness of Hegelian dialectic. In this way, I reconstruct Hegel's thought so that it may become acceptable to us—readers of the twentieth-century—as intelligible and coherent as possible. I conclude that dialectic, **as a logic of human reality**, has to be grasped and expressed from the viewpoint of the particular historical individual, in constant interaction with the cultural environment of his or her time.

Using this approach, I investigate the questions at issue from Hegel's *Logic* point of view. Historically speaking, it seems to me that so far the *Phenomenology of Spirit* has been widely studied, whereas the *Science of Logic* remains relatively unknown. Systematically speaking, I agree with Hegel that the ***Logic*** , and therefore the logical dialectic, is the structuring essence of the System, the animating drive that explains its internal movement at least in part. I believe that an examination of the *Logic* may enable us to test the reliability of the entire building and to bring out its various flaws.

This work is a critical reconstruction of my dissertation. Composed of **relatively independent studies**, it may be examined in the order that suits the reader. There is, however, also **an internal movement**, from abstract to concrete, from simple to complex, each study presupposing those that precede it ; thus, it may be noticed that the different Parts become progressively longer as the research progresses. Each study also presupposes those that succeed it, so that a certain circularity takes shape.

In this way a **quasi-organic** approach is developed, attributing to each Part, in interaction with the rest of the work, a conception of the whole from its particular view-point. It has been impossible to prevent apparent **repetitions** which, however, at each new elucidation state more precisely the perspective in play, corroborate some established results and anticipate others. Strict **consistency** can neither be demanded, for the questions raised and the answers obtained acquire their particular meanings in the context of the ongoing discussion : at every new level a new vision leads to new interpretations. As for the content of each study, each issue examined is in general enlightened from different angles. Hence I develop a conclusion that is more and more specified, better and better grounded, within the limits of validity of the discussion, of course.

The work opens with a Foreword describing the projected task : to propose a modern version of dialectic that is capable of contributing to the elucidation of the problems with which we are concerned nowadays. In the Introduction I specify the point of view I adopt. In Part One I deal with the fundamental categories of Hegelian thinking, maybe of any human thinking : the categories of identity, difference and contradiction. That inquiry determines the structure of all subsequent discussion. Afterwards, proceeding with systemic investigations, I explore the problems of the beginning (Part Two) and that of the end (Part Three), the 'circular' Hegelian dialectic being brought to confront them. Since Hegel affirms that he has 'overcome' the Other, the sensible given said to be "not-true, not-rational", I examine the cogency of that affirmation (Part Four), and discover that the pretensions pass far beyond the results actually attained. A similar difficulty is addressed in Part Five, where I discuss the problem of language: in order to communicate, 'pure' thought must be embodied in a sensible, non-adequate medium ; from that an unceasing fight against the 'Other' follows. Finally, in Part Six, I ask whether dialectic might have a certain explanatory power. Inquiring that issue, I elucidate Hegel's mode of 'explanation explaining itself', at first in a general manner. Then, going into details, I close the book with a concrete examination of a specific dialectical movement. Hence the following principle: 'Dialectical development is not organic genesis'.

Thus the work proceeds from abstract to concrete, from a categorial study to a systemic study and then to a detailed exposition of the dialectical movement, passing through material and linguistic investigations.

Eventually, I would like to ask the reader to show indulgence and goodwill in matter of style : in my view, complex subjects lead to complex discussions! When I have had to opt between elegance and precision, I have given preference to the latter, in so far as I have understood. Moreover, I beg the reader not to blame the form when the content is at issue.

1st May 1992
Auvers-sur-Oise

FOREWORD

Dialectic was not born yesterday, and some scholars refer to Zeno as its inventor. After a brilliant beginning in ancient Greece it fell into discredit, being accused of dishonesty, malevolent manipulation and, at best, of futile chattering. Although dialectic played a part in the birth and development of formal logic, it was nevertheless first used to defend any cause, bad as well as good ; hence its name of 'sophistry'. Consequently, dialectic disappeared for centuries as a subject-matter studied for its own sake, and turned into a curiosity to be known only to guard better against it.

Hegel is correct when he attributes the renewal of dialectic to Kant. Indeed, from the first pages of his *Critique of Pure Reason* , Kant distinguishes between the **transcendental** logic of truth and the **dialectical** logic of appearance. This does not prevent him, at the same time, from calling the criticism of dialectical illusion 'Transcendental dialectic'. Already with Kant, and perhaps without his being aware of it, dialectic shows itself in its double meaning—both negative and positive, destructive and creative. I understand that this ambiguity is due to Kant's 'architectonical instinct' which leads him to join each 'Analytic' with a corresponding 'Dialectic'. However, at least explicitly Kant denies the existence of a dialectical organon that would be able to test a form as well as to produce an informative content.

As Kant's pupil, Hegel developed some potentialities that lay hidden in the various texts of his master. Thanks to this, we are at last in possession of a vast system, said to bring to light dialectic in all its richness and quasi-unlimited power. After Kant's shyness, then, came Hegel's imprudence : everything had to become either dialectical or nothing at all. The status of the amorphous, inert and sensible matter was settled! In the presence of such excessive spiritualism (or 'panculturism') which reduced nature to a moment of Spirit, Marx aimed to reverse the situation by replacing dialectic on its feet. This time the concrete world—of labour, production and consumption—would be explained dialectically. Engels went so far as to write a 'Dialectic of nature', even though Hegel had talked of nature in terms of non rationality, seeing it as a realm where extreme contingency and diversity reign.

In this way the circle closed, and dialectic knew again the dark days of contempt and diaspora, rejected everywhere, even by those appealing to it. In effect, like their colleagues in the Western world, Soviet physicists seem to be satisfied in their work with the formal or empirical thinking of the natural sciences, leaving the dialectical speeches to the concerned specialists.

In order to confer again significance to dialectical research, I have had to draw important distinctions in the field of knowledge. From the outset, indeed, if dialectic is to constitute 'a grasp of itself and the other, each of them being both same and other', it must learn to respect the fellow in his work and to honour his domain of research : did not Socrates ask his fellow-citizens to mind their own business? Therefore, with all due modesty, I endeavour to keep as well-informed as possible of the works of the so-called natural, experimental or exact sciences—and no matter what their names may be. I understand that in the physical domain, dialectic should not be revisionist ; at best it is just able to learn and interpret *post factum* . The devil take those alluding to a 'Dialectic of nature' : self-respect begins with respect for others! Thus I distinguish between two fields of reality, each with a proper mode of explanation and verification : natural reality *qua* physical, chemical and biological reality, and human reality *qua* social, cultural and historical reality. To that division of reality corresponds a parallel division between the natural and human sciences. It seems to me that a fruitful work requires of each of these branches to regard first of all its own domain and, moreover, to co-operate, despite the specific differences in ways of perceiving, conceiving and explaining. For I understand that the experimental sciences remain perplexed when confronted with the cultural object ; and, on the contrary, dialectical thought stammers when confronted with the quantifiable sensible given. If this is the case, neither the intention to trespass on a neighbour's preserves nor the precipitate exclusion of the other will lead to development and creativity. I therefore propose a **coexistence principle** that recognizes the respective essentiality of each of the two domains. For, after all, for the human being reality forms a totality, both empirical and *a priori*, physical and cultural ; in short, a multidimensional reality.

So far I have characterized my research field : social, cultural and historical reality. Now, I expound the approach I have adopted.

In this work I do not intend to expose, summarize or comment on what Hegel said or wished to say ; neither, speaking by the book, do I intend to set one quotation against another. As is well known, anything can be proved in this way : one has only to isolate the subject under discussion and to take it out of context—indeed hardly a productive manner of argument... I believe that today it mainly matters to grasp Hegel's topicality, namely, to see what we, as modern thinkers, are capable of drawing out of his voluminous work. In other words, the reader will not find here a historical investigation dealing with a text of the past, but an 'interested' project asking what is the meaning of Hegel's work **for us today**? What is it able to contribute, despite the important changes that have occured in our social, cultural and scientific backgrounds during the last two centuries?

Of course, that question is posed from a logical point of view. In this book I examine the System, or Hegel's thought, not so much for its own sake but rather for its structuring dialectic : how can the latter help us better understand our sociocultural environment, our technical world, our present history?

To this end, taking inspiration from the British scholar G.R.G. Mure I have had to prune Hegel's work. I have retained Hegel the dialectician and rejected Hegel the 'spiritual monist'—the one who perceives the Absolute as a movement of Spirit exteriorizing itself as Nature and returning to itself through the mediation of History. For, in my opinion, the thinker does not contemplate from the outside but is himself carried along with the movement he is observing. Consequently, this book is concerned with **dialectic or history in the making**, and every point of view is regarded as both universal and particular.

Now, what is dialectic? To answer that question, following Mure I have concentrated on the middle level of Hegel's *Logic*, the level of the Doctrine of Essence, looking there for the structure of a living and productive dialectic that is still human. By remaining within the limits of human experience, one can avoid the Hegelian apocalyptic tone of the 'high levels', characterized by widely optimistic and seemingly unfulfilled expectations.

All things considered, therefore, I have drawn closer to Kant, relating to the Hegelian absolute with a reflective mind, just as, conversely, one can read Kant in a Hegelian way, thus bringing out a latent dialectic structuring his work. Indeed, if dialectic is basically a logic of ambiguity and paradox, in Kant's work these latter are not rare either. To mention only some of them : the existence of an *a priori* sensible intuition, of an *a priori* productive imagination, the scheme as both intellectual and sensible, the object of knowledge as phenomenon or representation, time and space as forms and intuitions, matter as sensation and source of sensation, human beings as citizens of two worlds, Kantian thought as transcendental idealism and empirical realism, etc. Aiming at more coherence, in order to defuse the tension inherent in Kant's system, Hegel was led, at the high levels of his central works, to display much too perfect and quite abstract circular movements. There are accordingly good grounds for reading Hegel through Kant and Kant through Hegel, without expecting the formation of any 'total' synthesis: every human look, partial and particular as determined, brings about the partition of the examined object and thus its movement. In my view, it is advisable to grasp human reality at once from a Kantian and a Hegelian standpoint ; namely, to grasp human experience through a double dialectical movement, both *a priori* and *a posteriori*, respecting the ambiguities and paradoxes encountered, and avoiding suppressing them univocally, as Hegel does when he affirms that the sensible matter is nothing, being "not-true, not-rational"—for dialectic lives on **that** tension (between matter and thought), and when it denies the other *ipso facto* it denies itself.

In short, in this work I have tried to set forth an intelligible dialectical movement. My first question was to understand what the subject at issue is. Being unwilling to repeat words that I could not grasp, I set out in quest of some **criterion for intelligibility and acceptability**. In consequence, I divided reality into two distinct ontological realms treated by two different sciences (physical or human sciences), looking in human reality for some double movements, double beginnings and double meanings. Thus I found the whole as a moment and the moment as a whole : for us, I think, the absolute is at once a movement that carries us along and that we constitute, always being split up and rebuilt afresh. Ready for an 'all-thinking anew' to the best of my ability, I have subjected to criticism the encountered fixed ideas and 'irrefutable' principles, examining if necessary the criticism itself—whence the open circle of knowledge, or better still the developing spiral that still drags me along. I have done so, with no fear of pointing to the univocal limits of dialectic itself, to its flaws and own contradictions, since the matter is not just to see the mote in one's neighbour's eye. Otherwise, how could so-called 'self-consciousness' be achieved? Should not a dialectician learn 'to think against himself'?

I hope that such a dialectic reconsidered in this way—remaining within the boundaries of human experience, respecting 'the other' as physical and biological matter and as the corresponding sciences—will arouse some interest in the formal or empiricist thinker, for the aim is precisely to convince the non-dialectician.

That is the reason why I have tried in this book to enter into dialogue with that other, to expound in intelligible language the manifold movements of Hegelian logic, adopting his relevant objections raised to the dialectical argumentations, illustrating them when it is helpful. Thus the reader will find, presented in detail, the dialectical movements of

Being and Nothing (Part II par. 3.2., Part V chap. 4), of Identity, Difference and Contradiction (Part I), of Form, Matter and Content (Part IV chap. 1) ; more briefly, the dialectic of Means and End (Part III par. 3.1.) ; and, finally, a detailed and concrete study of the dialectic of Something and an Other (Part VI chap. 5). When exhibiting the complex movements of dialectic, I may sometimes have failed to remain intelligible. Undoubtedly, that failure proceeds from my personal limitations, but also from the need to translate into a 'clear and distinct' language a fundamentally ambiguous, paradoxical and 'tense' thinking.

To end with, I apologize to the reader for not having solved in this book important problems encountered on the way. The blame lies certainly in my obtuseness, but likewise in the nature of the case. Among the issues I have not addressed, I want to recall the relationship existing between the two ontological domains : natural reality (physical, chemical and biological reality) and human reality (social, cultural and historical reality). Because these two are interdependent, it is difficult to determine their borders. In particular, dealing empirically with a specifically human world, economics appears to partake of both. In addition, I understand that, at the boundary, psychology and psychoanalysis are to be met, belonging at the same time to experimental sciences (through their physiological dimension) and to human sciences (through their relation to sociology). I believe that an elucidation of this issue might bring us to a better understanding of the sense in which the external world can be said to be 'external' and also 'internal'.

Furthermore, the question 'What is thinking?' remains to be discussed. Yet, this book as a whole provides an indirect answer to it. However, if, according to Hegel, 'Man thinks', nowadays, in consideration of contemporary technical calculus and electronic robots, it is important to explore the case carefully. For even if, so far, what thinking is was known, it can still be forgotten...

INTRODUCTION

FOR BOTH A FORMAL AND A DIALECTICAL READING OF DIALECTIC

Sometimes we are told: 'If you want to know what Hegelian dialectic is, nothing easier : read Hegel's *Logic* [1]. There you will find, properly exposed, the principles of that new logic and the necessary immanent movement of the categories of thought and being'. These words promise that the assiduous reader, after overcoming the difficulties proper to Hegelian language, will take possession of the famous 'Hegel's secret'— that which should explain the power of speculative philosophy.

This language seems very appealing. Yet, from the outset one must ask : which is the right approach to the *Logic* ? How should we read it and understand it? Several ways can be considered.

1. As far as **formal thought** is concerned, there is no special problem. The book lies there before our eyes, in cold print. Moreover, if need be, we can get a **'dictionary'** of Hegel's language[2] that will help decipher the difficult text of the *Logic*. With such an instrument we should be able to progress, step by step, towards the comprehension of this complicated, obscure and problematic work. If, however, in spite of all our efforts and goodwill we realize that the results obtained are not satisfactory, it is then up to us to blame this badly written, badly conceived book, and reject it outright.

I frankly acknowledge that a formal reading of the *Logic* is of a certain interest and, in my opinion, the dialectician has to take it into account[3]. But the matter cannot rest there for, in reality, things are not so plain and simple to the point of permitting such a decisive conclusion to be reached. Consequently, to this formal thinking I propose the following.

First of all, is the *Logic* such a work as to admit a 'dictionary'? In other words, is it fitting to investigate the terms of a dialectical work when they are detached from their original context and then attached to formal thought? It seems that this approach is not fitting, since Hegel suggests a new doctrine of signification summarized by the words : **"The True is the whole"**[4]. Here I understand this proposition in the following way : 'The True' is the meaning of the element when grasped in its proper context—if so, a formal and abstract 'dictionary' proves to be quite inappropriate.

Accordingly, the answer to the question is negative[5]. Indeed, when we read the *Logic* —where content and form are said to be one—a deciphering key cannot satisfy since, as the text unfolds, it is supposed to become self-illuminating. In other words, we find there the Spinozian view concerning the truth as a "norm of itself and of the false". The aim is really to expose the System as autonomous, self-explanatory and self-grounding.

If that is truly the case, then, in order to decipher the text an external help will not do. Only by basing ourselves mainly on our own knowledge, our experience of the subject, can we give sense and life to this orderly set of propositions, paragraphs and chapters that form the *Logic*.

On this account, if the assiduous reader fails to grasp that difficult work despite the help of his 'dictionary', he is yet not justified in concluding what about it. Rather, he has to ask himself whether, when deciphering the text, he did not set off on the right foot, in consequence of which he made no progress? An alternative way is therefore suggested : **to understand things differently**. This means, here, putting aside that abstract 'dictionary' of a relative importance only, and setting out to address 'the thing itself' *(die Sache selbst)*, in order to consider it with a view to bringing out its sense and structure.

2. Thus, we can see that from the outset the **problem of the beginning** arises. Or, according to Hegel's formulation, from the very first there are good grounds for asking : "With what must the learning begin?"[6] If one cannot be satisfied with an external tool such as a prearranged 'dictionary', would it be more suitable to study the book with the help of an introductory summary? Hegel himself seemed to think so : effectively, his famous Preface to the *Phenomenology* constitutes a sort of introduction to the System.

Thus, it can be said that an important step has be taken towards an adequate reading of the *Logic* : on our way to achieving understanding we shall not open with 'the truth', the well-prepared dictionary ready for use, but with the non-true ; for example, with the Preface to the

Phenomenology, 'untrue' in so far as it sums up in a few pages the whole of a dynamic and living thinking, which does not bear any summary. For, *qua* summarized, truth here becomes non-truth[7].

The willing reader will therefore start with the Preface to the *Phenomenology*, followed by a reading of the *Phenomenology* itself, the 'first Part of the System'[8]. From there he will pass on to the *Logic*, the work supposed to contain the essence, that is, the original exposition of the Hegelian dialectic.

But **how will the reader be able to move from the non-true up to the true?** Is not the context a basis for the clarification of the element? To put it another way, the work itself is needed to an adequate understanding of its Preface! Thus we find ourselves in an hermeneutical circle, where part (the Preface to the *Phenomenology)* and totality (Hegel's philosophy) are supposed to explain each other[9]. Moreover, the *Phenomenology* itself only becomes intelligible against the background of Hegel's whole system.

How can the reader escape this difficulty? Can he count on a 'pushing' instinct of truth or on an 'appealing' true Idea? Maybe. In any case, as suggested by Plato's *Meno*, he may progress thanks to his inherent power of recollection. For, as an educated and cultured human being, the reader contains within himself the whole richness of Western culture, that which is said to be explained by Hegel's philosophy[10].

Indeed, he does not start with nothing. Aroused by Hegel's text, he simply has to follow his intellectual impulses and reconstruct the 'immanent and necessary' dialectical movement of the categories. Again, working in a 'true' cultural context—that of his time—he will discover in it the guiding strand, leading to the real and the concrete. In this way, as he progresses in his study of the *Logic*, step by step the intended *Sache* will reveal itself to him.

Consequently we see that, in order to gain an understanding of the Hegelian propositions, they must be read 'over and over'[11]. Starting always anew, working out and reflecting on his logical studies in the context of the System, the willing reader—with the help of the Preface to the *Phenomenology* and of the *Phenomenology* itself—will manage to appropriate the text gradually and to discover the genuine meaning that Hegel intended to express.

3. If this was really the case, the problem would have been solved and 'Hegel's secret' disclosed a long time ago. Yet, things seem to have turned out otherwise : up to now, idealists and realists have argued continuously about the meaning proper to the *Logic*. Does the work deal with the movement of human thought and its different categories, or with the movement of reality and its essential characteristics, or with both together? Do we have to do with the presentation of "God... as he is in his eternal essence before the creation of Nature and Finite Spirit", or is nature (*qua* externally determined) also treated in that work seemingly quite hermetic? Finally, is the content of the *Logic* a moment of the System or the whole System itself?[12]

If the views on all of these issues are so disparate, the reason must not be sought in the lack of research or publications, nor in the scarcity of scholarship and ability. It must lie elsewhere. I suggest here the examination of the following question.

As already noticed, the dialectical way presupposes that "Truth is the criterion of itself and of the false". If so, the task is only **'to follow the thing itself'** through an immanent procedure. But what is this thing? At the beginning, how can we grasp it when it is still unknown? For a dogmatic Hegelian there is no problem: immerse yourself in the text while 'forgetting yourself', he advises, and *die Sache* (the thing in question) will reveal itself to you[13]. However, a critical regard immediately shows that the matter is far more complex than it seemed at first sight. For the *Logic* was written by a rational yet limited human being, for the sake of other human beings, rational and limited as well.

What does it mean 'to be rational and limited'? Among other things, it means that in his search of truth and reality (as rational), every man belongs to his time and his near surroundings, and that he is the subject of a particular experience (as limited). A man can free himself from all these influences, only at the risk of losing his own sight, in so far as **reading is interpreting**.

At first, a book is in fact nothing but something made of sheets of paper, on which some signs of diverse forms are printed ; and it is up to the reader to endow them with sense and life. Therefore, from this point of view it can be said that man finds in its (cultural) object only what he himself puts in it. Although that judgement gives voice to an idealistic principle, in this context I will defend it : when reading a text, every man re-creates for himself (and for his audience if he has one) the signification of that text—its original meaning having disappeared for ever with its author, if that author has gone.

To read, then, is to invent anew[14]. But how can we invent anew, if not by drawing from the materials at our disposal, such as our personality, our education, our background? In this respect, self-denying or self-forgetting really seems to be impracticable[15].

Thus, **the way to dialectical comprehension** not only **proves to be dialectical** in the indicated sense, but also reading itself constitutes a re-creation, **from the standpoint of a certain historical and cultural position :** there are no disembodied, non-cultured men, as we can learn from anthropology ; everyone brings with him his epistemic store and his own prejudices. As a criticism of determining prejudices, self-criticism therefore remains a main principle of the real grasping of dialectic, namely of a dialectical understanding of it (it seems that, in his *Critique de la raison dialectique*, Sartre had an inkling of this principle)[16]. In this way, the deciphered work becomes a certain mirror of our time and our experience ; and, actually, reading a work becomes also the deciphering of the corresponding time and experience.

Yet, the aim is to grasp Hegel's thought and not our own. That is, a 'dialectical' work of appraisal, criticism and re-appraisal is needed[17], applying to all categories, all propositions, **with no possibility of reaching a final stage of interpretation**, in any sense of the terms. Every period, every culture and every individual will always grasp Hegel's dialectical thought from its own viewpoint, a universal as well as a particular one. This remark applies equally to my own work[18], naturally.

Thus, at the outset the reader is really in possession of two different 'leading strands' : the common culture of his milieu in his own time, and his personal experience formed by everyday contingencies. Necessarily, his understanding of the examined work will reflect his own 'objective subjectivity' or 'subjective objectivity', in so far as universal and particular constitute for him a concrete and meaningful whole[19].

Is this personal 'deviation' to be seen as an error, a defect? I do not think so since, in the opposite case, the other alternative would be asserted : a universal reading by a universal spirit, that is, a divine reading of the work. Yet, a human being is not God.

It follows that the debate between specialists, cultures or periods does not lead to an ultimate thought, a definitively proved truth : *stricto sensu*, there is no 'true' Hegel nor 'one' authentic comprehension of his philosophy[20]. Nevertheless, to the extent that the different exposed points of view are dialectical, that is **plastic**, flexible and ready to undergo necessary transformations, to that extent the ongoing, endlessly open dialogue may show an enrichment for each of the involved sides, while each one internalizes its own particular interpretation of the 'general' dialectic, which is abstract in that sense. Indeed, according to its very principles, every 'general' grasp of the dialectic constitutes only an abstract and dead schema, which is therefore a non dialectical one : the essential moment of the particularity should be added[21].

4. Almost two centuries have passed since Hegel thought, wrote and published his philosophical works. Are we still able to understand them nowadays? In a dialectical context, is that question itself meaningful? The *Logic* was published after the *Phenomenology*, that is to say, 'at the end of history'. Yet, in so far as history has progressed (as may safely be asserted), some of Hegel's ideas are no longer ours. Thus successively Kierkegaard, Marx, Heidegger and Sartre, among other thinkers, have completed Hegel's philosophy, complementing it not *qua* ill conceived or badly reflected, but *qua* a philosophy that appeared earlier, when reality had not yet disclosed some of its important characteristics : **Hegel did not know Hegel and had not read him** (as an Other)! However, coming afterwards, we are in a privileged position to add to his work, without showing thereby any genuine wisdom.

Accordingly, I understand that today Hegel would not be an orthodox Hegelian, but rather a critic suggesting a more adequate presentation of dialectical logic. In this sense, it is precisely the orthodox who proves a bad disciple, he who tends to forget the subsequent development of human reality : doesn't the understanding of dialectic require also the grasping of one's own period in thought? Ours is undoubtedly a different time. In consequence, there are good reasons for a new comprehension of the matter.

At the present time, I believe, Hegel's logic has to be interpreted also from the standpoints of Kierkegaard, Marx, Heidegger and Sartre (to mention only a few). In his *Thus spoke Zarathustra*, Nietzsche observed that his true follower was the one who could manage to lose him [22]— here, the matter 'to lose' (and find anew) is the comprehension of Hegel's dialectic.

A dialectical reading of dialectic is therefore also a reading that takes into account the progress achieved in the fields of reality and cognition[23].That is why **we cannot merely rely on the various judgements passed by Hegel on his own work**[24].

5. At the beginning of this discussion, I talked about formal thinking that intended to grasp the Hegelian logic **from the outside**, with the help of a prearranged 'dictionary'. On the contrary, the strict Hegelian, attending to his Master's works, demands a pure **inner** comprehension of the text. For he believes that only the thinker who is working 'within' can succeed in seizing the matter in question. Actually, it seems to me that, in their one-sidedness, both are wrong.

5.1. I suggest again that a quasi-organic work, which is both conceptual and historical (such as Hegel's work), requires the active participation of the reader, and thereby cannot be understood from the outside alone—in that context, **to understand is to constitute**. As Hegel showed in his Preface to the *Phenomenology*, a 'detached' reader supposed 'to survey freely' the treated subject would miss it : he would find himself involved in an abstract reading, devoid of any content. For the act of 'sinking into' or the 'close consideration of' the thing is essential in order to distinguish something, to obtain **a content** because something is produced : the point at issue is the thought of the reader, moving and grasping itself in the course of its own moving[25].

A purely 'external' reader will therefore see nothing, not even his own prejudices. For instance, he will complain that Hegel does not respect the fundamental principles of thought and language, such as the principles of identity, non-contradiction and excluded middle. In his opinion, the whole work deals with an 'illogical' logic that must be entirely rejected, for it represents a fruitless or even harmful bundle of twaddle and chattering[26].

5.2. A reading only 'from within' would, however, lead to a problem well known to Hegel : how can a good-willed individual manage to penetrate that well kept 'mystery', without the help of a ladder that allows him to climb up to the level of truth? For the time being, no matter what the ladder is like, but it should be given to him. Moreover, as Hegel does not forget to add, this ladder is essential in the sense that the achieved result has to contain it and not to throw it away. Otherwise, the discussion would be about Wittgenstein's atomistic thought and not about Hegel's dialectical thought[27]. **Dialectically speaking**, then, **the**

external point of view constitutes a starting point which, in one form or another, **must be found again within dialectic itself**.

It follows that an adequate reading of dialectic should be, at once, **inner**, since the subject-matter is a living thought formed by the reader himself, and **external**, since dialectic is said to contain not only the truth but also the way to this truth, that is the not-true. So it is no wonder that we are to meet the external reflection within the *Logic* itself, as an essential and constitutive moment of dialectical thought[28].

For all these reasons, in so far as the previous discussions are valid, the **'reading of Hegel'** remains a problem and **'Hegel's secret'** is forever concealed from us. However, as is known to every serious dialectician, the opposite also is true : we are able to read Hegel, partly understanding him and so partly penetrating his secret. In that sense, there is no secret at all : every willing reader has his own Hegel, his own understanding of the dialectic and, from this viewpoint, it is advisable to regard every serious thinking as worthwhile.

Of course, that "the True is the whole" should be recalled ; namely, that the 'objective' truth would be formed by the gathering or synthesis of the different adopted standpoints. However, we have seen that things are not so simple. In fact, every concrete truth must be a truth for someone. On this point, Kierkegaard's thinking is entirely justified[29] : that is, as existing for a subject, the 'objective' truth itself becomes subjective, at least in part.

Ambiguity is thus a fundamental principle of dialectic[30]. For not only is the dialectical thought said to grasp the universal "which includes within itself the particular", but the way we understand that proposition is itself both universal (as based on our common Western culture) and particular (as based on our own experience). Eventually, it is by and for the individual that human reality exists and thinks itself[31]. And, from that angle, Existentialism and Hegelianism not only seems to exclude each other, but also oppose each other dialectically ; that is, they call and complement each other too. **Understanding Kierkegaard** is then understanding an isolated and lost individual at sea in existence, who nonetheless belongs to a certain time, a certain culture. And **understanding Hegel** is perceiving too, behind the self-confident scholar, the restless and anxious individual who is destined for an inescapable death.

It may be asked whether we have to stand up for Kierkegaard or for Hegel, to opt for the particular or the universal? 'Avoiding dichotomous divisions, dialectic is the grasping together of the opposites', should be an appropriate answer, I believe. Moreover, the understanding of dialectic itself is both particular and universal with no possibility of separating, like oil and water. Concrete truth, still abstractly expressed, lies then in the tense unity of these two opposites, apprehended in turn from each of the two points of view.

Finally, I understand that we have to read Hegel's dialectical thought **dialectically**, remembering to give place to a **formal** reading, which is equally essential since the abstract and 'not-true' external dimension constitutes a necessary moment of the dialectical movement[32].

PART ONE

DIALECTICAL IDENTITY, DIFFERENCE AND CONTRADICTION

According to Plato's *Parmenides*, Zeno said that "If things are many, they must be both like and unlike. But that is impossible"[1]. If that is the case we can assert that, from the dawn of Western culture, the problem of identity and difference has already been posed. As everyone knows, in the world things are identical from a certain point of view, and different from another[2]. What is more obvious than this general statement?

Nonetheless, the great thinkers of the past did not leave it at that. In particular, Plato questioned the possibility of stating a proposition concerning reality, which would be neither tautological as 'The One is One', nor contradictory as 'The One is Being' (if the One is Being, then it is two : One and Being)[3]. In other words, every identical proposition would be devoid of any real signification, and every proposition about reality would contain some contradiction—a thesis that is reminiscent of Antisthenes' Cynical School[4].

On the plane of being, then, identity and difference seem to constitute basic determinations. And, **on the plane of language**, every proposition can be said to be either tautological (when subject and predicate are identical) or contradictory (when subject and predicate are different), in any case with no signification—a necessary conclusion when the predicative proposition, whose structure is 'A is B', is considered[5].

Hegel poses the problem **on the plane of thought** : for him, identity and difference become categories of reflection, the 'determinations-of-reflection' with which the second part of the *Logic*, the Doctrine of Essence, begins. As for truth, Hegel presents it as the movement of the Concept said to determine our thought, our language and human reality in which we are living[6]. But what is the Concept itself?

In this Part I do not intend to deal with that question in particular, although it is important. In chapter 1, I merely briefly expose the development of Hegel's logic into its three levels : the Doctrine of Being, the Doctrine of Essence and the Doctrine of the Concept, and that in order to elucidate the movement of Essence, which I believe is essential. In chapter 2, I present the fundamental paradox of the 'relation of the independent terms', a presentation that leads, in chapter 3, to the exposition of the dialectical determinations of identity and difference, *qua* identity of the differentiated terms and difference of the identified terms. Finally, I attempt to show that the Hegelian contradiction must be understood as 'identity and difference of the determinations of identity and difference'[7].

1. THE MOVEMENT OF THE DOCTRINE OF ESSENCE WITH REGARD TO THE MOVEMENTS OF THE DOCTRINE OF BEING AND OF THE DOCTRINE OF THE CONCEPT

Superficially speaking, the Hegelian **Concept** may be understood as a **'unity of Being and Essence'**. In this context, 'Being' means the world of phenomena as opposed to its Essence, while 'Essence' means the world of thought. That is, the Concept must not be grasped as an abstract thinking, but as a thinking of human reality (logic is concerned), of an *'erinnert'*, or interiorized and recalled, human reality : the Hegelian logic is also an ontology[8]. Moreover, at the beginning of the Doctrine of Essence Hegel says that Being, in the course of its previous movements, has internalized itself *qua* Essence ; and Essence, for its part, first develops, according to the moments of reflection and determinations-of-reflection *(Reflexionsbestimmungen)*, the categories of identity, difference and contradiction with which we are dealing[9].

In another language, Being may also be conceived as the world of exteriority, and Essence as the world of interiority. Now, the Concept shows itself as an interiorized external totality, that is, as an experienced and learned one, or as an exteriorizing internal totality, 'recollecting' and determining human reality in its movements of thought and action. We saw that the Doctrine of Essence considers interiorized Being to be thought (Essence), which is why it opens with the movement of reflection and determinations-of-reflection. Let us throw some light on these terms.

The **reflection** of the Doctrine of Essence is to be understood according to the two meanings of the word. On the one hand, in the meaning of a **form of thought**, 'reflection' calls to mind the most basic determinations of human thought[10], and therefore also of human reality, in so far as the Hegelian logic is an ontology. On the other hand, the **etymology** of the word indicates that it also means a 'reverberation' of Being having interiorized itself as Essence, namely as thought.

Hence the following conclusion : the movements of thought (Essence) 'reflect', in the double meaning of the term, the movements of human reality (here apprehended as Being). But we should be careful : at this level of logic, we are allowed to talk neither about parallel development (as with Spinoza's Thought and Extention), nor about correspondence between entities (as if considering ideas could suffice to discern the characteristics of reality)[11]. Indeed, in his *Logic*, Hegel affirms several times that the **transition** of the Doctrine of Being has become the **reflection** of the Doctrine of Essence[12]. The discussion is thus about reflection into Otherness, according to the double movement of the Hegelian logic : Being (the Phenomenon) reflects or expresses the determinations of Essence, and Essence (Thought), in its turn, reflects or interiorizes the determinations of Being. This mutual reflection or mediation is said to constitute the movement of the logical Concept.

In this way, we have obtained a more concrete explanation of the nature of the Concept than before, where the Concept was merely exposed as a unity of Being and Essence. Consequently, studying the movement of the determinations of identity and difference amounts to studying the movement of Being. If so, the Hegelian logic is really an ontology.

We now consider the differentiation of the *Logic* into its three consecutive levels.

In his discussion of the dialectic of identity and difference, Hegel expounds it as a self-developing movement of **Essence** positing itself as its **Other**. At the starting level of **Being**, on the other hand, there is a transition to the Other that shows rather as a leap ; that is, at that low level, human thought does not recognize itself in that Other, the realm of the finite and of the quantitative determination that seem alien to it. As for the level of the **Concept**, Hegel claims that at this apex of development, difference *qua* difference is no longer to be found, since there is an entire self-recognition in the Other, with no residue left. At this high level, then, the human being would fully feel 'at home' (*bei sich*), in a world where he finally recognizes his own determining activity[13].

From this point of view, the movement of the Doctrine of Being may be regarded as an **external** process marked by abrupt transitions and leaps, as a process that advances only because of an external look that 'pushes' whereas, on the contrary, the movement of the Doctrine of the Concept seems far too **harmonious**. Hegel thinks that, at this summit, the universal thoroughly particularizes itself and, in its turn, the particular thoroughly realizes itself as universal. Hence a double movement producing the Individual, as the subject supposed to be the truth of human reality.

Therefore, with regard to the abrupt movements of the Doctrine of Being (lacking any immanent mediation) and to the far too flowing movement of the Concept (lacking any real driving opposition), **the level of the Doctrine of Essence alone seems truly dialectical**. On the one side, the tension existing between the opposites explains the need for a development that should bring relief ; while on the other side, at least in part, the human thought recognizes itself in its Other, which is supposed to be its own reflection. Reciprocally, we can also say that, in so far as thought partly recognizes itself in its Other, a pressing tension arises—whence the consecutive development[14].

As pointed out by Mure and Clark, it is just at that level of the Doctrine of Essence that a certain explanation of concrete human experience may be found[15]. For, according to that conception of things, **human experience** is not atomistic, as the level of the Doctrine of Being lets it suppose to be, when it exposes its determinations as leaping from one category to another, and when it explains reality with the categories of quality and quantity only. Neither is human experience this perfect harmony, this 'well-rounded' being that Parmenides talks about, as a subject-Concept that would no longer be substance, a Spirit that would no longer be matter, an Idea that would no longer be a thing—constituting thus a mere *Gedankending*, just a 'thing of reason'.

At that level of the Doctrine of Essence, where the dialectic of identity and difference develops, the basic paradox of the 'relation of the independent terms' comes to light. We now examine this paradox.

2. THE PARADOX OF THE 'RELATION OF THE INDEPENDENT TERMS'

To sum up, at the level of the Doctrine of Being the different categories appear to be **independent** of one another, while at the level of the Doctrine of the Concept, where the movement is said to develop harmoniously, all would be **linked** and united. Now, at the level of the Doctrine of Essence we find a posited category that is not its Other, as was the case in the movement of Being. However, since we are dealing with a middle level, where every determination reflects itself in its other and vice versa, we can also find there a relation between the opposites, hence the **'internal relation between the independent determinations'**, which I merely call the 'relation of the independent terms'[16]. In effect, the opposite categories of the Doctrine of Essence are **identical**, like those of the Doctrine of the Concept, and **different** like those of the Doctrine of Being, thus producing tension and paradox. At that level, neither external as the level of the Doctrine of Being, nor harmonious as that of the Doctrine of the Concept, we shall perhaps discover a somewhat satisfactory account of human experience.

The basic paradox of the 'relation of the independent terms' which, in my opinion, confers a certain intelligibility to the dialectical movement, may also be exposed in a more concrete manner. For this purpose, we have to consider dialectic as a movement developed by **human spirit**, grasped at once as **Understanding** and **Reason**[17].

Qua **Understanding** *(Verstand)*, human spirit apprehends the different characteristics of reality as being independent of one another, as data linked through a third only, a *tertium quid* , the external intelligence that considers them. Here, the discussion is about the empirical or formal thinking collecting particular facts, classifying them and drawing from them general laws by means of various procedures.

From the idealistic point of view adopted here[18], however, Understanding itself constitutes the source of the determinations it believes to find as 'data'. For Hegel grasps **reflection as determining**[19]. That is, in opposition to the corresponding Kantian thesis, Hegel asserts that judgement is not only reflecting but also producing—hence the 'determinations-of-reflection' of the Doctrine of Essence. Therefore, if every judgement is both reflecting and determining, as Hegel asserts, then the 'data' of the Understanding cannot remain isolated, fixed and with no immanent movement. That is, in addition to the Understanding itself, we must attribute to the 'data' a proper life, i.e. the dialectic illuminating the movements of transition, reflection and development from one logical category to another. Now, from that standpoint we may consider **spirit as Reason** (*Vernunft*), which connects what spirit as Understanding first separated.

Concretely exposed, according to Hegel, the human spirit cannot adhere to the conception of a finite that is independent of the infinite, of a form isolated from matter, of a particular external to the universal, and of an object not existing for a subject. For Hegel believes that, sooner or later, human spirit has to link up the opposites.

At that point, the paradox of the 'relation of the independent terms' appears, a paradox whose signification is as follows : that which thought to be a univocal, fixed and ready-made determination, actually acquires a content only when set in its proper context, which here may be called 'its Other'[20]. In this sense, every logical determination necessarily relates to its Other ; namely, it obtains an adequate signification only when related to this Other. For instance, one cannot merely be finite if not *vis-à-vis* the infinite that shows the limits ; a form is such only as a form of a structured matter ; every particular relates to a universal it may deny, but not ignore[21]; and subject and object do exist with regard to one another only, as Kant showed in his first *Critique*. In Hegelian terms, this idea can be expressed as the necessary **mediation of opposites**, an activity wherein every determination is itself only through its Other, its opposite.

Consequently, the paradox of the 'relation of the independent terms' can be said to be constituted by the pretension (or the illusion) of independence of each of the opposites which, on the one hand, believes to exist for itself, as indifferent to its Other, and, on the other hand, is obliged to relate to this Other in order to acquire some content[22]. The contradiction lies there : spirit as Reason has to link what spirit as Understanding had first separated.

In this way, the dialectical determinations of identity and difference produce the paradox that 'relates' the independent terms or, in other words, that reveals them as acquiring content and adequate meaning only when they interact—this paradox is of prime importance in the context of this investigation. From this the result is that, for Hegel, identity *(Identität)* is an **identity of the differentiated terms** and difference *(Unterschied)* is a **difference of the identified terms**[23].

3. THE DIALECTICAL DETERMINATIONS OF IDENTITY AND DIFFERENCE AS IDENTITY OF THE DIFFERENTIATED TERMS AND DIFFERENCE OF THE IDENTIFIED TERMS

The paradox of the identity of the differentiated terms and of the difference of the identified terms may be expounded from two points of view.

3.1. From the viewpoint of the very meaning of these expressions, I propose the following thesis : **only terms differing in a certain respect may be said to be identical.**

Let us look at the formal example 'A=A'[24]. In this example, the relation of identity is asserted about two different 'A' : one lies on the left of the equality sign, the other lies on the right. These two signs differ :

a) **intrinsically,** in that they cannot be the same, since the inked or chalked figures are not strictly identical[25]; and

b) **externally, from the standpoint of space** (or of time, if the signs are thought), in that the two signs necessarily appear in different parts of space (or of time). Although the relation of identity does not apply to the **signs** themselves but to the **signified**, formally speaking, this relation can be neither expressed nor thought without the help of these different signs, inscribed in space (or aimed at in time). Thus, **the moment of difference proves indispensable**. In consequence, if we consider the actual totality (the relation sign/signified), the paradox becomes apparent.

Undoubtedly, the formalist does abstract and separate, thus proving not to be concerned with the sign itself but only with the signified. On the other hand, the dialectician does not abstract but deals with totalities ; hence the paradoxes[26]. In his view, the identity of the signified or **content** is built only through the differentiation of the signs or **form**. The dialectician claiming to think 'the identity of form and content', that is, the concrete and living totality, we find again the 'identity of the differentiated terms' (regarding the content), which also constitutes a 'difference of the identified terms' (regarding the form).

This also holds for the converse relation : a difference can exist only between terms that are identical in a certain sense—here 'identical' is to be understood as 'belonging to the same genus'. So, in dialectic the difference is considered as a relation developing between 'proximate' kinds ; hence the expression 'difference of the identified terms'.

For instance, two tables may be said to be different to the extent alone that they belong to the same genus 'table' (we are really dealing with a relation connecting universal genus and particular species). It seems to me that this familiar assertion proves pertinent, for it expresses the following thesis : in human reality, we may assert that 'all is identical' or 'all is different' in so far only as we can speak significantly of a highest genus, comprehending every wordly entity. Although Aristotle was opposed to such a conception, Hegel adopts it unhesitatingly, Absolute Knowledge, Absolute Idea or Absolute Spirit constituting that all-encompassing highest genus.

Of course, rejecting this point of view is legitimate. But in that case, when comparing various entities of various areas, strictly speaking one will not be allowed to talk in terms of identity and difference. For example, the dualist will not be allowed to say that 'matter and spirit are different'.

Therefore, according to the paradox of the 'relation of the independent terms', in dialectical ontology the identity will be understood as identity of the differentiated terms (of different kinds) and the difference as difference of the identified terms (of the same genus).

Obviously, we cannot leave off there.

3.1.1. In this inquiry, the word 'identical' has been employed in two meanings. First, the discussion was about **formal identity** expressed by the formula 'A=A', and then about the dialectical identity that exists between different kinds belonging to the same genus. In this connection, in contradistinction to W. Becker who complains about the strategy, the cunning or the trickery peculiar to Hegelian logic[27], I propose to regard **ambiguity** as a fundamental characteristic of dialectic[28].

Indeed, if dialectic consists in the movement of human spirit unifying as **Reason** what it had separated as **Understanding**, then we have to distinguish—without separating—the **abstract meaning** of a term (corresponding to the **analysing** activity of Understanding) from its **concrete meaning** (corresponding to the **synthesizing** activity of Reason). Thus, the formal-analytic meaning of identity is opposed to its dialectical-synthetic meaning. In the former case the matter is about a given, univocal determination independent of any context in principle ; whereas in the latter case a determination is posited by human spirit, one acquiring adequate meaning and content only when related to its Other, here the complementing determination of the difference : from the viewpoint of the 'proximate' genus, the kinds show both as identical (to) and as different (from one another). The double meaning appears again, the 'dialectical meaning' comprehending now the synthetic moment (of Reason) and the analytic-synthetic whole (of Understanding and Reason). Out of that tension, unavoidably the impulse to develop arises[29].

3.1.2. In so far as ambiguity characterizes dialectic, the determination of difference must also be grasped according to its double meaning : **difference of Understanding** versus **difference of Reason**[30]. As far as 'difference of Reason' is concerned, the discussion is about the **relation** that is said to exist in a rational world between all the linked terms, thus constituting a continuous reality devoid of any impassable gap. On the other hand, when dealing with 'difference of Understanding', we must bear in mind the **discontinuity** that exists between the various points of reality, this time considered as atomistic. In this way, we pass from the identity of Understanding as **tautology** to the difference of Understanding as **leap**. And, in the 'syllogistic' movement at stake, between the extremes of tautology and leap we can discern, as a middle term, the dialectical identity which is also dialectical difference.

We should not be surprised at this movement : we are merely concerned with an activity of self-constitution through the mediation of the Other[31], while the middle term, grasped at once as identity and difference, contains a double meaning—we really encounter the 'fault' of the *quaternio terminorum*, signifying that, in the concerned syllogism, the middle term is not univocal[32]. I understand that in this case the 'fault' is conscious and deliberate, for to me the development seems possible under this condition alone : that the identity of the differentiated middle (differentiated into dialectical identity and dialectical difference) and the difference of the identified meanings (identified as formal and dialectical meanings of each opposite) should produce a driving tension that seeks relief ; hence the producing movement.

This new paradox, according to which reality is regarded both as **continuous** (from the standpoint of Reason) and as **discontinuous** (from the standpoint of Understanding), may be explained in another way.

So far, the discussion has been about the 'relation of the independent terms'. However, since in dialectic the logical proposition is always convertible directly, let us examine the symmetrical and complementary paradox of the **'independence of the related terms'**. From this new point of view we can see the opposites, which appear related as both identical and different, as also existing by themselves : thus, in the 'immanent' movement of the Concept, a certain discontinuity emerges. At this point it may also be noticed that, in the Hegelian logic, the continuous and discrete magnitudes are constitutive moments of the development (they form the number)[33]. That is, in so far as my reading of the *Logic* is acceptable, according to Hegel the real is truly continuous then rational, and also discontinuous then atomic.

If the paradox of the 'relation of the independent terms' expressed the activity of spirit which, *qua* Reason, connects what it had first separated *qua* Understanding, the new paradox of the 'independence of the related terms' indicates that spirit *qua* Understanding has also to separate what it apprehends, *qua* Reason, as connected and continuous. Dialectic again shows as a game of reciprocal and alternate reflections between opposite points of view, those of Understanding and Reason, a game where each one, in turn, unties what the other has tied and reconstructs what the other has destroyed. Hegel discerns there a circular movement whereas, in that context, I would prefer to talk about a bidirectional, **double movement**. In my view, it is that unceasing fight between opposites, here human spirit as Understanding or as Reason, that constitutes the motor of dialectic, its driving soul.

3.2. The expressions 'identity of the differentiated terms' and 'difference of the identified terms' may also be explained from the viewpoint of the movement exposed in Hegel's *Logic*, that is, from the viewpoint of the Doctrine of Essence.

Now, the discussion is about Essence (or Thought) developing itself and positing the determinations of Being (or Phenomenon) it had first interiorized. At that level we may talk about the **identity of the differentiated terms**, since we are dealing with Essence positing itself as an Other—the original determination has divided into 'the Same' and 'the Other'. This activity constitutes the movement of **mediation or reflection**, which can be summed up as follows : to posit the presupposed determination, to display it as an object *vis-à-vis (Gegenstand)*, and to connect itself with that Other grasped as the proper reflection of the original essence, which thus developed further.

In the idea of an Essence positing **itself** as its Other, we again meet the 'identity of the differentiated terms'. But since a relation to the **Other** is also concerned, when we think it over we may also call to mind the **difference of the identified terms**—it is merely a matter of point of view.

Indeed, the poles reflecting on one another necessarily differ from one another, since a relation can exist between differentiated terms only. Moreover, we should not overlook the moment of Understanding, namely that of separation and independence : in order to connect with each other, the opposite poles must also exist by themselves, at least to a certain degree. In the dialectic of the determinations-of-reflection, that moment comes to light as Variety, a level at which an external reflection compares the opposites from the standpoint of Likeness and Unlikeness. Thus, the moment of Reason, here of immanent opposition between the Positive and the Negative, is not sufficient[34].

This is why we find, at once, the identity of the opposite poles mediating themselves reciprocally, and their difference *qua* separated. Translated into the language of our discussion, we can say that the matter is about the **identity of the identified and the differentiated terms** and about their **difference**[35]. In other words, identity and difference are identical (according to the activity of the synthesizing Reason) as well as different (according to the activity of the analysing Understanding) : the identification of the differentiated terms and the differentiation of the identified terms take place between **connected** and **separated** terms, that is, between 'tense' terms.

In that constitutive tension we can recognize Hegel's dialectical contradiction *(Widerspruch)* : in my view, this category means nothing more than the relation of identity and difference holding between connected terms that are also separated[36]. As soon as the Hegelian contradiction is undersood in this way, it seems to me that the dialectical movement becomes intelligible, at least in part.

4. THE DIALECTICAL CONTRADICTION AS THE IDENTITY AND DIFFERENCE OF THE IDENTIFIED AND DIFFERENTIATED TERMS

4.1. In the context of the Hegelian dialectic, the exposed paradox of the 'relation of the independent terms' signifies that human spirit connects as Reason what he had first separated as Understanding.

From this point of view each opposite, when grasped in isolation, constitutes a fixed and formal determination lacking in content. In order to gain some content, it must be connected with the context from which it had been abstracted, that is with its Other through which it develops its particular meaning. Hence the following paradox : it is when human spirit tries to grasp the determinations posited as a pair of independent data that it is brought to connect them with each other. In that very activity, Hegel sees the 'return into itself' of each category out of its 'moment of exteriority'. In that sense, then, the dialectical identity (of the opposites) constitutes the **identity of the differentiated terms**—another formulation of the paradox of the 'relation of the independent terms'.

As regards the paradox of the '**independence of the related terms**', it expresses an inverse movement : it means that external reflection is a constitutive moment of the dialectical movement, and that synthesizing Reason comprehends within itself the moment of analytical Understanding. Indeed, Reason *'aufhebt'* that moment, but the Hegelian *Aufhebung* has to be grasped in its specific ambivalence: it negates and affirms, at once repelling and maintaining[37]. The repelled Understanding is also maintained.

The same holds for the **difference of the identified terms** : empirically, the connected terms must also exist by themselves, so that they may appear before a look apprehending them—we should not forget that moment of independence and isolation that is essential to the being of the finite[38]. For, of course, without the finite there would be no world, and accordingly no thought, no determination and no constituting movement either.

Thus, in accordance with the conversion rule proper to the logical proposition, the relations of identity of the differentiated terms and of difference of the identified terms have just been expounded ; that is, **the**

identity and difference of the identified and differentiated terms. And, in my opinion, Hegel's dialectical contradiction can be understood in this way : it comes out when we grasp, **at once** and **from the same point of view** (the human spirit) the identity of the differentiated terms and the difference of the identified terms. This paradoxical activity produces a tension that seeks relief and reconciliation and, in that very meaning, dialectical contradiction proves to be creative and not destructive[39].

This kind of contradiction requires the apprehending together of the paradoxes of the 'relation of the independent terms' and of the 'independence of the related terms' ; or, in other words, **the reflection that connects (determines) is also that which excludes**—another Hegelian expression of the thing in question[40]. Indeed, on the one hand every term is itself only through the relation to its Other (from the standpoint of Reason) ; here we find a mediating reflection connecting the opposites, and hence a movement of **attraction** that links them to one another. On the other hand, for the analysing reflection every term also exists by itself, as independent : it is what it is, identical to itself and to nothing else (from the standpoint of Understanding)—hence a movement of **repulsion**, through which every term excludes its Other.

In consequence, the contradiction also consists in this double movement of attraction and repulsion that develops between the dialectical opposites : every determination at once **repels and attracts** its Other ; namely, it is both independent and dependent, possessing by itself an abstract meaning and, moreover, a concrete meaning through the mediation of its opposite[41]. Generally speaking, here we are dealing with the **identity and difference of the opposite determinations** that exist for themselves, written on paper or aimed at by thinking[42], and that also exist in a context that provides them with sense and content : their constitutive Other.

For example, Hegel thinks that **finite and infinite are identical and different** : different according to their empirical or formal meaning, and identical in so far as the **finite** contains the infinite (its essence) within itself—hence its movement of self-realization—and as the **infinite** contains the finite (as moment) within itself—hence its movement of self-particularization[43]. Likewise, form and matter are identical as materialized form or formed matter, universal and particular do exist only as related to each other—separated, they become respectively abstract universal and particular devoid of any signification. As for the identity of subject and object, it means that the object is posited by a subject, and that the latter constitutes itself through an interiorization of the object *vis-à-vis (Gegenstand).* Moreover, that double movement implies that each of the concerned poles also has its proper mode of existence, as spontaneous subject and inert object.

Some will ask whether **the discussion is about contradictory or contrary opposites?** The Hegelian answer is that it is **about both, at once**[44].

Human spirit *qua* Understanding produces the contradictory opposition, signifying that each term excludes its Other and exists independently of it. As for spirit *qua* Reason, it connects the opposites, hence the contrary opposition—like white against black—and, in addition, it grasps together contradictory and contrary determinations[45]. In effect, Understanding sees every determination as **univocal**, as excluding any relation to the other, except, of course, the relation of cancelling negation, constitutive for the formal contradiction ; whereas Reason, for its part, posits a necessary relation and a reciprocal mediation between the opposites—hence the formation of a **double meaning**.

If this is really the case, then in dialectical logic the moment of 'attraction', together with that of 'repulsion', form the basic ambiguity that alone explains the possibility of a progressive movement. Here each term is both itself and its Other ; and in so far as a synthesis is realized, it does not produce an 'average', like grey with regard to white and black, but a new category containing again contradiction, paradox and tension. In this way, from one synthesis to another, a movement is engendered, until some totality is constituted[46]. Yet dialectic cannot stop there, for it is life and spontaneity. Consequently, the totality itself posits its own presupposed opposite, that is, the part that begins a new cycle of development. Concretely speaking, the totality acquires signification with regard to the part only ; that is, totality and part also 'attract and exclude each other', forming at once a contrary and a contradictory pair[47].

4.2. Now, **is the Hegelian contradiction thinkable or not?** We saw that this kind of contradiction must be understood in its double meaning. Namely, if 'contradiction' means '**excluding contradiction**', then the answer is obviously 'no'. For, from the standpoint of human spirit as Understanding, the opposites 'repel' each other and exist for themselves only as independent, univocal and with no immanent movement. But since the Hegelian contradiction also has to be considered as **linking contrary terms**, from that angle the answer becomes 'yes' : in my empirical example, black and white can be grasped together as giving a grey. Therefore we may say, from the point of view of Reason, that the opposites 'attract and mediate each other' and thus produce a new determination, their dialectical unity. It can be asserted, on that account, that Being and Nothing produce Becoming, Infinite and Finite posit the For-itself (a still abstract subject), Form and Matter constitute the Thing, and Universal and Particular give rise to the Individual. Conversely, the thinking of Becoming signifies a thinking of the passage from being to nothing and from nothing to being, every Subject is both finite and infinite (i.e. is self-reflecting), the Thing can be analysed into form and matter, and every Individual is both particular and universal, its essence consisting in the universal relation[48]. In this sense, then, **the Hegelian contradiction is both thinkable and not thinkable**, depending on the adopted point of view : that of the connecting Reason or that of the excluding Understanding[49]. In place of ambiguous language, a somewhat vague expression may be used: the Hegelian dialectical contradiction is **thinkable up to a certain point**[50].

Given the problematic nature of dialectical synthesis, the point at issue may also be presented in another way : when we try to grasp directly Being and Nothing together, or Finite and Infinite, what are we really thinking?

The answer does not seem easy. How can we then justify the dialectical development? I propose the following : to conceive it as a movement **both positing *a priori* and reconstructing *a posteriori*** . Indeed, dialectic does not work in a vacuum but, among other things, it presupposes the history of philosophy it claims to reconstitute in the form of a dialectical system. Therefore, when Hegel *a priori* 'deduces' the category of becoming from the categories of being and nothing, he uses the works of past philosophers as a **leading strand** ; hence there is a certain validity that corresponds to the validity of these works of the past[51]. For instance, it may be said that Parmenides and Heraclitus justify the dialectic of Being, Nothing and Becoming ; Plato justifies that of Essence, Show and Reflection ; the dialectic of Form, Matter and Thing originates with Aristotle; and many critics have discussed Hegel's debt to Spinoza, Kant and Fichte[52].

So, even from that point of view the dialectical contradiction seems **thinkable** as a movement reconstructing *a posteriori* and **not thinkable** as a movement positing *a priori*.[53] The dimension 'thinkable' reminds us that dialectic presents itself as a movement of thought which 'reflects' the past and the actual movements of historical, social and cultural human reality. As for the 'non thinkable' dimension, it refers to the impossibility of foretelling *a priori* the course of things: before the works of Parmenides and Heraclitus, it could not be known explicitly that being and nothing produce becoming. Indeed, empirical (here: historical) determinations truly affect 'pure' thinking[54].

From this it follows that, in human experience, facing the dialectical we find the non-dialectical, which dialectic fails to *'aufheben'*, that is, to resolve and move.

At this point we may mention, for example, the contingent existence of such a philosopher or of such a philosophical school. This 'non-dialectical' dimension constitutes an insuperable obstacle, an irreducible residue that ensures that the history of philosophy will never come to an end. Dialectic will always be confronted with an uncompleted task, which plays the part of an incentive as well as that of an univocal limit. For ambiguity characterizes the very validity of any dialectic : curiously, the exposition of the 'necessary' movement of thought has to ground itself on the opaque fact. In that perspective, the horizon of research remains open for ever.

4.3. The dialectical contradiction may also be presented in another way.

According to Hegel, it is possible **to perceive** either the identity of things or their difference, independently of each other[55], as we do in our everyday lives. Yet, we are not able **to think** these determinations in isolation without relating them to each other. Thus, the contradiction appears afresh : thought compels to connect what perception presents as separated, and that from the same point of view, the knowing subject.

In other words, if sensation reveals to us an atomic world made of isolated impressions and irreducible data, thought, on the other hand, as one living activity, offers to us a certain totality, living and active itself, which relates the different determinations to one another. Now, the human being is both sensation (or perception) and thought, a unity the dialectician does not divide[56]. Hence the contradiction in the Hegelian sense of the term, always present and creative since a 'solution' has to be provided to the unveiled tension—a 'solution' which, in its turn, will be in need of a further development.

I understand that the non-dialectical thinker, for his part, separates what is formally thought from what is empirically perceived, and has difficulty accounting for logic and mathematics as applying to reality. At this point re-emerges the paradox of the **relation of the independent terms** in which, according to Hegel, every dualism becomes entangled, in that it is not able to unify what it first separated[57].

As for the paradox of the **independence of the related terms**, its understanding may correct a certain interpretation of Hegel that I think is inadequate : Panlogism[58]. For, in my opinion, the concrete dialectician does respect the empirical element present in human experience, regarding it as the other of dialectic, partly irreducible : involving contradiction, paradox and tension, dialectic moves too through the opposition holding between the perceived and the thought elements—the separating barrier *(die Schranke)* truly constitutes a necessary condition of a movement that partly goes 'beyond'[59]. Therefore, while rejecting any reductionist theory, the concrete dialectician is supposed to comprehend human reality at once as empirical (*qua* perceived) and as dialectical (*qua* thought)[60]. Within that reality he also has to distinguish the moments of relation and independence, of continuity and discretion, of organic development and leap, of the *a priori* and sensation[61]; hence his discernment of an actively present contradiction. Accordingly, this dialectician cannot avoid coming up against the non-dialectical, the empirical 'residue' that, on the one hand, is an obstacle to the movement of the Concept and, on the other hand, represents a *sine qua non* for that movement. As Sartre noted, for dialectic, nature really constitutes a threat as well as an opportunity[62].

4.4. In Hegel's *Logic*, at the beginning of the Doctrine of Essence the dialectical movement unfolds itself according to the moments of Identity, Difference, Variety (*qua* Likeness or Unlikeness), Opposition (between the Positive and the Negative) and Contradiction.

In this Part of the work, I do not intend to expose and elucidate in detail the various phases of that development[63]. On the one hand, such a task seems tedious and rather uninteresting ; and, on the other hand, in any case a certain basic structure comes to light at the different levels of development—a structure that can be expressed in the form of the '**identity of the differentiated terms**' and the '**difference of the identified terms**' (or in the form of the 'relation of the independent terms' and the 'independence of the related terms'). Indeed, since this movement may be qualified as 'quasi-organic'[64], that very structure recurs at each new level with different degrees of realization and explication—so far, that structure has been the subject-matter of our discussion. In Becker's language, I see in it 'the structure of the dialectical argumentation'[65].

However, more interesting in my view is the **movement** exposed in the *Logic*, **from Contradiction to Ground**, a movement said to bring the 'solution' : the dialectical contradiction should find there relief and reconciliation[66]. In order to understand that 'solution', we should recall the ambivalence that is characteristic of the Hegelian logic : the enriching **progressive** movement of the categories also constitutes a **regressive** movement towards the basis. This is why Hegel can affirm that, in the circle of the logical development, "the first is also the last and the last first" ; namely, that the last, from the standpoint of the order of exposition—Concept or Idea—proves the first from the standpoint of the order of grounding. From that angle, we can say that the Concept or Idea self-particularizes into its structuring moments, the different categories of the *Logic* : Being, Nothing, Becoming, etc.

If this is the case, it is no wonder that the dialectical contradiction has **to return to the ground from which it stemmed.** Originating in Essence—a mere self-identity in a process of self-differentiation—the contradiction must return to Essence, but this time in the form of Ground, that is, of Essence enriched with the different determinations-of-reflection. In this way a 'mere identity' is again formed, the human spirit (here as a still abstract thought) having to recover a content it has first set aside. From here up the Ground will develop further, according to the essential determinations of Form, Matter and Content[67].

I want to explain this point with a brief recapitulation of the logical movement in question.

In the *Logic*, following the exposition of the external determination of Being (as Quality, Quantity and Measure), the Doctrine of Essence exposes the movement of the immanent determination developing towards subjective totality (the Concept) becoming also objective totality (the Idea). In that movement, Essence first shows itself as pure unity devoid of any determination, as mere self-identity. As self-identical, it is also self-different, the relation of identity necessarily existing between poles differentiated from a certain point of view. Hence the production of the determinations of Identity, Difference, Variety, Opposition and **Contradiction**, the determination of Contradiction consisting in **the**

identity and difference of the identified and differentiated terms.

Now, Hegel says, the contradiction *'zugrunde geht'*, according to the two meanings of that German expression : passes away and passes over into Ground. Namely, the contradiction goes back to its foundation from where it stems—let us not overlook the fact that the dialectical contradiction expresses just a certain level (Hegel says: a moment) of human reality self-differentiating and recovering its original unity. That is why the 'solution', in the double meaning of the word (cf. Note 66), lies in that returning to the unity, which is said to be the source and basis. However, in that movement of explication from potential to actual that constitutes the Hegelian logic, the moment of exteriority and independence is not only negated but also preserved[68]. Therefore the new unity obtained, said to be 'immediate', will keep on developing in order 'to resolve' the emerging new tension[69]. In this way identity again turns into self-contradiction and, conversely, reconstructed contradiction returns to its original identity. A double movement is thus shaped, moving in two opposite and complementary directions while positing the presupposed determinations.

We saw that by 'dialectical contradiction' we had to understand 'the identity of the differentiated terms and the difference of the identified terms grasped together'. The formalist sees that 'unity' as unrealizable. As for the dialectician, he discerns there a pressing tension demanding a movement, since **the dialectical contradiction proves to be not abolishing but producing**.

Ontologically speaking, in his experience man becomes entangled in contradictions he himself has created[70]. In order to escape from them, he is obliged to rise up to a level where he might find rest and relief, a level that Hegel calls 'Ground'. The truth, the ultimate that hides 'behind the things', that which is the 'cause' of everything and constitutes the appeasing explanation, all that should be reached at the level of Ground. However, to the extent that a certain validity may be ascribed to Hegel's thought, it seems that even on this ground the human being will experience new contradictions, and hence new developments. To that extent, self-contradiction constitutes the *'Bestimmung'* of the human being, in the two meanings of the word : determination and destiny.

PART TWO

THE PROBLEM OF THE BEGINNING

Philosophy as metaphysics aims to provide a certain explanation of human experience. To this end, it has to construct different argumentations founding and enlightening that experience, that is, to turn obscurity into clarity, vagueness into obviousness. The philosopher thinks. His thought, as discursive, progresses from concept to concept, from judgement to judgement, from syllogism to syllogism, constituting thereby an explanatory set intending to disclose the essence of reality. In this explanatory activity, the philosopher produces some sequences of arguments that ground and imply one another. Therefore, from the outset the following question arises : **in that production of successive arguments, with what can philosophy begin?**

If truth only exists as the result of some research, then philosophy cannot begin with it. At the beginning, it has to be content with the non-true, if not with the probable : as in Aristotle, the demonstrative syllogism, which is certain, must be grounded on the dialectical syllogism, which is only probable.

If philosophy, here considered as a serious and basic business, is not able to manage with the uncertain, it must then **begin with truth immediately**[1]. But how is that possible if, at the beginning, the philosopher is ignorant? Under these circumstances, referring to the history of philosophy I suggest an opening either with **a naive way**, without asking too many questions, like the first Presocratics, or with a **revelation** like Parmenides, or with Plato's **intuition of Ideas,** turned in Aristotle into **rational intuition**.

Indeed, as Aristotle noted, demonstrating everything is not possible. At the outset, we seem compelled to regard certain basic propositions as 'true'. In the opposite case, we would be brought either into the **infinite regression** of a ground needing itself to be grounded, or into the **circle** of a ground grounding itself[2].

In order to escape from that imbroglio, the Prekantian philosophical tradition chose to rely on rational intuition. This procedure permits the philosopher, from the beginning, to lay down some definitions and axioms and thus to deduce different truths said to explain human experience—like Spinoza in his *Ethica More Geometrico Demonstrata.*[3]

In itself, that rational intuition seems quite mysterious. Must we really believe that, in his essence, man is able to come into contact directly with the 'source', the 'absolute', so that he might immediately start with truth? However attractive that solution may be, it becomes quite strange when thought over : if the human being were able to know the truth directly and without more ado, why should he construct complex theories that demand so much work and struggle, and bring to the pioneer so many disappointments before the promised land of confirmation and recognition is attained?

Accordingly, I understand that things happen differently. Kant showed in his philosophical inquiries that man cannot rely on 'evidence', which constitutes at best an end and not a beginning[4]. On the other hand, he is able to perceive empirical data, a fact making human experience possible in that sense. However, in possession of a discursive intelligence only, the human being has to build, thought upon thought, the edifice of knowledge by means of hard and sometimes unfruitful labour, which certainly does not start with the expected result : truth.

Under these conditions, then, how is a beginning possible? The problem proves complex and, for the Postkantian thinker, the following alternatives are available :

(1) He may content himself with a starting-point 'that suits him', with no other justification. If so, he will not be allowed to refer to knowledge, only to **opinions and common ideas**. For, since Parmenides, Western culture has learned to distinguish Being from appearance. Now, what is aimed at is knowledge of Being, this being in need of justification and grounding.

(2) The thinker may also start in a **dogmatic** way, deciding that a certain proposition, which appears as fundamental, is to be considered as a basic postulate. Yet, since Kant, we know that this procedure is inappropriate. Human beings lack intuitive intelligence, as modern culture admits in general, so that such a beginning with 'a pistol' cannot be accepted, for it amounts to stating as true immediately what, in fact, proves to be only arbitrary.

(3) Between the infinite regression of a ground needing itself to be grounded, and the circle of a ground grounding itself, the perplexed thinker might find himself paralysed. In these circumstances he may

become **sceptical**, rejecting any possibility of a grounded science or of an explicative theory.

(4) Certainly, the sure and plain way of common sense remains : **to base ourselves on tradition,** acknowledging our forefather's discoveries and creations, noting that these have 'functioned' well and have yielded good results. However, with all due respect to tradition, a scientific mind cannot accept without putting it first 'through the riddle' of criticism. In that immense body constituted by tradition, he has to separate the wheat from the chaff. But how is this possible, without a safe criterion of truth?

It thus appears that, in his conceptual research, the thinker faces a discouraging dilemma : because the naive way is no longer available today, he must choose between **dogmatism** relying on immediate rational intuition, and **scepticism** appealing to the absence of any real justification. Under these conditions, with what can the beginning be made? With truth itself? We saw that a sure criterion is missing. With the non-true? But how can the true follow from the non-true[5]?

That question arises in all its acuteness and demands an answer. In this Part of the book I examine how Hegel dealt with this problem. In so doing I investigate, from its various sides, the solution he provided, reinterpreting it when necessary.

Philosophical questions generally do not lead to conclusive responses: various paths bring us nearer to the purpose, defining it with the help of more or less tortuous, more or less enlightening ways. We now turn to the way taken by Hegel.

1. THE HEGELIAN SOLUTION: ACTUALLY, PHILOSOPHY HAS TWO BEGINNINGS, ONE SENSIBLE AND THE OTHER INTELLIGIBLE

Hegel's intention was to develop a philosophical system that would be capable of giving to human experience a comprehensive and detailed explanation. Moreover, as a pioneer tending 'to think everything anew', he aims at a 'philosophy without presupposition'. From the first, then, he comes up against the problem of beginning : how can he embark on a task of elucidation that intends to be 'absolute', that is, total and unique, and that cannot content itself with immediate, rational intuition?

It is true that Hegel accepts the inheritance of the philosophical tradition, acknowledging its works and discoveries. However, he regards it rather as a material requiring reconstruction and interpretation in the form of his own system[6].

Under these conditions, thinking 'all anew', working 'without presupposition', Hegel seems to have had no choice but to start with arbitrariness or dogmatism. Nevertheless, we shall see that, in spite of diverse difficulties, the dialectical Hegelian can clear a way for himself (can find a met-hod) and escape the deadlock[7]. In order to grasp that original solution, let us go a way round with traditional thinking.

According to tradition, the task of philosophy is to explain reality. To this end, it has at its disposal two main sources of knowledge : Sensation and Reason. So, in the course of its inquiries, classical philosophy draws its first knowledge either from sensible or from rational intuition. Yet, as early as antiquity, the Greek sceptics showed that **sensible intuition** proved to lack any regularity, that is, any universality or necessity; for in that domain, according to the Heracliteans, "all things are in flux". In other words, what is concerned here is the world of generation and corruption, which lacks firm ground upon which to found a sure knowledge[8].

Shall we find a better answer when referring to **rational intuition**? Maybe. If in the sensible world "all things are in flux", we can say, in so far as thinking is concerned, that the flux itself is permanent : all things are in flux, necessarily and forever[9]. Expressed in Heraclitean language, the point at issue is the **Logos**, which represents the required constant principle, the universal measure that ensures the harmony of the opposites in everlasting struggle—like the musical harmony proceeding from the stretching between bow and lyre.

With that **Logos**, the rational intuition is supposed to have found the law that permits us to obtain a valid knowledge. If that were really the case, the problem of the beginning would have been settled, but, alas! things happen otherwise. In particular, in Western culture man is living in a certain world and experiences a certain reality. Now, how shall we understand the fact that this one world, which is common to all human beings, is explained by various theories, each of which proposing its own exclusive truth? In front of this mutual conflict of systems, what criterion will help us to discern the 'true' rational intuition? Certainly not a new rational intuition... Again, in the face of these School quarrels, the involved spectator will try to find shelter in a dumb or sour scepticism. So, rational intuition has deceived too ; for how can we believe the dogmatics when they claim to be in possession of 'the' truth, while they fail to agree with each other?

Thus, the two main sources of human knowledge prove to be insufficient ; but we must not conclude from this that scepticism is the only 'rational' solution. There is at least one other solution, proposed by the Hegelian dialectic. Let us proceed to its examination.

The question is explicitly raised by Hegel in the following terms : **"With what must Science begin?"**[10]. For the dialectician, the answer is included in the question itself : **we only have to consider the idea of beginning 'in itself'**[11]. For instance, earlier I mentioned a beginning either with sensible or with rational intuition. The dialectician does not think that a choice has to be made, in an exclusive fashion, between a sensible and a rational beginning since **a third way** is available : in the development of the philosophical explanation, the one necessarily leads to the other and vice versa ; that is, a reciprocal mediation between the two opposites takes form. Thus, in the course of the movement the sensible beginning develops until the intelligible concept, and the consideration of the beginning of thought brings back to the presupposed sensible experience[12].

Indeed, it may be said—even if it is surprising—that the Hegelian philosophy is not merely either empiricist or rationalist, but truly both at once[13]. That assertion becomes comprehensible when it is pointed out that, according to Hegel, concrete thought is nothing but the negation of experience, and experience itself a certain expression of thought[14].

Although uncommon, this thesis is plausible to me. Actually, when we consider closely the system Hegel bequeathed to us, we can distinguish there two beginnings, purely and simply: the *Phenomenology* starting with sensible intuition and the *Logic*, with rational intuition[15]. And, in my opinion, 'thinking together' these two beginnings renders the system intelligible to a certain extent.

In the *Phenomenology*, **Sense-Certainty** develops up to Absolute Knowledge, which appears at the beginning of *Logic* as Pure Knowing. As for the *Logic*, it opens with **Pure Being**, which is empty thought or empty intuition, and which develops up to Absolute Idea, the totality of concrete thought presupposing empirical reality. At this point, accordingly, the Absolute Idea "resolves to let the moment of its particularity... go forth freely as Nature" ; namely, it posits the presupposed sensible experience. In my view, the circle closes through the mediation of *Encyclopaedia's* ***Realphilosophie***, which, at the end, posits the Syllogism of the three Syllogisms said to unite the *Phenomenology* with the *Logic* [16]. It is in this exposition of the double movement, which leads from sensible to intelligible and vice versa, that the power of the Hegelian thinking can be discerned[17].

Over time, Western culture has become discouraged with the sensible or intelligible starting-point and, from blind dogmatism it has passed to a desperate scepticism. Hegel, however, like Kant before him, puts forwards a third way standing between ancient dogmatism and modern scepticism[18]: a certain conception of an organic system grounding itself. The paradox of the beginning, which might give rise to infinite regression, is here settled by the circular progression, or rather by the progression in a double movement.

If really so, then philosophy possesses two beginnings that mediate each other : a sensible beginning exposed in the *Phenomenology*, and an intelligible beginning exposed in the *Logic*. In that sense, Hegel is entitled to affirm that his philosophy 'comprehends' the principles of empiricism and rationalism, which are insufficient when regarded only 'in themselves', independently of each other. This assertion may help grasp in what sense Hegel may be said to *'aufheben'* or to comprehend the history of philosophy[19].

After that discussion in principle, constituting a first approach to the topic at issue, I first examine the beginning of the *Phenomenology* and later the beginning of the *Logic*.

2. THE SENSIBLE BEGINNING OF THE *PHENOMENOLOGY*

In the Preface to his first *Critique*, second edition, Kant affirms that, before developing any metaphysical system, we must "**prepare the ground ... by a critique of the organ**", that is, we must first examine the power of pure reason, enabling it to know itself *a priori* [20]. However, in this connection Hegel notices that to begin with criticism is absurd, for it comes to "seek to know before we know"[21]. In effect, this mode of beginning founds itself on different presuppositions that themselves require elucidation and criticism. According to Hegel, this beginning presupposes that knowledge stands on one side, separated from the thing-in-itself (which is real in the full meaning of the word) that is supposed to stand on the other side. Now, how is it possible to talk about knowledge, when the knowledge in question is presented as being unable to grasp reality[22]? Moreover, the conception of knowledge as an instrument proves inadequate, for the use of a means does not leave the examined thing indifferent but alters it. And how can we determine the influence of that means? With the help of another means?

As the Introduction to the *Phenomenology* shows, we can therefore start directly neither with truth itself in need of grounding, nor with a 'true' criticism that requires to be criticized[23]. We must then start with the non-true : we already saw that the dogmatic solution is not satisfying, and that the naive solution is no longer available.

At the outset, then, when it comes on the scene, Hegel's philosophy, like any other system, presents itself as 'the Philosophy' *par excellence*. Effectively, for itself every system is certain to grasp and expose the truth. According to Hegel, this general and reciprocal conflict between the various pretensions shows that we still are at the level of **appearance**, of the non-true—a point of view that has to be overcome. For we cannot accept all these various 'truths' that exclude one another. Consequently, in order to resolve the ongoing conflict Hegel suggests the dialectical way: to examine appearance 'in itself and for itself'[24], as he does at the beginning of the *Phenomenology*, when he opens with the presentation of the inner contradictions of Sense-Certainty—the starting-point[25].

The *Phenomenology* thus begins with the sensible consciousness containing within itself the impulse to know, but deluding itself about its own power—it is still the level of appearance, we said. Lacking any self-reflection, that consciousness is naive ; it believes it knows and has not yet apprehended its own limits. However, as soon as it attempts to test the validity of its knowledge, it discovers its error and, accordingly, loses all self-confidence. In this connection, the *Phenomenology* talks about the "pathway of doubt, or... the way of despair"—what brings to mind the questioning and paralysing Socratic method[26].

What is the origin of this disappointment? Hegel refers it to the very nature of consciousness, which is dual : on the one hand, it **distinguishes itself** from its object, and, at the same time, **it relates itself** to it ; hence the contradiction in the Hegelian sense of the word[27]. When consciousness discovers that moment of inner contradiction, it tries to overcome it by rising up to a higher level, turning then into a new pattern that relates to a new object—in this sense, Socrates is also a midwife.

Nonetheless, when reflecting on the nature of its knowledge, at every new attempt consciousness rediscovers the gap between what it aims at—to know the object in itself—and what it actually achieves—it only knows the object as it is for it ; hence its new disappointment[28]. Thus, forever pursuing self-certainty, consciousness is forced to advance from disappointment to disappointment, until the point where, finally, its knowledge of the object 'for itself' corresponds to the 'in itself' of that object[29].

At this point, Hegel says, where objective truth and subjective certainty coincide[30], the delusion vanishes and consciousness finally learns what its true nature is : its object only consists in its own essence it has exteriorized, an activity it was not aware of at the outset. That is why, Hegel continues, it must pass through this 'Calvary', a way re-vealing to it its own truth, through the mediation of desire, fight, work, language, alienation and suffering.

In this way, the *Phenomenology* exposes the 'education' of consciousness, delivering itself from error and delusion and gradually attaining truth : we really do start with the non-true. However, the novice may ask, "Why bother with the false?". Is it not possible to dispense with that moment of non-truth and to begin with truth directly[31]? The previous discussions have shown that this is not the case.

For the novice, at the beginning, Hegel's philosophy stands beside other modes of philosophizing, like them putting forward its own claim to truth. Under these conditions, how could that ignorant novice distinguish a fulfilled promise from a broken one, if not arbitrarily? Moreover, a direct leap to truth, "like a shot from a pistol", should not be made, for fear of finding a dazzled and paralysed novice—like the prisoner of

Plato's cavern (*Rep.* 516b) who must first become accustomed through successive stages, and only then pass progressively from shade to full light.

In Kierkegaard's words, we "first of all" have "to find" the novice "where he is" and to talk with him in a language he can understand, since in fact he is the one to convince[32]. That is, we must take his hand and lead him, step by step, like Diotima guiding the young man in Plato's *Symposium* (209e), towards the world of Truth (and Beauty), so that he can learn and comprehend.

In opposition to the **deductive-linear** method that starts with truth directly and has no use for the non-true[33], the **circular** method at stake here regards as essential the ladder having allowed to take possession of truth. In consequence, the achieved result contains within itself the way that leads to it : in Hegel's terms, **"the way to Science is itself already Science"**[34]—here the non-true proves to be a constitutive moment of truth.

In concrete terms, the result becomes meaningful for us provided that, at least in thought, we are ready to traverse the whole way already passed by reality. In other words, systematic reconstruction here takes the place of mechanical deduction[35]: appearance must really be considered first, for example in the form of Sense-Certainty with which the *Phenomenology* begins.

In any case, the non-true recognized as a moment already constitutes a relative truth[36]. For dialectic rejects any dichotomous thinking that tends to grasp the world in 'black' and 'white', that is, to separate the totality of knowledge into 'truths' and 'errors'. Now, instead of excluding oppositions, there is the vision of a reality that gradually discloses itself to a consciousness which, step by step, learns about things and itself: here the opposites are supposed to unite, each one positing its presupposed other[37]. It is the same for the opposites we are considering, such as subject/object, thought/being and produced/given[38]. According to Hegel, the same holds for the opposites 'true' and 'false', each being regarded as constitutive for the other : *qua* starting-point towards the true, the false already contains some truth, however poor it may be—in human reality, there is no absolute false[39].

This new comprehension of the relation of opposition (not as excluding but as connecting) throws some light on the need to begin with the non-true, said to contain potentially the whole development to come[40].

I believe that human experience approximately starts in this way. If this is so, a certain explanation of that experience may be given by dialectical thought.

Thus, we saw briefly how the *Phenomenology*, written before the *Logic*, opened the Hegelian system with sensible intuition. Caught in the contradiction arising between its high pretension and the emptiness of its actual content, that sensible intuition is finally obliged to undergo a conversion and to turn towards the realm of thought, towards Absolute knowledge which opens the *Logic* in the form of Pure Knowledge[41].

We now proceed to an examination of this new point of view.

3. THE INTELLIGIBLE BEGINNING OF THE *LOGIC*

In the *Logic*, at the beginning of the Doctrine of Being, Hegel raises the question : **"With what must Science begin?"** And, in the course of the discussion, he provides an answer in three different forms.

3.1. The first answer is formulated in terms of 'immediate' and 'mediate'.

Let us first present a truism : every beginning must be either immediate or mediate. **Every beginning *qua* beginning is indeed immediate**, that is, not reflected, appearing to the regarding look as a given determination. Therein lies the problem : from this viewpoint the beginning has no justification, no ground ; in short, it is an arbitrary idea selected from many by the thinker. Now, Hegel intends to develop an absolute philosophy explaining human experience, that is, a total and detailed one, to which no serious alternative could be found. Yet, an absolute philosophy requires an absolute beginning, which is why the arbitrary beginning of the immediate determination does not suit.

On the other hand, **if the beginning is regarded as mediate**, this means that it has already undergone a development out of a 'lower' Other. In that case, we have to deal with a result which, as such, cannot be called 'a beginning'.

We thus see that, if we look for a justified beginning, as science demands, then we are not allowed to consider it as a beginning since, in fact, as 'deduced' it already is a result[42]. But if we decide to regard the matter as a beginning *proper*, then we will not find what is required : a necessary, grounded beginning. Of course, the deductive-axiomatic system is quite inappropriate to dialectical thinking, as the previous chapter showed.

That was the presentation of the problem of beginning from the standpoint of Understanding, the latter thinking it in terms of dichotomous oppositions : for it every beginning is **either** immediate **or** mediate, and no third way is available. From that angle, there is no scientific beginning, and the only remaining solution is the act of choosing between arbitrary axiom and dogmatic postulate.

We therefore proceed to the point of view of Reason, which grasps the two opposites together: **the dialectical beginning proves to be both mediate and immediate.** According to Hegel, "there is nothing in Heaven, Nature, Spirit, or anywhere else, which does not contain immediacy as well as mediacy"[43]. That is, in our experience we always begin at a certain level, which is both a starting-point and a result[44]. In effect, we are born in a certain culture where a certain tradition prevails. When we begin pondering on our human condition, we do not start from nothing : from the outset, our reflection takes place in given surroundings we did not choose and which determine us—the cultural environment in which we are living, acting and thinking. In a Platonic language, we can say that opinion precedes knowledge[45] or, in other words, knowledge always implies a first opinion that we find 'already there', present to our reflection—we must begin with some thing.

If, as Hegel asserts, human thought constitutes the source of any determination existing for us, then every determination—though seemingly immediate, a fact or a datum we should submit to—is in truth the product of a long development, in the course of which man has formed his natural, social and cultural environment. Accordingly, when we attempt to reflect upon that immediate beginning, we then discover it as mediated, namely as implying other determinations constituting it. As for the mediate and 'deduced' determination, as soon as it is grasped without reflection it appears as immediate, that is, as an obvious fact taken for granted and requiring no explanation.

As we saw above, however, according to dialectic every determination is at once immediate, a datum 'for-itself' imposing itself upon the perceiving subject, and mediate as a result 'in itself' of a previous development. And if, as Hegel affirms, truth is 'in-itself and for-itself'[46], then **an absolute beginning of philosophy cannot be found**, in so far as by 'philosophy' one understands a total and detailed account of human experience, and not a system dealing with possible worlds : our subject-matter is the real world alone[47].

Therefore, in spite of Hegel's sometimes imprudent language, **any beginning is relative to the culture considered**[48]. The idea of an absolute beginning in the sense of a sure, univocal Archimedean point upon which philosophical activity could rely unreservedly—that idea is a mere chimaera.

What the dichotomous thinking requires from the outset, the dialectician proposes it as an end. For, according to Hegel, the absolute or truth exists as a result of cultural development, and not as a given or an immediately discovered determination[49].

3.2. In that first discussion, the dichotomous standpoint of Understanding has been distinguished from the standpoint of Reason, which relates the opposites. However, the problem of the beginning can also be exposed **from the standpoint of speculative thinking**, said to produce a third determination : the unity of opposites[50]. For that purpose, let us reconsider the concept of the beginning with Hegel's help, and bring out the determinations it contains.

At the beginning there is no single thing, only Nothing[51]. This beginning must not be regarded as an empty or abstract Nothing, but rather as absence and privation : Hegel thinks that every logical determination proceeds from here. In this way, the beginning is Being as well since, although Nothing, in a sense or another it truly must be—if a certain determination has to emerge from it. That beginning, therefore, Nothing as privation, is also Being as potential of that which will develop out of it.[52]

Consequently, when considering the concept of beginning we discover within it the determinations of Nothing (we are at the beginning only, and not a single thing has yet come out) and of Being (the beginning must be, at least as the beginning of something). The beginning, therefore, is both Being and Nothing.

However, dialectic is essentially concerned with the movement of the Concept or of human thought[53] and, accordingly, the 'grasping together' of the two opposites brings the following result : **the beginning is** as Being, **it is not** as Nothing ; therefore **it becomes**[54]. In Hegel's language, it reads : Nothing, as Being, becomes Some-thing.

We saw that the examination of the concept of beginning elucidated the categories with which Hegel's *Logic* opens: Being, Nothing and Becoming. The following 'syllogism' may then be posited: **the beginning is and is not, it becomes**[55]. In that syllogism 'to become' means 'to pass over from Being into Nothing' as Passing Away, and 'from Nothing into Being' as Coming To Be—in the Hegelian logic, the determination of Becoming is itself ambivalent[56].

The point at issue is therefore the speculative beginning of philosophy, namely a beginning which grasps the opposites Being and Nothing together and thus produces a third determination, Becoming. In this connection, Hegel presents the principle of all speculative thought in the form of **"The identity of identity and difference"**[57]. In these very terms, I expounded above the identity (here Becoming) of the identity (of Being and Nothing) and of their difference.

In other words, at the start of reflection Being and Nothing are declared to be identical, as empty determinations with which philosophy begins. Afterwards, in the course of a first development, these determinations show up as different, constituting Coming To Be (a movement from Nothing to Being) and Passing Away (from Being to Nothing). It seems to me that these few lines disclose one of the 'secrets' of the Hegelian dialectic[58]: **the grasping together of the identity and difference of the opposites**. Of course, that act implies the double meaning of the considered terms, which are 'same and other at once'—in this way, the structure of the dialectical contradiction has been exposed anew[59].

Given the speculative beginning of philosophy according to Hegel, we therefore should not ask whether that beginning is immediate or mediate, subjective or objective, formal or material. It is all these at once: **immediate** and **mediate**, as we saw above ; it is **subjective** as the beginning of a movement of thought, and **objective** because it constitutes the unique intelligible beginning of philosophy—at least in Hegel's view ; it is **formal** in that, at the start, the determinations are still abstract and deprived of any content, and **material**, in that the exhibited movement of thought is supposed to reconstruct a certain movement of reality (from a logical point of view).

We have been led to these results by considering the idea of beginning 'in-itself and for-itself'.

3.3. I now discuss the 'unique intelligible beginning' of philosophy mentioned above.

The *Phenomenology* presents the development of natural consciousness towards Absolute Knowledge, and Absolute Knowledge appears at the beginning of *Logic* as **Pure Knowledge**. According to the structure of dialectic, the result of the movement shows itself as a new immediate. Namely, at the end of the development Absolute Knowledge is said to have *'aufgehoben'* (overcome) all oppositions and become a mere identity, devoid of any differentiation[60]. If that is the case, a solely abstract, contentless unity remains.

In my opinion, that metamorphosis may become intelligible only in so far as a new point of view is referred to : in that transition the point of view of the *Phenomenology*, which thinks human experience, has risen to the point of view of the *Logic* which thinks that thinking of experience[61]. Consequently, from that latter viewpoint we can discern the formation of a new immediate, undetermined or lacking any relation to an Other[62]: Pure Knowledge that, as deprived of any determination, ceases to be a knowledge at all—for every knowledge must be a knowledge of something, that is, a determined knowledge.

Thus, it is but a mere unity, Pure Being, with which *Logic* is to start. Indeed, that 'undetermined determination'—to use paradoxical language—seems not very informative since, in a certain sense, we can say that everything is, without any difference : table, poem, boredom, imagination, or even mud and dirtiness[63]... In that new unity of being and thought, a product of the phenomenological development, **Pure Knowledge shows itself as pure Being or pure intuition[64].**

The issue here is the unique intelligible beginning of philosophy, since the successive consideration of, first, the nature of any determination, second, the concept of beginning, and third, the transition from the *Phenomenology* to the *Logic*, always leads to the same result—

and, in Hegel's view, there can be no other. In so far as we live a unified experience, there is only one intelligible beginning of philosophy (of Hegel's type), absolute in that sense, revealing itself to the thought that examines experience ; and any alternative will have to be rejected as imaginary or arbitrary[65].

Hegel has expounded the intelligible beginning of philosophy as immediate and mediate, as Being and Nothing and Becoming, and as Pure Knowledge which is also pure Being. That multiplicity of arguments seemingly points to the presence of a certain difficulty : the Hegelian system, which claims to ground itself, also has to ground its own beginning, that is, in Hegel's words, **to philosophize with no presupposition.** Is that really the case?

To see whether that is the case, we proceed to the study of this primordial issue.

4. THE HEGELIAN PHILOSOPHY, SAID 'TO GROUND ITSELF', IS HOWEVER UNABLE TO DISPENSE WITH SOME PRESUPPOSITIONS

We first return to the problem of beginning.

In the empirical or formal sciences, the appropriate field of research and method are supposed to be known. The scientist, in the first steps of his work, lays down some definitions and universal laws that are acknowledged by the community of researchers ; from there he progresses according to the accepted rules of work.

However, in the Hegelian philosophy things are not the same. Here, nothing can be taken as granted at the outset, since what is looked for is precisely the grounding of the totality of being and knowledge. As a systematic reconstruction of the different cultural products of Western civilization, that philosophy can rely on itself only—that is the price that has to be paid for the Idea of a comprehensive and detailed science that 'grounds itself'.

This ambitious plan renders the starting-point of the 'absolute philosophy' quite problematic, as saw above. If the project is really **to think all anew, to ground everything** and leave nothing to arbitrary will or private intuition, then how can the scientist start work? For indeed, according to Hegel **the concepts of method and system are immanent in the System itself**[66]. That is why these concepts can come to light only in the course of the development of philosophy. Previously, they are unknown and undetermined. In particular, the *Phenomenology* was written from the point of view of the *Logic*, which comes afterwards[67], and the 'method' reveals itself explicitly at the end of the *Logic* only, as a result and not as an *a priori* tool prior to it. If this is the way things are, beginning seems very difficult!

The problem proves complex. Let us look at it once more from this new angle.

On the one hand, it is not possible to rely on the traditional rules proper to other sciences—in which method and system are already defined—since, as Hegel says, "the familiar, just because it is familiar, is not cognitively understood"[68]. On the other hand, a beginning must be made, one way or another. Consequently, the philosophical reflection will focus its attention on precisely this 'familiar' thing, an opinion that has to be criticized and thus turned into a grounded knowledge.

For instance, the *Phenomenology* opens by unveiling the nullity of the pretensions of natural consciousness which, therefore, is compelled to rise to a more adequate level. Hence the following conclusion : if in dialectic everything has to be justified, yet this is not possible without some presuppositions, like that of the existence of a natural consciousness, with which the *Phenomenology* begins!

In other words, the expression 'philosophy with no presupposition' may now be interpreted as follows : the Hegelian philosophy grounds itself on some presuppositions that are indispensable to the development of the system but, on the way, these presuppositions in turn come to be grounded, that is, criticized and developed. They therefore become **inner presuppositions**, that are legitimate in this sense : the circular structure of Hegelian thought realizes itself through the reciprocal mediation of propositions that are presupposing one another.

On the other hand, a presupposition will be regarded as non valid if it is unnecessary to the development of philosophy, and not 'deduced' in the course of that development. For example, Pure Knowledge, with which *Logic* begins, acquires its signification from the movement of the *Phenomenology*. As for the natural consciousness presented in the opening pages of the *Phenomenology*, it is 'deduced' in the *Encyclopaedia* in the Subdivision 'The Subjective Spirit'[69].

We saw that Hegel plans 'to think anew' the basic determinations of human culture, such as the concepts of beginning, method and system—in this respect the approach proves **revolutionary.** Moreover, that 'radical' thinking is directed towards the very productions of culture, productions it 'digests' and reconstructs in the form of an organic system—in this respect, the Hegelian approach proves **conservative**[70]. Furthermore, Hegel affirms that the need to overcome the cleavages of human culture by the unifying power of thought (that is, the 'need for philosophy') implies a long stretch of road already traversed by that culture. If we add to these arguments the pretension of 'summit' and 'end', then we see that a **developed culture**, the **history of culture** in general and the **history of philosophy** in particular, constitute the essential presuppositions of the Hegelian sytem.

In particular, at the beginning of the *Logic* Hegel frankly recognizes that the forms of thought, studied by the Greek philosophy, are necessary conditions (as external material) to the construction of his own system. As for the *Phenomenology*, among other things, it exposes the history of human culture and that of philosophy closely intermingled with each other[71]. That is why that work has to show that "now is the time for philosophy to be raised to the status of a science", that is, as a self-grounded organic system[72]. If now is really "the time", then Hegel may truly assert that human reality, after having reached such a summit, is only waiting for the moment of self-consciousness—a moment that Hegel claims to have developed in his own philosophy.

In the discussion of this issue, I have discerned the **logical presuppositions** that allow us to undertake the construction of the system ; for instance, the phenomenological way in its different movements presupposes the viewpoint of the Hegelian Concept as already known. However, some **historical presuppositions** have also to be taken into account, to ensure that human culture has accomplished itself to such a point that the achievement of Hegel's plan proves not only possible, but even necessary.

In this context, we can say that the *Logic* grounds the logical presuppositions (such as the movements of categories, the concepts of beginning, method and system). As for the *Phenomenology*, its task is to ground the historical presuppositions (we should thus be 'close to the end'). If that is the case, then the Idea of a philosophy grounding itself immanently becomes feasible through the reciprocal mediation of the two dialectical roads—the *Logic* and the *Phenomenology* conditioning each other.

Indeed, the young Hegel expressly states that the *Logic* and the *Phenomenology* ground each other[73]. And if the mature Hegel mentions but a certain identity between philosophy and its history, in the context of that discussion this modification does not seem to be significant, since the history of culture and that of philosophy continue to be essential presuppositions of the system[74]. It only remains to notice that in Hegel's *Lecture on the History of Philosophy* the problem of the beginning is no longer raised, not because it has been solved, but rather because it no longer comes into consciousness—theses *Lectures* being not dialectical, but empirical or formal[75].

When studying Hegel's system, we must not overlook the important *Encyclopaedia.* Accordingly, we have to add the **different empirical and formal sciences** as essential presuppositions. In effect the Hegelian philosophy, said to be presented in the form of a comprehensive and coherent system, claims to give a total and detailed account of knowledge and being as a whole. For that purpose, Hegel borrows from the diverse sciences of his time ; if not the method, at least some results he interprets, incorporating them in the so-called 'organic unity' of the System. In that sense, the System comprehends an *Encyclopaedia of the Philosophical Sciences*, whose structure depends on the degree of development of the contemporaneous sciences ; and, along with these sciences, it has to progress and correct itself in consequence. Now, the philosopher ought not to interfere in scientific work ; but, *in retrospect,* he may attempt to present the achieved discoveries and inventions in an 'organic' form, in so far as that Idea itself is legitimate[76].

We have thus learned that, at the outset, the Hegelian philosophy based itself on some presuppositions: an advanced culture, the sciences of the time, the history of culture in general and that of philosophy in particular. The Hegelian philosophy therefore comprehends a *Phenomenology* that has to justify the historical presuppositions,a *Logic* that justifies the logical presuppositions, and an *Encyclopaedia* incorporating the achieved scientific results. Thus, in opposition to the linear method which opens with data, axioms or postulates, here the different presuppositions are 'deduced' in the course of the development of philosophy, acquiring thereby their proper signification. And talking about an 'organic system grounding itself' proves meaningful in so far only as that circular paradigm realizes itself[77].

However, in the beginning there is only opinion, which involves within itself the impulse to self-explication. In Hegel's paradoxical language, we should say that we do understand in the end only...

CONCLUSIONS

In the *Logic* Hegel talks of an 'absolute beginning', an expression that I believe must not taken literally. For, in rejecting any conception of an immediate rational intuition or of a substantial sensible given, the Hegelian philosophy is unable to find such a beginning, in the univocal meaning of the term. Instead, in the circular movement of self-grounding or, better still, in the movement of reciprocal mediation, we have discovered Sense-Certainty and Pure Being (which is empty thought), the phenomenological and the logical beginnings, conditioning each other and thus forming a system of concrete thinking, *'ratio sui'*, so to speak.

We have also learned that the 'philosophy with no presupposition' consists rather in a philosophy that relies upon inner presuppositions. Now, the question is to know the extent to which these presuppositions of the system are really 'deduced' in the course of philosophical development. Is now really the time for philosophy to be presented as 'a science'? Does the movement of the Concept truly structure the development of the *Phenomenology* ? Can we find in the *Logic* the ultimate *ratio* of human experience? Do the results of empirical and formal sciences suit a dialectical reconstruction? Or is there an insurmountable gap between being and thought, experience and dialectic, dividing the Hegelian absolute into a non-intelligible reality and a non-concrete philosophy? To answer these various questions would require going beyond the scope of this Part of the work, but nevertheless I think that they need to be asked[78].

H.F. Fulda, after studying the problem of the introduction to Hegel's philosophy, defends the idea of an absolute beginning[79]. Thus his thesis asserts that the *Encyclopaedia* closes on itself through the mediation of the Syllogism of the three syllogisms, thus forming a circle and accomplishing the 'identity of the beginning and the end'. In consequence, the *Phenomenology* would no longer find a place in Hegel's system.

Yet, Fulda acknowledges that it is indispensable, for it alone presents the dialectical moment of the difference between truth and non-truth, a moment that belongs to the natural consciousness with which one has to begin, in one way or another. In that perspective, the *Phenomenology* would be **internal** as necessary to the development of the System, and **external** as taking no part in the circle of the *Encyclopaedia*. Indeed, Fulda concludes with the assertion that the *Phenomenology* is "transcendent to the System"—an expression that, in the context of his work, does not seem very clear[80].

Again, in the Hegelian system Fulda gives precedence to the *Logic*, and only a secondary, yet essential, status to the *Phenomenology*. On the one hand, Fulda's argument may be justified in so far that, for Hegel, every philosophy is idealistic : Hegel sees the Idea to be all reality and, moreover, he sees thinking as man's essence and philosophy as logic.

Now, attention must be paid to the fact that, as far as the *Phenomenology* is concerned, it is about a concrete philosophy exposing human reality in its genetic and historical movement. In my view, therefore, by displaying that rich movement in the shape of a wide variety of patterns, the *Phenomenology* also deserves a primordial status, without which Hegelian thought would be reduced to just one more dogmatic metaphysical system.

I do not intend to look in Hegel for a 'well-rounded coherence' nor for a symmetry out of place, for the goal is 'to think the thing itself', that is, Hegel's work, and not to imagine a perfection or a perfectibility that is not human, to the best of my own experience. All things considered, however, I acknowledge the essential role of the *Phenomenology* in providing the system with a substantial content, for without that Odyssey of the consciousness at once divided and eager for plenitude, the system would remain formal. If so, a certain inner tension proves to be the law of the Hegelian system and, perhaps, of all human reality.

That is why, between ancient dogmatism founding itself on immediate rational intuition, and modern scepticism going so far as 'epistemological anarchy', I suggest a third way : the dialectical way that criticizes and reconstructs[81].

The examination of the idea of beginning, from the viewpoint of **experience** and **thought** of that experience, has thus brought me to assert the existence of two beginnings, one sensible according to the *Phenomenology*, and another intelligible according to the *Logic*. If, as

Kant claims, the human being really belongs to two worlds[82], a material one and a spiritual one, a natural realm and a cultural realm, then human experience must reflect that basic duality—just as the Hegelian explanation itself, within the limits of its validity, of course.

Up to this point we have been led by Hegel's way, by his met-hod.

PART THREE

THE PROBLEM OF THE END

In Part Two, I examined the problem of the beginning in Hegel's philosophy[1]. From the same point of view I now deal with a corresponding problem, the problem of the end.

Indeed, throughout his experience man cannot help asking: 'Where are we going? At the end of the road, what might be expected?'. Sometimes, man also wonders in the following terms: 'Where did we come from? What are our roots, our origins?'. The questions of the 'before' and 'after' worry him, and that in the one framework of life, thinking and culture. In principle, then, the problems of the beginning and the end are raised together and cannot be separated, only distinguished.

If, as I suggested, ambiguity characterizes man in his thought, language and action, then human reality itself must reflect that fundamental ambiguity, that is it must be dual[2]. In Part Two I showed that, according to Hegel, **the beginning** can also be regarded **as a result**. In this Part I present **the end as a new beginning**[3].

On the other hand, when reading through the central works of the System, one may find there a univocal 'pretension of the end'[4]. This expression signifies that Hegel curiously affirms, without the slightest hesitation, having studied reality from the viewpoint of the end, both in principle and in time. **In principle**, the Hegelian philosophy is supposed to accomplish the synthesis of all the great systems of the past. And as a total, comprehensive and coherent philosophy, it is said to leave no room for any other important discovery. **In time**, that 'absolute' philosophy would also be the last, in so far as all the principles of reality have been disclosed in the course of history. Even from that angle, no essential addition would still be left to be made. Hegel summarizes these audacious and imprudent statements in asserting that philosophy is identical to its history[5].

With Hegel, we would thus have reached the end of history and that of philosophy itself, human reality being realized and truth finally grasped and expressed.

And yet, since Hegel a lot of water has flowed under the bridge. **Philosophy** has gone on developing, first in the form of Kierkegaard's existential thinking and of Marx's social, economic and political thought. Then Phenomenology and Hermeneutics made their appearance : with Heidegger, we seem to have discovered a philosophy of world-wide importance that modifies our way of perceiving and conceiving. Moreover, in the English-speaking countries several Schools have arisen, such as the Analytical Philosophy and the Logical Positivism, rethought by the second Wittgenstein. Indeed, one cannot seriously think of reducing these different movements to 'moments' of Hegel's philosophy, for the excluded 'residue' would be much too important to be renounced.

As regards **universal history**, it may safely be said that, since the last century, various industrial and political revolutions—including two world wars—have largely contributed to change the face of the world. On the way, Europe seems to have lost its dominant position, and the New World has become more important (as for the Third World, that is another question...). It is therefore legitimate to affirm that reality has essentially changed. In concrete terms, the rapid progress of modern technology, together with different ideological reactions it has aroused, has transformed our world to such a point that some fear it will soon become unrecognizable.

Should we now conclude that Hegel was a naive or a deceitful thinker? The answer to this question is not univocal, so that in this Part of the book I use the following strategy. In chapter 1, I try to show in what sense the univocal pretension to philosophize from a standpoint of the end proves, for Hegel, at once necessary and yet indefensible. In chapter 2, relying on different texts of the *Logic*, the *Phenomenology* and the *Encyclopaedia*, I propose to reinterpret the concept of the end so that it may become acceptable while still remaining Hegelian. In chapters 3 and 4, 'thinking anew' the determinations of end and time, respectively, I complement the conclusions of the previous studies.

With this strategy I do not try to save Hegel at any cost ; rather I attempt to show that, in principle, a genuine dialectical thinking cannot lead to such nonsense, as the univocal ideas concerning the end of philosophy and that of universal history—I understand that, sometimes, Hegel has not been faithful to his own principles[6]. In short, I intend to show, in chapter 5, that every end constitutes a new beginning, and that philosophy itself develops through **successive revolutions**. Accordingly, for man there is no final solution, the mere thought of a 'final' solution producing the most unpleasant associations...

In the course of this study, I also propose a new approach to the problem of transitions in Hegel's philosophy.

1. A BASIC PARADOX: FOR HEGEL, THE UNIVOCAL PRETENSION OF PHILOSOPHIZING FROM THE STANDPOINT OF THE END PROVES NECESSARY AND YET INDEFENSIBLE

1.1. The pretension of philosophizing from the standpoint of the end proves necessary

If, as Hegel asserts, the absolute is a result, truth is the whole and philosophy is identical with its history, then the exposition of the 'absolute' philosophy implies a standpoint of the end univocally.

From that standpoint, as it were, reproducing the total movement of reality through an 'immanent and necessary' approach, the thinker finds himself in a position to grasp and present also the different principles explaining that very movement. If so, he is capable of constructing that 'last' philosophy in the form of an **organic system**, supposed to be the genuine manner of exposing the truth[7]. In this way, Hegel thinks that the identity of form (the organic system) and of content (the thought of the developing living reality) can be achieved, thereby ensuring the validity of his above-mentioned theses. Having completely manifested its essential determinations, Hegel claims, at last reality is about to reach self-consciousness, and that through the absolute philosophy Hegel's system is said to be—indeed, for Hegel the actual *(das Wirkliche)* is the self-conscious[8]. If that is the case, no wonder that the Hegelian philosophy stands **'at the summit'**, not only as the absolute knowledge of a completely developed human reality, but also as the self-consciousness of that reality, a reality that is supposed to recognize itself thoroughly in Hegel's dialectical 'science'[9].

On the contrary, if Hegel did not claim to philosophize from an ultimate cultural-historical point of view, dialectically speaking this would signify that not everything was unveiled in the realm of being and thought. Consequently, the comprehensive and coherent organic system would prove impracticable, on account of the gaps due to the still unknown. Like Schelling before him, Hegel would have to construct 'his' philosophy, in place of 'the' philosophy *par excellence* out of reach.

However, Hegel thinks that there is only one true philosophy developing in space and time, exhibiting its different principles in the form of various philosophical Schools tending to constitute the unique, total and ultimate system—his own naturally[10].

1.2. **Nevertheless, that pretension proves indefensible** Today, it is clear that Hegel has failed in part and that his promise—to grasp and express absolute knowledge—has not been kept.

First, the System is **not comprehensive**, since human reality continues to develop in its theoretical and practical aspects. Although Hegel seems to have regarded art, religion and philosophy as lacking any future, the diverse civilizations and cultures proceed on their way, notwithstanding some cruel regressions—this concerns the future of humanity.

As for its past, the Hegelian system gives no satisfaction either. In order to account for universal history, Hegel is constrained to reconstruct it in an over-simplified fashion[11]. Furthermore, in the System the great classical philosophies themselves are *aufgehoben* (comprehended) only at the price of a most doubtful work of 'adaptation'[12]. Non-Western cultures, which are better known now, are quickly sketched or even ignored ; and, as pointed out by Kierkegaard, in that reconstruction the German philosophers, contemporaneous with Hegel, occupy a place better explained by the lack of distance than by their real importance[13].

Second, the System is also **short of coherence**. In particular, the *Phenomenology* is characterized by a quite problematic inner structure, the System has two versions of the 'Phenomenology', two of the 'Logic', three of the 'Encyclopaedia', and the way the *Phenomenology*, *Logic* and *Encyclopaedia* are related to one another is not very clear[14]. Moreover, in his writings from youth to maturity, Hegel himself develops philosophically, and it is not easy to tell which thesis really expresses Hegel's thought ; strictly speaking, the promised organic system truly remains a non-achieved ideal[15].

Actually, then, Hegel did not carry out his ambitious plan, and even **in principle** that plan is impracticable. Like everyone, Hegel was a child of his time[16]. He spoke the language of a particular people and thought human experience from the viewpoint of a certain historical epoch. As a member of a certain socio-economic group and as a member of a certain family, he grasped the world according to his own understanding of things, which did not fail to be subjective too[17].

To be sure, my criticism is not aimed at refuting the System univocally ; for every author is but a man, even if he is writing in the context of dialectical thought, whose business is said to be self-consciousness. Hegel too perceives reality through glasses proper to him, without his being able to take them off[18]. Notwithstanding the wealth of his interesting and enlightening work, we should not forget the boundaries inherent to every human endeavour.

In consequence, together with the vanity of an 'absolute' knowledge, we also find the **vanity of a univocal pretension to philosophize from the standpoint of the end**. Thanks to the revolutionary character of his time and his wide encyclopaedic culture, Hegel was able to achieve some impressive results ; but not to reach the absolute which, to human thinking, reveals itself as relative[19]. We may leave the absolute 'absolutely understood' to the all-powerful gods. As for the 'relative *qua* relative', I discern in that expression a thesis that denies to man any ideel content and reduces him to a unidimensional being. According to a concretely grasped dialectic, however, man is endowed with a double nature and stands in an ambivalent position in the world, with his feet on earth but tending to heaven. If that is the case, a **basic ambiguity** characterizes every theoretical or pratical human undertaking, and even the 'summit' of human reality is stamped with it.

In the following chapters, I explicate further this assertion.

2. ACCORDING TO THE VERY PRINCIPLES OF DIALECTIC, EVERY END CONSTITUTES A NEW BEGINNING

In Hegel's central works, the passages dealing with the problem of the end are to be found mainly in the various Prefaces and Introductions and in the last pages. In this connection, with regard to the exposition of Hegel's philosophy I find it suitable to talk about a **high level**, Prefaces and Introductions having generally been written after the completion of the work itself. As the reader may notice, the point at issue concerns a vision that goes far **beyond** any real human experience[20]. Yet, in my opinion, dialectic, which presents the *a priori* movement of thought, is also *a posteriori*, the historical and cultural experience serving here as a leading strand[21]: the 'pretension of the end' is thus expressed in texts of quite a doubtful validity. At that high level, the researcher does not reconstruct the movement of reality but seems rather to make predictions, in opposition to any dialectical principle.

In order to remain within the boundaries of a valid study grasping and expressing the movements of human reality, I therefore think it advisable to turn to the **middle level** of the philosophical exposition, a level that proposes different dialectical developments, problematic perhaps but instructive too. As for the **low level** of the beginning of philosophy, it is unable to provide the necessary distinctions, since it consists in potential determinations that realize themselves only later. We can find there, ineffable in that sense, the sensible Consciousness of the *Phenomenology* or the dialectic of Becoming of the *Logic*[22]. Now, my intention is to throw some light on the problem of the end, and to exhibit the general structure of the dialectical movement according to the middle level of Hegelian discourse : from that angle the middle is equally an extreme, since it is the best[23].

Dialectic may be presented as a movement of the **immediate** given which, when regarded 'in-itself and for-itself', reveals itself as **mediated**, that is, as the result of a first development. For instance, let something appear before us. This thing is what it is only in relation to an Other, its opposite or context determining it as this Something *(Etwas)*. As for the Other *(Anderes)*, which is different from it, it is the negation of Something and is mediated as such[24]. Immediate and mediated thus relate as opposites.

The immediate term being actually mediated, and the mediate term being 'deduced' from a first immediate, each one presupposes its Other. In this circle, or rather double movement, a contradiction takes form and seeks a solution. This solution is reached in so far as the opposites, when 'held together', produce a **new immediate**—such as Being-for-self which, as qualitative Affirmative Infinite, contains within itself the movement to and fro between Something and its Other. Now the posited new determination, like any immediate, becomes the starting-point of a new development, and so on. In the quoted example, Being-for-self, as the One, reveals itself as positing its presupposed Other, the Many ; and the process goes on towards quantitative Infinite. In principle, then, the dialectical movement can never come to an end for, according to Hegel himself, thought apprehends every immediate as mediated, grasps together the opposites and produces a 'new immediate', mediated in its turn.

To put it another way, at the outset we have to do with a given, a **first positive** which, on thinking it over, reveals itself as **negative** too, since limited *qua* determined. Negative as being lacking, the positive develops. It posits the presupposed lacking and, when uniting with it, it forms a new determination[25]. Now, as such, that determination is itself limited, and thus negative—hence there is a new movement. At this stage the **negation of negation** appears, which, as a **new positive**, reconstructs the first positive. It is said to constitute 'its truth', since, as a new positive, it explicitly posits the presuppositions of the original one[26]: the dialectical movement really consists in a development.

For instance, we can say that the finite is not infinite. Consequently, it first posits the Spurious infinite that unfolds itself as an indefinite reproduction of the finite—here the Other of the finite is another finite which, in its turn, posits its own Other[27]. The Spurious Infinite, as negative, thus contains the positive finite within itself. But that infinite, being itself finite (the finite is opposed to it), posits what it is lacking and thus negates its own negative determination. This occurs when the finite *qua* finite recognizes itself in its other, another finite like it, and vice versa[28]. At this point the Affirmative infinite constitutes itself, as the circle of the determination relating to itself alone. As Being-for-self—a logical self-consciousness—it is a new positive, a determination limited with regard to its Other ; hence the 'driving power of the negative'. And the movement goes on endlessly.

As it were, the **totality** alone might constitute an ultimate determination, to which nothing could be added : it would contain all otherness. However, if we want to avoid the abstract totality thoroughly positive—thus not alive—we have to discern in it the activity of the negative.

In other words, the negation of totality yields the part without which that totality would be only an empty Idea—the part is truly the Other of totality, its categorial opposite. In concrete terms, the totality constitutes itself with regard to the part only ; and without the part, a negation of totality, the totality itself would no longer be dialectical[29]. Stopping at that totality *qua* 'new immediate' is then impossible, since reflection unveils it as mediated, so that the movement continues[30].

The dialectician is unable to consider the totality from the outside, as if it were a sphere laid down in front of him : in that case, he would find the Parmenidean One, devoid of any determination. On the contrary, as an immanent member of that totality, he is brought to think the dialectic of totality and part, a dialectic carrying him along with it. For him, therefore, the living totality (as human reality) continues on its way, for an ultimate determination would rather signify rest and death. Now, such a rigid result seems quite inappropriate to dialectical thought which, in principle, consists in the thinking of life, say of creative negation.

However, at the **high level** of the Hegelian discourse such a 'pretension of the end' is to be found, going far beyond any real human experience : although a genius, Hegel is not a god. In contrast, from the standpoint of the **middle level** we can perceive the part developing towards totality and the totality positing the part it presupposes ; then, through that reciprocal mediation, the movement perpetuates itself, progressing unceasingly.

In dialectic, therefore, a univocal end is not thinkable at all. This signifies that "in Heaven, Nature, Spirit, or anywhere else", there is no determination that does not contain the negation within itself, including the determination of totality (negative or limited as a certain determination)—this conclusion is drawn from the viewpoint of **human thought**, considered as reflecting and determining.

Spinoza has asserted that every determination is a negation, and Hegel, for his part, adds conversely that every negation is a determination[31]. From this angle, then, it is not possible to think of the living, creative, human reality as stopping to develop, if not by means of a natural accident[32]. A determination that did not contain the negative within itself would be an 'undetermined determination', similar to Being with which the *Logic* opens. Obviously, as an 'empty thought', that abstract determination is unable to constitute an end, in any sense whatever.

I wish to conclude, holding in opposition to **Kojève** that one cannot think of the death of human culture and, in opposition to **Fulda**, that one can no more think of an entire self-realization of being, wherein the lack would no longer act[33]. Ambiguity remains the necessity and destiny of human reality, the latter constituting an object to which the reflecting subject belongs.

The investigation of the general structure of dialectic has led us so far. The following chapters will corroborate.

3. A NEW PARADOX: IN A CERTAIN SENSE HUMAN REALITY IS ACCOMPLISHED, AND YET ON THE WAY TO ACCOMPLISHMENT

At first sight, the Hegelian concept of the end leads us to believe that reality is completely realized 'in itself', and that the moment of self-consciousness alone is lacking, a moment provided by Hegel's philosophy itself. This is not the case, however, for since Hegel's time, the history of philosophy and that of the world have continued to progress. Moreover, I do not think that an exclusive choice must be made between the idea of an **end of history**, signifying that reality might reach a full self-consciousness, and the idea of an **indefinite progression of history**, which denies any decisive stage and any possibility of attaining some essential purpose. In a 'speculative' perspective, is it not advisable 'to think together' these two opposing ideas[34]? In this chapter, that is what I attempt to do with the help of the following dialectical studies, referring also to the 'middle level' of Hegelian discourse.

3.1. The dialectic of Means and End

Hegel asserts that history is accomplished when spirit has achieved **its end** ; namely when, having exteriorized itself, spirit returns to itself **by means of** the temporal movement of reality[35]. To take him literally, I understand that a logical investigation of the categories of means and end may help to clarify the movement in question.

In his *Logic* Hegel shows how, in his opinion, external subjective teleology develops into inner realized teleology. As far as possible, I shall sum up that movement in intelligible language referring to human action[36].

When we want to achieve some purpose, we find ourselves in the following situation : facing us we can discern, successively, a **subjective end** to realize, a **means** by which we can try to reach that end, and an **external object** upon which we intend to act.

This situation is not so innocent as it seems, however, since it implies different presuppositions. To begin with, we grasp the world as objectively given and as indifferent to our activity. Secondly, in that rich and varied world, according to our interests we select this as an object to be changed and that as a means to attain our goal ; of course, other selections are always available. Thirdly, the three moments of action (end, means and object) appear as externally related to one another. However, Hegel remarks that if we examine these different presuppositions 'more closely', we can discover at their roots one unique source, the human thought that posits the various determinations—hence the inner tensions[37].

For instance, the means appears to be **independent**, as belonging to the external world. Now, as serving some definite purpose it is **dependent** (as a specific means for *this* purpose). Next, every realized end becomes a means for another end, and so on ; hence an **infinite progression** of end and means, never leading to any real satisfaction. Moreover, end, means and object are perceived as externally related to each other, so that they can be united in a third only. However, from that standpoint the third itself is regarded as external ; hence this time an **infinite progression** of uniting terms, the related couples (third-means, third-end, third-object, etc.) always being in need of a new third[38]—from a logical point of view, the realization of that end seems quite unintelligible. Subjective teleology is truly a 'nest of contradictions', and thus proves inadequate[39].

Indeed, in the reified world it imagines, everything may serve as a means for another thing. Accordingly, no real satisfying end can be found in it and, in view of that lack, there is no means either. The agent used to think that every end implies some corresponding means, and now he learns that likewise every means implies some definite end.

The infinite progression, devoid of any meaning, therefore turns into the circle of the **affirmative infinite** when, in the presence of these diverse *aporia*, the agent becomes aware of its own determining power—then, the external exclusive interrelation changes into an inner constitutive relation. This time, the agent perceives the different determinations at stake as moments of the **realized teleology**. That is, 'subjective' and 'objective', 'external' and 'inner', 'means' and 'end' finally reveal themselves in their truth : they are but determinations of human thought. However, we should be careful : at this point, we are no longer dealing with individual purposes but with the Idea of the Good, in the realization of which the agent takes part, at first without his being aware of it. Now, he sees himself as a moment of the whole[40].

In Subjective Teleology, the agent aimed at realizing some peculiar and contingent thing. But at present, Hegel concludes, at the level of

Realized Teleology, he understands that reality is self-actualizing, and that the intended end, the transformed object and the mediating means constitute the moments of the self-determining Idea[41].

That was the Hegelian exposition of the dialectic of Means and End, summed up in our usual language. However, the critical standpoint must be recalled : given the ambiguity of the Hegelian *Aufhebung*, dialectically speaking we can nevertheless find that in reality **Subjective Teleology and Realized Teleology are acting together**. That is, the negated determination of exteriority is not only cancelled but also preserved[42].

According to the very principles of dialectic, therefore, the described experience should induce the agent to regard the world at once as real and ideal, and to regard the means also as an independent and external thing. Exteriority as such proves to be an **essential** dimension of human reality too, without which there could be no history and, following, neither Concept nor Idea or Spirit[43].

Moreover, we know very well that, in any sense whatever, we are not just a 'moment': we are also these individuals, **essentially** irreducible to the universal[44]. Hence the following movement.

3.2. The dialectic of Individuality

The preceding critical examination showed that, in human reality, an inner-organic (the means as a moment) as well as an external-empirical (the means as given) dimensions are present—the discussion is really about the double nature of teleology.

From the standpoint of Subjective Teleology, that is, of the agent intending to realize a personal goal, there is an **infinite progression** of an end turning into a means for another end, and so forth. From the standpoint of Realized Teleology, however, there is the **circle** of a partly accomplished, then partly self-conscious reality—the image of the circle expressing the movement of self-reflection implied by any self-consciousness. From that standpoint, one may already talk about a certain end of history.

This double result can also be achieved through the examination of the dialectic of Individuality, exposed in the *Phenomenology* [45].

The *Phenomenology* presents the individual as **Virtue**, acting with the intention of causing the 'bad' **Way of world** to advance. Yet, according to Hegel, the Way of world is but the product of the individual's constitutive activity. If so, in his action Virtue is only fighting against himself. Therefore, when the individual discovers that contradiction, he becomes reconciled with the world and no longer attempts to change it ; from now on he will content himself with realizing himself within.

Indeed, in so far as the acting individual is seen as a moment of reality, Hegel is correct. In so far as that individual is also an independent, isolated and particular being, however, he has to exist in a world where being and ought-to-be are opposed and irreconcilable : he then continues to try to change it. From the viewpoint of thought, one can grant Hegel that the individual is the ground of human reality. But, as existing here and now, the individual must also protect his personality and satisfy his multiple needs. That twofold nature—where the inner contradictions of the finite, both real and ideel, can be recognized—forms the dialectical mover of progress, for it is the fate of dialectical contradiction to seek an appeasing solution.

We have thus distinguished between the ideel **universal** individual and the real **particular** one—hence the following double result[46]. For the individual *qua* universal who has realized himself and knows that, reality proves accomplished, therefore good. As a **'Sage'**, this individual is relatively satisfied, at peace with himself ; his main needs seem to be fulfilled and he does not aim at impracticable goals—in that sense, for him **history is actualized.** In contradistinction to that, as **rest-less** (in movement) the particular individual shows dissatisfaction. Consequently, fighting against the 'evil' he is still aiming 'to change' the world, to reconcile ought-to-be and being ; and for him **history continues**, forever producing some new creation.

Furthermore, in view of the dual nature of man—of any man, whether a great philosopher or a simple peasant—the above-mentioned opposition may also concern the individual Hegel himself. In fact, though gifted and educated, Hegel was not just a 'universal individual' making speeches on the end of history—Kierkegaard was right : such a metaphysical being does not exist[47]. Both as thinking and as acting, man is essentially twofold, so that to Hegel too reality appeared bidimensional : as the **circle** of a philosophical self-consciousness having gained a certain rest, and as the **endless progression** of a purpose which, no sooner achieved, already leads to posit another purpose, whereas the intended ideal remains forever out of reach[48]. In particular, as a father and an intellectual, Hegel had to maintain his professional position and appropriate means of subsistence. He also wished to have his philosophy recognized as 'the' philosophy *par excellence* . Thus, even for him, being and ought-to-be partly opposed each other ; and, in his vital practical action, Hegel certainly also behaved as a particular individual who aimed at subjective ends : is not every man **essentially** a theoretical and a practical being, sometimes a 'sage' universal thinker, sometimes a 'restless' particular agent?

This is why the universal thinker who talks about a univocal end, can but be an abstract being, fancying and not thinking concretely. For even to Hegel reality is at once accomplished and still accomplishing itself—it is only a question of point of view. Yet, at the 'summit' of his works, carried along by his enthusiasm for discoveries, Hegel could not help dogmatizing. Then, at this point he prophesied the end of Time...

This basic ambiguity will appear again in the determination of time.

4. THE DOUBLE CHARACTER OF TIME

In order to elucidate the problem of the end, studying the relevant logical and phenomenological movements is not enough : the real determination of time must also be examined[49]. The *Logic* does not deal with time at all, for the movement of the categories is regarded by Hegel as purely logical, that is, as supratemporal (beyond time). To find an investigation concerning time, we have to turn to the beginning of the Philosophy of Nature in the *Encyclopaedia* and to the end of the *Phenomenology* : of course, the ***Realphilosophie*** cannot ignore this dimension of reality, just as the short philosophy of history concluding the phenomenological way[50].

In the discussion referring to the 'high level' in Hegel's discourse, as far as possible I translate the 'absolute' point of view into an intelligible one. In particular, I understand the term 'Concept' as signifying 'human thought'[51].

At the beginning of the ***Realphilosophie***, Hegel presents time as "the being which is not in so far as it is, and which is in so far as it is not", as an **intuition of becoming**[52]. From this I understand that any real being is located both in space, its **static** dimension, and in time, its **dynamic** dimension. In other words, in space all persist but in time all come to be and pass away. Accordingly, Hegel observes, time is becoming itself, the movement of coming to be (from nothingness to being) and of passing away (from being to nothingness), 'Chronos devouring his own children'[53].

Moreover, *qua* limited everything is finished. That is, the determination inherent in it has also its origin outside, in the other, since being limited is having an other 'beyond'—here the infinite or Concept[54]. Whence the tension and rest-lessness that are time itself ; indeed, in its effort to overcome its own limit, the finite being exhausts itself, for 'beyond' it finds only itself, another finite being[55]. In this way a movement of temporalization is produced, as a movement of self-realization of the finite, which is also a movement of its own self-destruction—self-contradiction is truly the *Bestimmung* of the finite, in both meanings of the German word, determination and destiny.

In contrast, relative to the limited thing the self-determining **Concept** is supposed to be not in time, but to be the "power of time". Namely, thought, in determining the thing as finite, is said to determine it as moving in time. For Hegel, then, time implies the producing activity of thinking.

Consequently, for us, ***qua*** **natural** the finite thing moves in time, but ***qua*** **moment of thought**, as the very category of the finite it is supratemporal—the finite is truly real and ideel at once. On the other hand, Hegel says, as the source of the activity of temporalization, thought (or the Concept) proves supratemporal. From that, the relation existing between **nature** and **spirit** (or movement of the Concept) can be understood as corresponding to the relation existing between **time** and **supratime**.

Thus, in Hegel's philosophy the following movements can be found. On the one side, the ***Realphilosophie*** in the *Encyclopaedia* presents the development of spatio-temporal intuition (in the Philosophy of Nature) towards the 'divine' supratemporal Concept (in the Philosophy of Spirit), and this runs through universal history. On the other side, the *Phenomenology* talks about time as the Concept "being-there" *(der da ist)* [56], namely, as thought exteriorizing itself *qua* a determined, finite thing. A double movement therefore develops between the **experience of time** and the **thought of the Concept**, the latter being the "power of time"[57]. All of this holds, of course, in so far as human thought really determines the thing as finite, and thus produces the inner contradiction that brings about the movement of temporalization.

That double character of time—grasped both as conceptual supratemporality and as natural temporality—may also be explained as follows : let us distinguish time *qua* time, its essence, from time *qua* appearing, the real time that passes. With regard to essence, in Aristotelian language it can be said that 'time is not in time', i.e. that in itself **time is supratemporal**—it has neither beginning nor end[58].

However, in a Hegelian context the matter may also be considered from an opposing viewpoint, allowing us to talk about supratime as appearing, namely about its empirical existence : indeed, Hegel himself presents time as "the Concept being-there". From this it results that supratime, an original power, reveals itself to us in the form of the phenomenal time. Hence the conclusion that for us **supratime is temporal** (is in time), *viz.* essence becomes and necessarily appears in time. Grasping together the two propositions 'time is supratemporal' and 'supratime is temporal', we encounter an example of the activity of conversion, which is proper to the Hegelian logic[59].

Although time *qua* "intuition of becoming" is not regarded as a logical category, it seems possible to think 'the thing itself' *(die Sache selbst)*, especially as a determination of becoming[60]: for Hegel, any real becoming occurs in time. Say, the double character of time may also be exposed in the very term of becoming.

In effect, on the one hand becoming constitutes the movement of coming to be and passing away of the finite thing, an event that takes place **in time** ; on the other hand, that very becoming neither comes to be nor passes away : it persists as **supratemporal**. In a similar sense, the Preface to *Phenomenology* seems to describe the Manifestation as "the arising and passing away that does not itself arise and pass away"[61].

In the *Phenomenology* itself the double character of time may again be found, exposed in an indirect way. On page 558 (in English: p.487) it can be read that, **from the standpoint of the Concept** which knows itself as Concept, the developing moment precedes the intended full totality ; and, **from the standpoint of consciousness**, which is not self-consciousness, the totality precedes but at first is not grasped.

Translating afresh the text in intelligible language, I oppose thought (here Concept) and experience (here consciousness). Thus, according to Hegel, when we are grasping experience from the standpoint of thought, we can discern in it a **necessary development** where the part, through its movement, is supposed to participate in the constitution of the whole, 'pregnant' with the structuring determinations : that movement is not temporal. We look at a pattern of consciousness, discern the contradiction present in its attitude and thus discover the need to move up to a higher level, until we have entirely passed through the different constitutive patterns.

However, when we ourselves are the subject of experience (at a certain level of consciousness), then the movement develops, so to speak, "behind our back"[62]. In that case the whole precedes the part, thus producing, through its ideel presence, the driving tension we experience without our being aware of it. In that sense, the movement of experience appears as **contingent**, as "the pathway of doubt... the way of despair" ; and our limited point of view **changes** from one pressing contradiction to another, without our ever attaining self-consciousness—the whole remains unknown to us.

In so far as this interpretation is valid, it shows that human consciousness, while necessarily changing in the course of its experience, moves in a temporal development, for its viewpoint proves partial and limited : consciousness is thus submitted to the power of becoming. Thought, on the contrary, with its synoptic vision, grasps being according to all its constitutive moments—it proves therefore supratemporal.

If this is really the case, we can **distinguish, without separating**, two dimensions of human reality : the conceptual-supratemporal and the empirical-temporal dimensions[63]. Whence the concrete-synoptic vision that apprehends together the 'supratemporal' movement of the thought of experience (the abstract-synoptic one) and the temporal movement of that very experience ; namely, the **logical** exposition of reality (from the standpoint of the Concept) and its **historical** exposition (from the standpoint of consciousness)[64]. According to the first movement, we can discover a certain 'summit' from where all consideration is made ; and according to the second movement, history goes on, ever-opened in its essence.

Moreover, given the activity of mediation unfolding between thought and experience, we may talk about a **historical movement of the logical summit**, that summit itself developing as partial self-consciousness where the moment of consciousness is still acting ; and also about a **non absolute, logical summit of the historical movement**, for at every stage a certain self-consciousness is reached, however poor it may be[65]. For instance, it is now recognized that there are no peoples without a religion. In that meaning, then, reality is dual, accomplished and yet on the way to accomplishment. In other words, the end itself being on the move, **every end is a new beginning**. As for the 'absolute' end, we should leave that ideal to the gods and remain satisfied with our 'erotic' and limited human vision.

In this way, the ambiguity of the concept of the end has been examined from different angles. If I found in Part Two of this work that the beginning 'is and is not'[66], similarly, with regard to **the end** I suggest the following thesis : **it is** in so far as our culture has reached a certain degree of self-consciousness, and **it is not**, for history seems to be directed towards an ever-open future.

Actual history *(die wirkliche Geschichte)* is thus conceivable as a 'grasping together' of the progressive development of reality and of its circular movement as self-reflecting knowledge which, indeed, is not Absolute Knowledge[67].

The problem of the end has proved complex. Let us continue a few steps further with its elucidation.

5. THE CONCEPT OF THE END HAS TO BE THOUGHT ANEW: FROM A LOGICAL POINT OF VIEW, HUMAN REALITY SEEMS TO DEVELOP THROUGH SUCCESSIVE REVOLUTIONS

With the refutation of the idea of an Absolute Knowledge, we saw that the ideas of a univocal end of the history of philosophy and of the history of the world were also refuted[68]. In regard to philosophy, even if we were ready to accept the problematic thesis stating that 'all' the principles of reality have been discovered, the development of those different principles would still remain possible, if not necessary.

For instance, admittedly **Kierkegaard** studied the existential situation of the particular individual and, in so doing he explicated the category of existence *(Existenz)*. **Marx**, for his part, closely examined the genesis and structure of the Civil Society in its relation to the State, thus disclosing the basic contradiction that tears it. As for the **last Heidegger**, he pondered on a new way leading to "the thing", which is neither the Aristotelian external and separated thing nor the Hegelian thought and 'tensed' thing[69].

Positively speaking, it seems that the emergence of these different standpoints constitutes a certain unfolding of the System ; negatively speaking, we can say that the System itself breaks up. Accordingly, partial reconstructions have recently been attempted, in particular by G.R.G. Mure as a plan, and by Sartre as a synthesis of the principles of Hegel (as dialectical ones), Kierkegaard (as existential ones) and Marx (as materialistic ones)[70].

The history of philosophy is thus continuing or, to put it another way, **the end constitutes a new beginning** : in the Continental or Anglo-Saxon philosophical world, from all sides new Schools are arising.

If, at first, the very idea of an end seems shocking to common sense, would it not be possible to talk, nevertheless, about a certain end of geometry with the construction of the Euclidian system? about the end of classical physics as a result of the discovery of Newton's system? and about a certain end of formal logic after the publication of *Principia Mathematica* ? Of course, these different ends are also new beginnings, so that science continues to progress.

The same view may be expressed in another language : **science develops through diverse revolutions**. That statement is not new ; it was first proposed by Kant and more recently by Th.Kuhn[71]. As for the Hegelian *Logic*, it seems to show that any development is both continuous and discrete, i.e. accomplished at once in a continuous manner and by leaps[72] ; and in the dialectical movement, the nodal moments are not rare. These are generally understood as breaking points, giving rise to the well-known problem of the transitions in Hegelian philosophy. For my part, in this context I discern there a development through changes in points of view[73].

As regards the problem of the end, I would suggest that with the French Revolution, from a practical point of view, and with Hegel's system, from a theoretical one, a certain period of history did come to an end : in Western culture, the seemingly good days of monarchies and great systems have passed away. The period that has followed—our own—has been fruitful too, but differently. Later, another 'end' will allow its proper appraisal.

Nonetheless, at the high levels of his basic works, Hegel talks of the end in the following terms : **essentially** all is already accomplished 'in itself' and, to the full accomplishment of reality, only the moment of self-consciousness is missing, a moment developed in his own system[74]. On the other hand, the **contingent** history of the world continues indefinitely without revealing any decisive new principle, since 'all' of them have already been discovered.

To me, this presentation of the concept of end seems not very 'dialectical', in Hegel's sense of the word. Therefore, faced with the pretension of Hegel who claims having unveiled **'all'** the principles, I first ask the following questions:

a) Does the set of the principles really form a countable 'whole', whereas Hegel is known to see quantity in general, and numbers in particular, as determinations inappropriate to the world of spirit (to the not-material world)[75]?

b) How can a decision be made about the 'essentiality' of a new discovery, about its being 'in principle' and not of a secondary rank? Who can answer this crucial question decisively ? And with no univocal criterion for deciding, how can Hegel claim to hold in history an absolutely privileged position?

c) How does Hegel know that no important principles remain to be discovered? Is his system 'organic' to such a degree that no place is left for a new member? The answer is well known : the inner structure of the system appears to be somewhat broken[76]...

Secondly, to the essential end of history Hegel opposes its accidental continuation. Now, the dialectical determinations are not rigid or univocal, but fluid, 'plastic' *(plastisch)* and polysemous[77]. That exclusive opposition, affirmed between essence and accident, thus seems to be established from an inadequate, dichotomous point of view. Indeed, from the consideration 'in themselves and for themselves' of these logical categories[78], it follows that **the contingent is essential** ; namely, in the movement of essence, the accidental appears as the moment of exteriorization, essential to the essence that has to express itself. Accordingly, with no accident, no essence, and vice versa of course. That is, likewise **the essential is accidental** in the sense that, for **this** individual, essence reveals itself in this form rather than in that one[79]. In other words, essential movement and accidental movement do condition each other.

Must the conclusion be that, given the frequent presence of 'plasticity' and polysemy, dialectic has to offer only a blending of unintelligible determinations that are devoid of any real content[80]? For all that, I do not see this sceptical view as inescapable. Instead, I propose **a fluid and ambiguous concept of the end**, wherein essential end and accidental continuation prove hardly discernible—is it not human reality we are discussing, rich, various and ever-surprising?

On that account, we may talk about an essential end that is also accidental, the series of the different ends going on indefinitely ; and about an accidental continuation that is also essential, for with no essence still expressing itself, we would obtain Kojève's 'death of culture' and not a universal, living history[81]. Dialectically speaking, for a human being, at once real and ideel, there will always be a **telos** to realize, an essence to disclose.

In consequence, instead of confining oneself to dichotomous thought, it seems better to see essence and accident not as separable like wheat and chaff, but rather as present and related to each other to a certain extent[82]. In accordance with this basic duality (which is not a dualism), it is also possible to conceive this both essential and contingent end as a new beginning, which is itself also essential and contingent[83].

In short, I suggest a pattern positing a series of successive ends, which constitute successive beginnings too. Hence my reference to a theory of **development through successive revolutions,** each replacing one point of view with a higher one by a leap.

In my opinion, the Hegelian philosophy may be said to develop in such a way : from the standpoint of the phenomenology of experience to the standpoint of the logic of thought ; from the doctrine of external Being to that of inner Essence, then to the doctrine of the mediating Concept ; from the ideel standpoint of the logic to the real standpoint of the philosophies of nature and spirit, etc.[84] Through its problems of transition (except for the elevated discourses on 'absolute totality'), the System appears to reflect the movement of human reality to a certain extent—this thesis holds good, of course, within the limits of validity of my theory about development through successive revolutions.

CONCLUSIONS

Classical philosophy considers man as rational and yet finite. Indeed, only a being of flesh and blood is able to write a book, for the Concept has no hands. Thus the thinker who aims at 'rising to the heavens' still remains bound to the earth—man is really **bidimensional**, 'wise' and 'rest-less' at once[85]. I have found this very ambiguity at the heart of Hegel's thinking : dialectically speaking, reality, actualized in a certain sense, is still actualizing itself. To put it another way, human experience may be said to reflect the dual structure of human spirit[86].

That was the conclusion drawn from the standpoint of the thinking and acting individual. Now, from a logical standpoint, I suggest that human culture has developed **through successive revolutions**. Namely, according to Kuhn the historical movement is not linear and cumulative but proceeds by leaps : every new paradigm, when it appears on the scene, requires **the thinking afresh** of yesterday's problems, questions and responses—which is what I am trying to do in the context of the 'Hegelian revolution'.

Thus I conclude that the idea of an absolute end is devoid of any intelligibility. The thinker, standing in a certain social and historical situation, is really not in a position to explore from the outside that very human reality he intends to reconstruct as a system. There is no Archimedes' point on which to lean in order to lift the world up, for the thinker belongs to the very movement he is examining, and so find himself carried away with it. Therefore, to him the assumed enveloping totality always appears as 'broken'[87].

If, in Part Two of this work, **the beginning** came to light **as a result**, then in this Part **the end** comes to light **as a new beginning**, and thus a certain dialectical circle takes form. That is, to our mode of thinking, human culture proves to have no absolute beginning nor absolute end. And, from that angle, human reality manifests itself as progressing, without our being able to identify either a first leap or a final revolution[88].

About two centuries ago, Kant noted that science advances through revolutions and, recently, Kuhn has developed that idea. A similar idea has come out in the Hegelian logic, in the ambiguous form of a dialectical development that also progresses by leaps, from one level to another, from one point of view to another—in a process that, in principle, will never come to an end.

Unfortunately, Hegel was not always faithful to his own principles. Coming after him, we can understand that neither death, nor the absolute, may constitute a dialectical result of the social, cultural and historical human life.

PART FOUR

THE PROBLEM OF MATTER AND NATURE

At the end of the *Logic*, Hegel asserts that the Method constitutes an infinite force that penetrates everything. Nothing can resist it. This means that dialectic claims its ability to transcend any limit, to overcome every otherness and to move everything according to its own rhythm[1]. As a strategy intended to convince, this promising passage may seduce many naive readers. But as for its truthfulness, that is another question.

Indeed the System, which is supposed to be 'organic', also appears to be fragmented. For instance, Hegel wrote two versions of the *Phenomenology*, two of the *Logic* and three of the *Encyclopaedia*. Moreover, the relations holding between these different subsystems seem quite problematic. In particular, some scholars think that the *Encyclopaedia* renders the *Phenomenology* useless: accordingly, the 'way to Science', which 'is itself already Science', would have become obsolete. Obviously, if by 'system' one means '*Encyclopaedia*', then the matter stands. However, things are not so simple.

The System is in fact composed of a somewhat undetermined collection of articulated and detailed subsystems. Furthermore, it may be asked whether the System also includes the various *Lectures* that Hegel continuously changed and enriched over the years as he pursued his research and teaching[2].

On discovering Hegel for the first time, a willing reader is likely to be rather surprised, if not impatient or disappointed, by such a lack of clarity and distinction. Does the fault lie in the reader's ignorance, in Hegel's limited capacities, or in the so-called 'Spirit of the times'? In this Part of the book, I attribute the source of the difficulties to the **other of thought**, that is, to **matter** or **nature** which, when grasped from an 'idealistic' point of view, seems to be insurmountable[3]—without specifying, for the moment, what sort of idealism is concerned. Just as empiricism has trouble explaining the nature of soul, thought and culture, so any idealism sooner or later finds itself confronted with the **problem of matter and nature**.

It is this problem I attempt to elucidate, in the framework of Hegel's philosophy, without claiming to solve it, of course : in dialectic, the 'solution' lies in the process itself of elucidation, and there is nothing else to look for.

To this end, I proceed from abstract to concrete. First, I study the dialectic of Form, Matter and Content, with the aim of clarifying, from a logical point of view, the meaning Hegel attributes to the category 'Matter'. Second, I ask the philosophy of nature (*Encyclopaedia*, second Division) what signification Hegel attaches to the terms 'real matter' and 'nature'. Third, with the help of G.R.G. Mure I examine what really happens in a concrete experience, and I conclude that matter is, for dialectic, both a necessity and a univocal obstacle. Finally, I linger somewhat over Hegel's strategy regarding the question of matter, and suggest that it should be seen as a certain ambiguous answer to a basic problem of knowledge. In that sense, the conclusion of this Part of the work—dealing with the legitimate extent of dialectic—will itself be dialectical : in the presence of that inert and amorphous given, Hegel should admit that he succeeds neither in 'swallowing' it nor in 'spitting' it out. In metaphorical language, I conclude that matter is the fate of dialectic.

As concerns the examined object, Hegel, too Aristotelian for his time, failed to recognize the irreducible and yet essential reality of the particular sensible. As concerns the examining subject, it is within Hegel himself that nature shows itself, thus limiting univocally the possibility, for a human being, to develop a dialectic constituting a comprehensive and coherent system ; intermingled with the enlightening *Logos* we can find the opaque *Chaos*, a sign that matter was there.

1. LOGICAL MATTER AS A CATEGORY OR AN ARISTOTELIAN SUBSTRATUM: THE DIALECTIC OF FORM, MATTER AND CONTENT

Hegel regards the *Logic* as the essence of his philosophy. For here the content of human thought, as a *Denkbestimmung*, is said to be exposed in the form of the Concept, that is, of thought itself—hence the expression of the 'identity of form and content', serving as a guarantee of the validity of Hegel's logic. In my opinion, however, since the System claims to grasp concrete reality in thought, Hegel still must show that its content *aufhebt*— comprehends—human experience in all its bearings, and that with no essential loss. An examination of the dialectic of Form, Matter and Content, as it is presented in the *Logic*, constitutes a first attempt to elucidate that crucial problem[4]. I therefore ask the following : is the logical content really a unity of form and matter? What sort of unity is at stake? What kind of matter?

In various Hegelian studies, many logical movements have been examined. However, for reasons perhaps due to tradition, in the *Logic* the examination of the 'Absolute Ground' seems quite neglected. Such an examination nonetheless proves essential, for it allows a better understanding of this supposed 'identity of form and matter'—indeed, a basic identity of the System[5]. Moreover, when the 'content' of Hegel's philosophy is under discussion, it would be advisable to know what it is all about, and what is the meaning of that common term, which is not always well defined.

As for the general background against which this examination occurs, it must be noticed that we are at the level of the Doctrine of Essence, which consists in an interiorization of the movement of Being. That is, we find there a mediation developing between Essential and Non-essential, Essence and Show : as a determination of the dividing reflection, Essence is always related to an other than itself. At the outset, then, Essence posits the non-essential Show it presupposes[6]. That Show *(Schein)* is its own reflection ; or, in other words, Essence determines itself as reflection, for reflection is both an object *(qua Schein)* and a movement *(qua scheinen)*—attention must be paid to that double meaning, which constitutes an important help to the way of understanding the difficult movement presented here[7]. Having posited and unfolded the determinations of reflection (Identity, Difference and Contradiction), Essence goes on developing and positing the determinations of the 'Absolute Ground' : Form, Matter and Content, the categories we are dealing with.

In view of the complexity of the treated subject, I retain Hegel's tripartite division, and conclude with a discussion.

1.1. Form and Matter as determinations of Spirit

First let us see how Essence, or human spirit as producing, is supposed to divide into active form and passive matter.

At the beginning of the Doctrine of Essence, Hegel says, Essence has revealed itself as ground **in itself** or **for us**. But things cannot be left off there, of course. Essence also has to become ground **for itself**, i.e. it must determine itself as ground explicitly[8]. Yet, every ground is a ground of something ; which is why Ground divides itself within itself.

At that opening level, **Ground** then has a double nature : **grounding** as determining (*qua* active essence) and **grounded** as determined (actually, Ground just determines itself). That 'duality in unity' will characterize the entire subsequent development. In particular, at the outset that 'duality' reveals itself as a relation between form and matter, namely, as a self-partition of the original Essence.

In the course of the movement, then, Essence posits itself as Ground. And, with regard to that producing activity, it becomes the difference grounding/grounded. It is thus possible to distinguish Essence *qua* activity from Essence *qua* Ground—we again find the double meaning of 'reflection', which signifies at once movement and object. And, because Ground is both grounding and grounded, that determined difference differentiates itself from the unity of active Essence that becomes Form—that movement is for us, the examining subjects.

In Iena's Hegelian language, we meet here with **the identity of the identity** (of Essence as active Form) **and of the difference**[9] (of Essence as divided Ground against the one Form)—in this way, Form and Essence each become the totality as well as a moment. Consequently, as the self-reflection of Essence, Form relates itself to its different moments as positing and determining. In that activity I propose to see a movement of 'determining step back' *(recul déterminant)*. Form, the totality of the movement, *qua* mere moment, grasps itself as determining its other[10]. 'Form determines Essence' means then that Form is led to grasp the difference present in Essence (*qua* Ground) as posited by itself (by Form itself).

From that, the following contradiction or inner tension arises : *qua* total form, the human spirit grasps the present determination as an activity as well as an outcome of that activity. Again we meet the duality movement/object above-mentioned.

As usual, contradiction proves unbearable. Therefore, Form (our mode of thinking) relates to the determination it has posited (as it were) as to an alien other, in which it no longer recognizes 'its own activity' : consequently, pure reflection turns into external reflection[11].

From now on, that other is grasped as lacking any form, that is, as non active and non determined, an element in which the determinations of Form are said to subsist[12]. In that very element we can discover, *qua* identity devoid of any difference, the substratum *(hypokeimenon)* of the Aristotelian philosophy[13].

From this discussion we have learnt that, *qua* substratum, the logical matter may be said to be just a self-determination of spirit positing the determination it presupposes[14].

1.2. **Form as determining the material substratum**

However, we must be careful : the so posited matter is what remains after having abstracted from the form or structure proper to the thing. In that meaning, it is the opaque undetermined entity **of which we can get no sensation**.

Indeed, Aristotle has shown that, in our everyday experience, we always perceive a certain matter already structured by a certain form. In other words, Aristotle asserts that Form implies Matter that it determines, and Matter implies Form that confers to it reality and structure. Matter may thus be conceived as the **passive** element in our experience, relative to Form, which is **actively** present in it. Consequently, Hegel notices, Matter, *qua* determined by Form, is supposed to be 'receptive, sensible' to its other. In this case Hegel mentions *'die Empfänglichkeit'* (the sensibility)—this time the explication of a Kantian concept is at stake[15]. Hence the following logical conclusion : Matter has to form itself and Form has to materialize itself—a thesis that can be referred to Aristotle, in so far as the sublunary world is concerned[16].

In the course of that movement, then, the original unity—the spiritual Essence as starting-point—has divided into **essential identity**, the indifferent substratum, and into **essential difference**, Form that determines itself as Matter : a new contradiction comes to light. On the one hand, Form sees itself as independent (**it** is the determining source) ; on the other, it has to determine Matter upon which it is thus dependent[17]. The 'solution' consists in the *Aufhebung* , or overcoming, of the pretension of independence.

At this point, Matter comes to be grasped as an inner other, or a moment, posited by Form and in which Form subsists. Moreover, Matter also is a contradiction, since it contains within itself the determining Form which is said to be external to it. From this I understand that 'Matter' now appears in its double meaning : at one time, it belongs to the movement as a moment of the developing spiritual essence ; at another it is external to it, as an indifferent and independent Aristotelian substratum.

However, we should not be mistaken : this double meaning does not reveal a flaw, but rather the opposite. For, dialectically speaking, the **driving tension** is precisely produced by the opposition that arises between the movement of **unification of Essence** determining itself as an inner other (on Matter as a moment of spirit), and the movement of its **differentiation** in whose course it grasps Matter as an external other (on Matter as an indifferent substratum)—in **this** meaning, we can say that spirit *qua* separating Understanding is a moment of spirit *qua* uniting Reason.

We saw that, according to Hegel, the solution of the contradiction lies in the act of *Aufhebung* which posits the dialectical identity of Form and Matter: grasped as indifferent to Form, Matter is yet 'sensible' to it since it is supposed to be posited by that very Form.

Thus, dialectic appears here like a play of reflections developing between the two different points of view, the inner and the external. That double play proves possible, even necessary, to the extent that man is at once a dividing Understanding and a uniting Reason. If this is the case, then the logician (for he is the concerned one) grasps **matter as independent** (*qua* an Aristotelian substratum) as well **as dependent** (*qua* an Hegelian moment of the developing essence)[18].

1.3. **The content, formed matter or materialized form, as a determination of Ground**

Hegel thinks that it is impossible to leave off at that opposition between Form and Matter, since the 'tense' relations of identity and difference arise there. That is why a return to original unity has to be made, this time to a unity reconstituting itself as a **content.**

The movement unfolds as follows :

Form is finite, since it implies Matter as its other ; it is not Content but only determining activity. Matter too is finite, since it implies Form ;

it is not Content either but passive substratum. Namely, when separated, the two opposite determinations are not grasped in a true way—they appear abstract.

Therefore, only in the now emerging unity of Form and Matter can we meet again the original Essence which, in the course of that dialectic, has developed as Content. In other words, in that process we saw the unity positing itself as unity of unity (of original spiritual Essence) and of non-unity (of the differentiation between itself and Form, between Matter and Form)[19].

Hence Hegel's following conclusion : that movement of Spirit's self-realization, which contains within itself the moment of difference, shows that the determination of Matter by Form is nothing other than the **self-determination of Essence as Content**—the Content being here understood as formed (determined) Matter or materialized (subsisting) Form.

In that dialectical movement, therefore, Ground has become determined, while Content shows itself to be a determination of Ground. To put it another way, 'Ground is determined' means 'Content has a Form', a proposition in which Content proves to be essential. That Determined Ground is the self-identical, reconstructed Essence, enriched with the determinations of Form and Matter[20].

1.4. Discussion

1.4.1. In the examined dialectic, a certain **reconstruction of the Aristotelian theory of form and matter** has come to light. Aristotle sees substance as consisting in a matter structured by an active form[21]. Explaining change, he believes that the material substratum is the permanent principle that 'bears' it. In this way, different forms unite to a same matter in order to constitute the changing thing, whereas they themselves do not change. As is well known, that theory is affected by **some difficulties**.

First, in so far as we always perceive an already formed matter, a **primary matter** *(prote hule)* must be posited as a logical principle deprived of any form—this principle therefore cannot be real, in the

Aristotelian meaning of the word. At the other end of the 'scale of nature', we find the Aristotelian god as a pure form, a pure act lacking any matter—a concept that is also hardly intelligible. In sum, in this theory both the extremities of the ontological scale appear quite problematic.

Next, Aristotle explains reality with the help of the determinations of form and matter, but these determinations themselves require explanation. In the **linear** mode of explanation at stake, the issue refers to the problem of a **beginning** that stands unjustified : 'What are primary matter? pure form?' it may be asked[22].

Moreover, since the principles explaining change do not change, the cause of changing must lie outside. Yet, as regards the natural thing, the formal and final causes are expounded as inherent to the thing[23].

On the other hand, in the Hegelian doctrine of form and matter the inner tensions, discerned in Aristotle, find a certain 'solution'.

First, as moments of the developing essence, form and matter are both grasped as dynamic : now, while explaining change, these categories can also explain themselves. That is, **form and matter are said to mediate each other**, each realizing itself through the agency of its opposite. In that **circular** mode of explanation the 'extremes', primary matter and pure form, become partly intelligible—hence the way of an 'explanation explaining itself'[24].

Second, in Aristotle form and matter are external to each other, and are thus made independent. If that is the case, how can they unite in the thing to make up a synthesis[25]? In my opinion, Hegel's answer to this question lies precisely in his doctrine of the **double meaning of 'matter'** : for him, matter is a moment of self-developing essence as well as an indifferent, independent and external substratum. If so, one can understand, at least up to a certain point, in what way the **passive** matter is structured by the **active** form, since identity and difference of form and matter are both asserted. These are **identical** in so far as matter is posited by form as its inner other determined by it ; they are also **different** because, at the same time, matter appears as an external and independent substratum. What is really involved is, so to speak, first an act of determination, and then an act of self-alienation. And dialectic consists **in grasping**, as far as possible, **these two opposite thoughts together** : the determining identity of form and matter and their alienating difference.

We thus encounter again dialectic as a play of different viewpoints reflecting in each other, and requiring to think both oneself and the other, both simple unity (matter as the inner other of form) and separation (matter as an external other, within)[26]—hence the reconstruction of the creative tension. That is why human thinking does not leave off at that point : the Hegelian logic goes on developing, for instance as formal, real and total Ground.

Understanding is truly a necessary moment of Reason that can unite only the separated, just as in its turn Understanding can separate only the united ; in principle, the double movement proves inescapable.

1.4.2. This study has shown that, in Hegel, 'matter' can be understood in three different senses : as a moment of spiritual Essence, as an external, indifferent and independent Aristotelian substratum, and as the sensible given (which, in Aristotle, corresponds to the accident not treated by science). In other words, matter successively appears as a moment of the living thought, as its external other that lies within, and as its irreducible external other.

The **first signification** (on matter as a pure category) seems to express an extreme idealistic point of view, that solves the problem of matter by suppressing it, purely and simply. Undoubtedly, the coherence of thought is thereby safeguarded, but at what price? at the price of abstraction and lack of content! In that perspective, we will not find an account of human experience, but the pure and void Parmenidean Being.

The **second signification** refers to matter as the substratum of any determination, any formation—in his activity, man must indeed form something. However, now the problem of Kant's unknowable thing-in-itself again arises for, as we know, Kant too has trouble dealing with matter[27]. Hegel, for his part, claims to know what that substratum is : it is just the inner other that is inherent in spirit, posited by human spirit itself as an external other[28].

As for the **third signification**, it expresses an extreme materialistic point of view, considering the touched and perceived given as the object of any knowledge, of any concrete thought. In a certain sense, one can concede to the materialist that being is sensible. But Hegel asks us not to leave off there, for the perceived object, which is not the sensated atomic 'this, here and now', can also be thought. It therefore contains a form. If that was not the case, how could we know it and grasp it in our mind?

In consequence, Hegel may legitimately assert his philosophy as being at once **idealistic**—it gives explanations through thought categories—and **empiricist**, since it is just the present concrete experience that is supposed to be accounted for[29] except, of course, the 'sensible residue', an absolutely irreducible given. Through that productive tension, Hegel's thinking reveals itself in all its wealth, its ambiguity and univocal limits[30]. Thus, in concrete terms, by 'matter' one can understand either the perceived object containing a universal form—which is then explainable—or the sensible particular residue, a 'this, here and now' that is absolutely unintelligible[31].

The *Logic* therefore proposes an explanation of matter that appears to be either abstract or ambiguous. At this point, we can discern two basic problems in Hegel's philosophy.

With regard to the Hegelian explanation as **abstract**, we can observe that a **leap** is needed to reconstitute the concrete determination, the "universality which includes within itself the particular". For the moment of sensible intuition, that is 'set aside' in the exposition of the movement of logical matter, is restored in the *Logic* 'from behind' only, by a *tour de force* that goes far beyond the real possibilities of the 'necessary and immanent' movement of the Concept.

At the beginning of the *Phenomenology*, Hegel presents the sensible given as "not-true, not-rational,... merely meant"[32]. But, in my opinion, that is the case only in so far as a beginning is concerned. Likewise, the *Logic* deals with the 'ineffable' difference that exists between the first categories of Being and Nothing[33]. Nevertheless, the *Phenomenology* later recognizes the concreteness of any experience, which is always empirical as well as spiritual, that is, contradictory : in subsequent developments, the sensible given is really restored.

As for the logical movement, the second moment is generally that of the external, particular and accidental difference, and is thus the moment of the sensible dimension[34]. From this it follows that empirical intuition truly takes part in the logical development, such as in the form of the categories of Dasein, Finite, Variety, Real Ground, Properties of the Thing, Determined Causality, etc.[35]

With regard to the **ambiguity** that characterizes the Hegelian explanation, whenever Hegel talks about matter it is advisable to ask whether it is question of the category as a moment of thought, of the amorphous and inert substratum as an other inherent in thought, or of the sensible given that is external to it[36]. This many-sided viewpoint constitutes an aspect of the problem of language in Hegel's philosophy : the wealth of the exposition allows us to obtain a total and yet detailed account, even though this is entangled in disconcerting ambiguities that are likely to mislead the uninformed reader. In every case, the context alone indicates the relevant meaning.

1.4.3. At this point, it is time to ask **what is the meaning of 'content'** in Hegel's philosophy? According to the examined text, the concept of content may be apprehended in its double meaning : at one time with respect to form and matter which make it up, and at another time with respect to thought, the form that considers it—'form' also appears in its double meaning.

First, by **'content' as identity of form and matter** I understand the substratum (or matter) that is determined by thought (or form) in the dialectical way exposed above. Here the discussion turns on the essence of experience, given the fact that Hegel has 'set aside' the sensible particular, which is irreducible even to dialectical thought—in so far as the Content may be said to comprehend *(aufheben)* Matter, it is not real matter that is at stake but only its shadow, the Aristotelian substratum.

Second, **with respect to the considering thought itself**, Content is the object examined by thought and which is thus moved in its 'immanent and necessary' dialectical process. The sensible given remaining irreducible (as 'residue'), we encounter abstraction again—in that sense, *Logic* is really a "realm of shadows".

The discussion has thus brought to light two different dimensions of dialectic : one according to which human experience constitutes itself—its **ontological** dimension—and another according to which human experience is grasped and exposed—its **epistemological** dimension[37]. In the two cases the 'sensible residue' is discarded, not-structured and not-moved ; so that, in this context, the **'identity of form and matter'** can be understood as the unity of the considering thought with the considered experience, the latter being a formed matter or a materialized form.

This was the conception proper to the dialectic of 'Absolute Ground', as far as I have undestood it. But, as we saw, **in fact** the empirical intuition plays a constitutive part in the *Logic*. Therefore, by 'content' it is also possible to understand the apprehension of the whole of human experience which is really reconstructed, but only by leaps, thus forming a 'not-immanent and not-necessary' process. That is the moment of the **difference** between form and content, which is essential too. This result perhaps does not agree with some of Hegel's programmatic statements ; but, in any case, it proves quite faithful to experience. And, in dialectic, that is the main thing[38].

In this way, the term 'content' has been understood as meaning:

a) an abstract constituted totality : as a subject-object with no sensible matter;
b) an abstract considered object : the 'pure' examined thing;
c) a concrete considered object : as a real, empirical thing.

As usual, human reality proves complex and multidimensional[39].

When the point at issue is a concrete topic, referring to the *Logic* is not enough. We therefore now proceed to the examination of the problem of matter, from the standpoint of the philosophy of nature in the *Encyclopaedia.*

2. REAL MATTER AS MATERIAL NATURE

In Hegel's view, the explanatory power of the *Logic* consists in that it is supposed to realize the identity of form and content. But, obviously, in the philosophy of nature that is not the case.

Let us consider the *Encyclopaedia*. It is divided into 'Science of Logic', 'Philosophy of Nature' and 'Philosophy of Spirit'. The 'Science of Logic' is said to treat the 'immanent and necessary' movement of the Concept, unfolding its different determinations and then accomplishing itself as the Absolute Idea. When moving in its other (the determined category), thought would feel here 'at home'.

On the other hand, on the threshold of the Philosophy of Nature Hegel declares that the Idea "resolves to let the moment of its particularity... as **nature** go forth freely from itself" ; that is, as an immediate other. Since nature is the milieu of exteriority and contingency, the movement of thought will meet there the greatest difficulties. Indeed, for thought nature is the world of alienation where it is an other to itself, an **empirical representation** *(Vorstellung)*. Later on, in the Philosophy of Spirit thought will come back to itself, to find itself at the end of the road as a so-called "wholly realized" logic[40].

Consequently, the point is the following : how is the movement in the Philosophy of Nature to be understood, given that there is no identity of form and matter here, but rather the deepest alienation? If dialectic consists in the moving unity of thought, thinking-subject and thought-object—as I think it to be—now the movement of the thinking subject seems to differ largely from that of the thought object ; that is, the movement of thought has turned into a process of representation. And, with representation we have 'fallen' into the sphere of the empirical, a synthesis of thought and sensibility[41].

We have therefore moved to the level of the difference holding between form (or thought, for we are dealing with **philosophy** of nature) and content (or empirical representation, for we are dealing with philosophy **of nature** as well). Now, the development will hardly be made by an *a priori* movement of sinking and restraining, but rather by an *a posteriori* movement of reconstruction[42].

In other words, in the Philosophy of Nature, Hegel seems to organize the various determinations—drawn from the natural sciences—into a more or less coherent system, in accordance with the malleability of the 'given', which is mainly physical and biological. When reaching the level of the 'Organic', thought feels somewhat 'at home' again, at least as being itself an organic system.

However, in view of the gap that exists between form and matter, despite Hegel's explicit statements **any talk about a 'dialectic of nature' is to be considered as invalid** : here, we have mainly to do with a work of external reconstruction. I understand that the most to be done is to detect the 'traces' of a well hidden Concept[43]. For this reason, the external criticism has no difficulty in making fun of the movement of the 'solar system' determinations, for example, or of the so-called development of sensibility *(Sensibilität)* into irritability and reproducibility[44].

2.1. The empirical representation as the other of thought

The Philosophy of Nature deals with thought that does not think itself but thinks its other, the representation *(Vorstellung)*, as far as possible. We now see that the classical topic of the relationship between subject and object is shifted to the 'inward' aspect of human spirit, which both thinks and represents[45].

How is thought able to think its alien and external other? It is able to do so, solely because the representation is not entirely alien! In truth, since representation is a synthesis of thought and sensibility, thought can recognize itself in it, partly at least. In other words, the representation does for thought a preparatory work of formation, bringing about a first interiorization of the external object.

There is no direct thinking of nature, for, as sensible given, nature constitutes an irreducible 'residue'[46]! Therefore, thought grasps nature **indirectly** through representation, to the extent that the latter both interiorizes the external object and exteriorizes the thinking subject—at least, it seems to me so. In consequence, 'representation' may also mean 'synthesis of outer and inner' or 'synthesis of particular and universal'.

In this way, a **basic ambiguity** is again encountered in the Hegelian philosophy : if, from the standpoint of pure philosophy, thought is 'the best', from the standpoint of concrete experience it is representation that deserves such a title. For without it, thought would be only that "night in which... all cows are black"[47].

Actually, an indispensable interaction develops between representation and thought, since—we should not forget—representation is somewhat opaque (a sensible element resides in it) and needs thought to attain self-clarification. Translated in Kantian language, this assertion reads: 'Thoughts without a corresponding representation are empty, representations not grasped by thought are blind'[48].

If this is the case, abstract (pure thinking) and concrete (representation with its double character) dimensions complement each other, to constitute together a concrete philosophy that sets itself the task of giving a total and detailed account of human experience[49].

In principle, then, thought is able to think itself only[50]. Therefore, what matters to Hegel is to show that the other of thought is still thought, but as exteriorized. If so, in the representation, thought can discern the 'essence' of the thing, the **universal** present in the empirical given. However, as concerns the sensible given itself, which is **particular**, it remains an irreducible external. Neither thought nor even discursive language succeed in grasping and expressing it. That is why, in order to apprehend the object in question one also needs 'to look with one's own eyes'. In the philosophy of nature, direct observation (or recording) must then make up the discursive report, and concrete signification is only achieved with the unity of inner sense and external designation—physics or empirical science in general is really a presupposition of the philosophy of nature[51].

Hence the following conclusion : necessarily, the Hegelian system comprises at least a logic wherein thought thinks itself, a philosophy of nature wherein it thinks the empirical representation, and, in order to restore the concrete experience, a philosophy of spirit supposed to unite logical thinking and real thinking—actually, however, the sensible given appears here also as **an external yet essential residue**. As Kant used to say, to distinguish the 'conceived' from the 'given' is not enough : we also have to separate between them[52]. For by 'nature' or 'matter' we must at first understand the particular 'this, here and now', irreducible to human thought which grasps the universal only[53].

I now intend to clarify this aspect of the problem.

2.2. Material nature and formal nature

Some scholars frequently ask whether the philosophy of nature in the *Encyclopaedia* deals with the movement of nature itself or solely with the movement of the determinations of nature. To that basic question I propose two different answers.

Firstly, as is often the case with Hegel, Kant has to be recalled. In his *Critique of Pure Reason* Kant has shown that "The conditions of possibility of experience in general are also the conditions of possibility of the objects of experience"[54]. If this is the case, in a transcendental or dialectical study, **knowledge of nature** and **nature** itself cannot be separated—in that sense, 'nature' is equivalent to our own experience of it[55].

With a view to rendering this assertion intelligible, I think it worth mentioning the Kantian distinction between **natura formaliter spectata**, nature in the formal meaning of the word, and **natura materialiter spectata,** nature in its material meaning—in short, between formal nature and material nature[56].

In the philosophy of nature, **formal nature** is concerned ; viz. nature as conforming to laws, man being regarded by Kant and Hegel as constituting nature in that sense[57]: in that sense, then, the movement of nature *qua* formal is also the movement of the laws of nature, that is, of its necessary and universal determinations. As for **material nature**, the object of sensible intuition, it is given in our spatio-temporal experience and not constituted by us, and external observations alone allow us to grasp its inexhaustible richness. Namely, the philosophy of nature thinks the sciences of the time, that is, *inter alia*, the Newtonian laws of nature, and not the natural diversity it can only describe from the outside. Like every thinker, Hegel is unable 'to resolve' or 'to dissolve' the sensible given, and he can do so only with regard to the determinations posited by man himself[58].

In this way, we have again found the Hegelian content as the identity of form (the determining natural laws) and matter (the determined Aristotelian substratum)[59].

But we cannot leave off there and so we turn, secondly, to the text itself for a closer consideration of the 'idea of matter'.

In the *Encyclopaedia*, the Philosophy of Nature opens with the real determinations of **space** and **time**. Yet, remembering Kant, Hegel regards space and time as **pure intuitions**, and not (like Newton) as absolute realities nor (like Leibniz) as an order of appearance of the phenomena. However, in opposition to Kant, Hegel grasps intuition as a moment only, as the lowest level of self-realizing thought. This thesis is exposed in the *Encyclopaedia* par.445-468, that lead from sensible intuition to pure thought through empirical representation.

If that is really the case, then nature, as ruled by laws, is not as foreign an element for thought as it seemed at first sight : in nature thought meets with itself, but at a low level of development. In this sense, then, a certain dialectic is possible, and the substratum may be seen as a determination truly posited by human thought.

This result seems appropriate with regard to the forms of sensibility, space and time, but with regard to the **matter** of sensibility (nature *qua* sensated and perceived) this is not the case. If one tarries a while at the 'psychological movement' from intuition to thought, two important things can be pointed out. First, at the beginning of the movement (*Enc.* par.440 note), Hegel affirms that he abstracts from the phenomenal content of human mind ; later, at the end (par.465), he comes to the univocal conclusion that 'being is thought': the sensible given, set aside at the beginning, is now entirely denied...

From this I understand that, when the sensible given is likely to hamper the exposition of some movement, then Hegel overlooks it or abstracts from it, and the trick is played! And, of course, Hegel restores it when necessary, with the help of a 'little push', that is, by an unavoidable and unjustified leap. For instance, in paragraph B of 'Mechanics', the 'weighting' matter is suddenly reinstated. It seems that, here, Hegel solves the fruitful Kantian paradox of the ***a priori* sensible intuition**[60],by turning the intuition into a 'self-intuiting' one: the Other has become a Same, through an annihilating (or abstracting) reduction and not through a comprehending (or including) *Aufhebung*.

From this discussion I conclude that, in the *Encyclopaedia*, nature is understood as a problematic synthesis of abstract thinking and sensible intuition, constituting together the empirical representation, or as formal nature implying material nature[61]. Further, nature in the *Encyclopaedia* may still be understood as a spatio-temporal substratum structured by the determinations of logical thought, to which must be added the sensible given or matter of sensibility, introduced on occasion by unjustified leaps—we thus find again the Hegelian content as concrete thing[62].

Strictly speaking, in the *Logic* we have discerned the identity of form and content (at the 'Absolute Ground') and their difference (in the *Logic* in general). In the Philosophy of Nature of the *Encyclopaedia*, we have discerned the identity of thought and nature (as formal nature) and their difference (as material nature). Consequently, either we find there a 'flowing' but abstract dialectic dealing with a non material, thus a non real nature ; or the treated topic is real nature, the object of physical and biological sciences, but then all rigorist claims must be given up. In that case, we have to content ourselves with a systematic reconstruction of the results achieved by science.

I understand that, in general, Hegel has opted for a concrete exposition of the concerned thing, and no matter whether the movement of the

Concept remains 'immanent and necessary' or not[63]. For the leap and the external ordering according to the degree of generality, prove here inescapable. It is only to be regretted that sometimes Hegel gives way to self-deception and presents his work, yet so fruitful, behind an harmonizing mask.

In this way, nature has been reconstructed according to its ideel dimensions (as thought and formal nature) and also to its real dimensions (as representation and material nature). However, for all that our task is not finished. We have still to consider what really occurs in concrete human experience[64].

3. THE ANTINOMY OF HUMAN EXPERIENCE

In the *Logic*, we saw that the exposed 'pure thought' was in fact dependent on empirical intuition. As concerns the Philosophy of Nature in the *Encyclopaedia*, it is able to restore the sensible given at the cost only of unjustified leaps. This sensible residue thus seems to constitute a univocal obstacle to Hegelian dialectic. Matter, encountered in human experience, is then not only a determination of thinking or the substratum to be formed. It is also the unintelligible *Dasein*, a being-there that resists any attempt to reduce it. That is, it happens that the 'immanent and necessary' dialectical movement succeeds in rebuilding the human experience in all its richness, but at the price of becoming external and accidental. In these cases it seems that, paradoxically, precisely the quality of experience is not preserved nor reproduced—Hegel's promises notwithstanding. For this quality is an immediate sensation that cannot be grasped, moved or transcended by any mediation[65]: the concrete quest for signification will always require a return to the lived experience.

The rationalist postulates that, to the human being, reality is conceivable and expressible in discursive language ; and yet, the sensible given escapes him. Hegel gets out of the difficulty by deciding, at the beginning of the *Phenomenology*, that this given is "not true, not rational", and therefore a nothing. And yet, that **non essential** proves to be **of some importance**, since it is capable of limiting the realization of Hegel's plan : to provide human experience with a coherent and comprehensive *Logos*.

In this chapter we shall again find these different conclusions, by examining directly a particular human experience with the help of G.R.G. Mure. Accordingly, we shall see that in any experience, at any level, even the highest, the forms of sensibility (space and matter) and the matter of sensibility (as sensible given) subsist, irreducible to thought, so that the external is present within, yet *qua* external.

The result is that, despite Hegel's own statements, in the System the problem of matter and nature stands, unsolved—a 'thorn in his flesh'.

3.1. In any experience, even the highest, the determinations of space and time subsist

Let us consider a simple judgement of perception, at first from the standpoint of space ; for instance, the following judgement: **'I see a house'**[66].

This judgement implies that I perceive the different parts of the house as extending one next to the other, and that in the same one experience. That is, on the one hand this experience is **spatial**, since the various perceived parts appear in external relations to each other. On the other hand, this experience takes place within my consciousness, which is **supraspatial**. Indeed, my eyes or my sight are not a centre for me, for I do not see them in my experience. I am really present in that experience, but only as a non-spatial conscious subject who perceives a house.

A contradiction arises : this empirical experience is spatial at its **noematic** pole, and non spatial at its **noetic** pole. Namely, any experience of a spatial object presupposes a non spatial subject as a conscious source of this experience. With no subject no space either, for the determinations 'here', 'there', 'next' are only for a perceiving and thinking subject—this being said, of course, to the extent that man really constitutes his experience. To that extent, despite its numerous external relations, space does not form a system ; for, as an active subject, I alone am able to confer a certain unity to the present aggregate of sensations I organize as a totality.

According to Mure, from this it follows that any experience implies both a spatial object and a non-spatial subject. That is, any experience proves **spatial** as an experience of an object, and **supraspatial** as an experience made by a subject—hence the paradox. However, as we saw, this thesis presupposes the idealistic theory of constitution, according to which the active subject unifies the multiplicity of a given intuition into an ensemble that is significant to him[67]—what here is called an experience.

In this sense, then, **the spatial dimension of experience is not susceptible to being transcended!** And even at the summit of the dialectical development, at the level of the so-called pure thought, we again encounter, in the concept of the experienced object, the presence of space as the moment of external relation[68]. Thus, the thinking of experience leads to an experience of thinking[69].

From this viewpoint, therefore, Hegel's plan seems unrealizable : to reconstruct, in the element of pure thought, the concrete human experience that has been interiorized, then reminded *(er-innert)* . In concrete language, it amounts to saying that pure thought succeeds in building human experience anew, solely through spatial representations *(Vorstellungen)*, which persist as external relations present within.

We arrive at similar conclusions on examining **temporal** experience[70].

Let us consider again the judgement 'I see a house'. According to the adopted idealistic theory of constitution, this judgement signifies that, in my experiencing the house, I intuit its different parts, imagine the totality, judge this perception and infer about the possible tenants[71]. That movement is **temporal,** for the different perceptive acts are made one after the other. But that movement is also **supratemporal** since it concerns different events occuring in the same one experience.

In other words, my experience develops in time for there is no visual judgement without a previous vision—in the process of vision, this is the moment of **cancelling** where an 'after' empirically implies a 'before'. However, I also intuit, imagine, judge and infer in one and the same act, for neither is there a vision without an *a priori* corresponding judgement—now this is the moment of **preservation**, where a 'before' in principle implies an 'after'[72].

Yet, we have to distinguish carefully between the temporal and the conceptual order. In Kantian language this means, following the temporal order of the event, that visual experience begins with vision; but, following its conceptual order, that it does not entirely arise out of vision. In Hegelian language this time, let us say that the posited—the judgement succeeding the vision—is presupposed (by vision) as a necessary condition.

The term 'order' has thus acquired two different meanings : it means the **temporal** order of a process, in which every stage follows the previous one ; and it also means the non-temporal, **logical or conceptual** order of an activity, in which the different moments are supposed to mediate one another—whence, as Mure dares to suggest, the 'horrible' expression of the **'logico-temporal'** order[73].

In so far as the proposed description is valid, in every lived, exposed and thought experience, we encounter that logico-temporal order. For instance, in the ***Realphilosophie*** of the *Encyclopaedia*, the development advances from Movement in space and time towards Logic as the philosophy itself. Yet, at the very summit of the System, as far as it is intelligible, we can find the temporal dimension tainting the purity of the philosophical Idea : as was shown by Kant, man thinks in time.

Consequently, in every experience, even the highest, it is not possible to separate, but only to distinguish, between the **temporal process** of the genesis of the determinations and the **logical activity** of thinking : as far as time is concerned, there is no overcoming—hence the **coexistence of opposites** that do not really merge into a dynamic unity. At any level of human experience, temporal and supratemporal relations are always present. That is why the Concept lacks transparency and, strictly speaking, the required self-consciousness remains out of reach.

This paradox is well-known, and formalistic thinking tries to solve it by separating the opposites : it considers the logical relation as non-temporal, and the empirical given as existing in time. However, as Mure points out, if in the visual experience **the temporal process is isolated**, then the outcome is not an ordered development but only some change ; and, for the thinking subject, the sequence 'I intuit, imagine, judge and infer' do not constitute a united experience : he then perceives an atomic world in which unintelligible events are happening. On the other hand, **if the logical activity is isolated**, then we do not understand the necessary order that characterizes the genetic movement—intuition, imagination, judgement and inference—an order that temporal experience alone can teach us.

In order to apprehend the concrete experience we therefore should not separate, but grasp together, temporal process and logical activity—in our experience or thinking of it, this condition constitutes a basic *sine qua non* of dialectical intelligibility. Say, the so-called pure thought thinks itself in time, and does so through the mediation of spatial representation. Such is the case even at the level of logical thought[74].

Thus, in the sphere of 'pure' thought, space and time, the **forms of sensibility,** subsist—a fact that can be called the '**antinomy of experience**'[74a]. We now proceed to an examination of this problem, this time with regard to the **matter of sensibility**.

3.2. At any level of experience, the sensible and the *a priori* dimensions are also present[75]

In the previous discussions, we understood experience as a total *Erlebnis* (a lived experience), including a subject-pole and an object-pole. According to the adopted idealistic point of view, however, it was pointed out that the difference subject/object is posited within the experience itself and not prior to it, since the **self-identical I** synthesizes the multiplicity of a given intuition into a structured object. Hence the following categorial distinction : this multiplicity facing me here, organized in space and time, is not me who is one and the same consciousness grasping and synthesizing that multiplicity.

Thus, experience is not only **impression**. It is also **expression**, for an ***a priori*** element is necessarily present within it[76]. And, on the contrary, in the highest experience a **sensible** element subsists, as a necessary condition of intelligibility.

Let us elucidate this problem, first from the viewpoint of the opposition **universal/particular**.

In the empirical experience, corresponding to the abstract universal (the concept) stands the particular individual (this object). And, since what is looked for in knowledge is the adequacy of the concept to the given, the development advances from the mere given to its conceived determination—we must not forget that, in Hegel, the explaining concept is understood as the essence of the explained thing (here the particular object). From this it follows that, in order to become concrete, the universal as a species is in need of the particular ; for a concept that was not a concept of a particular example would be reduced to a mere *Gedankending* , an empty 'being of reason'. Consequently the particular subsists, **not overcome**, for it proves indispensable to the existence of the universal.

In this way the relationship has been reversed, and now the intelligible universal proves to be dependent upon the opaque particular[77].

Or, generally speaking, in every human experience the presence of the sensible element can be discerned, as the milieu of realization of the universal.

Hence the following surprising results. On the one hand, the particular is **necessary** to the activity of thinking that finds in it its realization, its *Dasein* ; indeed, thought attains self-knowledge through its determined object alone[78]. On the other hand, as a sensible given, that very particular **hinders**, since it is an opaque and contingent residue : it is the 'this, here and now' that may be pointed out with a finger, but not expressed in discursive language. In principle, therefore, the unity of subject and object is not achieved in human experience ; and we always have to deal with a non entirely explicated object, that object having been interiorized in part only.

Moreover, on the side of the subject we can meet a non expressed inner, for the particular 'I' proves richer than the universal concept that is supposed to grasp and reveal him[79]. Hegel, for his part, exhibits **the concrete universal as the genus that thoroughly particularizes itself into its species.** However, this thought is formalistic to the extent that it overlooks the example, the particular sensible given that proved indispensable. Yet, as we saw, **the particular that enables the realization is also the one that hinders it**. Consequently, in human experience, beside the contentless and still abstract universal stands the independent sensible particular. And, in addition to a certain identity of the opposites (as formed matter or materialized form), we can also discern their **coexistence**—or again the identity and difference of the opposites.

The same problem also appears in the form of the opposition existing between **sensation** and **thought**. Hegel claims that thought 'comprehends' sensation and that, of course, with no essential loss. Nevertheless, in his exhibition of the 'psychological' movement from intuition to representation and thought, by 'content of the sensibility' Hegel means the feelings of spirit structured by the forms of sensibility, space and time. As for the irreducible matter of sensibility, he does not mention it at all[80]. On that account, at this point 'thought comprehends sensation' means that the sensible given is overlooked: it is not 'put up' but rather 'put aside', put apart.

Hegel occasionally expresses the same idea ('thought comprehends sensation') by affirming that thought contains its other within itself. Let us call that other 'outer'. It may be noticed that the word 'outer' has two different significations : 'outer' signifies either the general category of the

Logic opposed to 'inner'—after the **meaning** of the word—or the external thing **denoted** by that word. It is thus possible to understand the other of thought in these two significations. Namely, that other is itself either internal, a determination that is intelligible to thought (as is 'outer'), or an external other, the non thought 'this, here and now' intended only. That other, with which the *Phenomenology* opens, cannot be the category of Being with which the *Logic* opens—since it is said to be "non true, non rational" (while *Logic* deals with truth alone). Thus, the universal discursive language does not express the difference existing between the philosophical thought of the category and the corresponding concrete experience of the external sensation—yet, that difference should be 'comprehended'. In this very meaning, Mure suggests to talk about the non explainable **'sensuous-empirical'**[81].

Within human experience, the presence of the sensuous-empirical element sets a univocal limit to the dialectical approach, which draws its concepts from the empirical object, without its being able to claim that 'no essential is lost'. One dimension, which is essential in my opinion, remains irreducible to thought, in that it can be seized by direct perception alone : in the face of the wealth of the immediate **lived**, the 'divine' **Concept** is powerless[82]... On the side of the object, that 'non digested' opaque element appears in the form of leaps or 'accidents' that are necessary to the dialectical movement ; on the side of the subject, it appears within the thinker himself engaged in a lived experience conditioning him—the 'absolute' philosophy proves subjective too[83]...

4. HEGEL'S STRATEGY IN THE FACE OF THE IRREDUCIBLE SENSIBLE GIVEN; A PROBLEM OF KNOWLEDGE MUST BE SOLVED

As has been observed, the sensible given constitutes the Achilles' heel of all idealistic philosophy wanting, as Hegel says, "to think that which is". In order to overcome this obstacle, Hegel has recourse to different strategies, depending on the moment and the context.

First, sometimes Hegel purely and simply decides to consider that **irreducible given as nothing, as not-being!** To do so, he uses two opposite approaches, one indirect and one direct. Following the **indirect** approach—which I discuss first—Hegel declares that to be is to be thought, that **Being is Thought**[84]. Consequently, as non seizable by universal thought, the particular and sensible given comes to be denied all real existence.

From the standpoint of classical philosophy, that stratagem is understandable[85]. But when the issue is Hegel's system, which claims to expose the "universality which includes within itself the particular", then it becomes problematic. What 'particular' is he speaking about ? Like Aristotle in his theory of science, it seems that Hegel deals essentially with genera and species[86]. As for the particular individual, in his view it is only a contingent given, and therefore negligible. Indeed, the particular, as a universal and thought determination, is very present in Hegel's philosophy ; but as an external sensible thing apprehended by empirical intuition, this is not the case[87]. So, in this sense, **to be is to be for thought**, and the sensible, immediately lived experience, would prove to be of no importance...

That **tyranny of the universal** over the particular seems to me rather undialectical. Hegel often repeats that man differs from animals in that he is a thinking being. This is true. But man is also a natural being—as the originator of the 'concrete' philosophy, Hegel should not forget that. Does the Concept have hands to write? a brain to think? Hegel knows quite well that it does not, and his notion of **mediation** appears to be most fruitful. As a matter of fact, there is no philosophy without the mediation of philosophers, and no disembodied philosophers. When Hegel's soul and body separated from each other, they both disappeared for us and Hegel ceased philosophizing and writing. In that sense, Mure's 'empirical-sensible' proves essential. Like everyone, the philosopher is composed of matter and spirit. I therefore think that, instead of talking about an entire subordination of one dimension to the other, supporting their **co-existence** is more adequate. Thus, if need be, Hegel suddenly remembers the great discovery of the modern age, namely the **principle of subjectivity**[88].

A reality in which the Absolute, the Idea and the Concept alone would be subjects might perhaps constitute a world, but certainly not ours—a fact that all so-called concrete philosophy should know. It seems that Hegel, Greek and modern at once, failed to reach a coherent view on that issue. Sometimes, he is too Aristotelian and thus pushes away the sensible given as "non true, non rational". At other times, he proclaims the universal value of the particular individual who, in his action, is entitled to expect some satisfaction.

Whence my conclusion : **theoretically** the particular is sacrificed, for man would feel 'at home', *bei sich*, in universal thinking alone ; but, **practically**, the particular is in principle recognized and, in this respect, a certain balance is established. As for the realization of that principle, Hegel appears to be still traditional, still Aristotelian, since he advocates the primacy of the universal in the State. Does that wavering stem from the fact that Hegel had not yet interiorized the new principle of subjectivity? Or was the time not sufficiently ripe?

The second way of denying the real existence of the sensible given, consists in showing its non-being **directly** : here, Hegel claims that the sensible given is not, since it proves **ineffable** in discursive language[89]. As a matter of fact—except the proper name for which the use can only remain restricted—the language expresses solely the universal or the particular as universal, for instance as species. As for the atomic impression, it resists any immediate expression by scientific discourse. That is, for Hegel **to be is also to be communicated or communicable**.

However, it may be asked, are there no other languages? Cannot **art**, for example, work as a medium through which the particular may find a certain expression? Does this not explain its success with the general public, who sometimes finds it difficult to recognize itself in the universal institution? If so, art may be said to represent a refuge, a solace and a vehicle for the self-expression of the alienated individual[90]. A similar idea is suggested by Heidegger, who saw in the artist, and especially in the poet, the true philosopher, the ontologist *par excellence* who unveils, in his own language, a dimension of Being that is forgotten by scientific thinking[91].

Likewise, might not the particular individual express himself in direct or indirect, normative or deviant **action**[92]? Does not his action constitute the concrete support of 'Objective Spirit'? Therefore, even if expressibility is accepted as a criterion for existence, in my opinion discursive language should not be the unique decisive touchstone. There are other languages, and a concrete philosopher ought to know that.

Hegel's second strategy consists in **reducing** the sensible given **to the category of quantity**. In this connection, I wish to mention a certain **Hegelian formalism**. Indeed, while fighting against the formalistic way of thinking, Hegel also abstracts from an essential dimension of human reality, sometimes attempting to reduce the sensated quality into a thought category. In particular, in grasping **the given as posited**, he seems to interpret the Kantian expression of 'the multiplicity of given intuition' as mere 'multiplicity', finding thus in that determination a quantitative category. In this way, Hegel describes the sensible particular as 'Something' that is 'Various', or as a 'Particular, External and Accidental Individual', all of which categories express some characteristics of the purely quantitative determination[93]: in this Hegelian doctrine of knowledge, Hume's 'impression' has disappeared, and only Leibniz's 'expression' remains.

In this sense, maybe that Hegel saw himself as 'elevating' sensation to the level of thought, this latter preserving the concrete experience in its essentiality—I am of course referring to the act of ***Aufhebung***. However, as Mure has shown, this trial ends in failure. Or, in Hegel's own terms, the 'impotence of Nature' turns into the 'impotence of Reason'—that is, of Hegel himself[94].

In this respect, Hegel keeps from the lived experience only the **forms** of sensibility, space and time, at first conceived as determinations of spirit. As for the **matter** of sensibility, in contrast it disperses like smoke. Only traces of it remain, as quantitative categories.

This brings us to the third strategy. After claiming that the sensible given is nothing but a negligible quantity, Hegel reintroduces it surreptitiously when needed ; hence the different **leaps**, acknowledged in any serious account of Hegel's dialectic. In so doing, Hegel shows that he neither 'knows what he is saying'[95]...

The concrete Hegel rightly restores the sensible matter that disappeared for a while. However, the dialectical Hegel cannot convince us that he is still dealing with the 'immanent and necessary' movement of the Concept, a movement he claims to reproduce in the form of an argued discourse. As Hegel's critics, we admire him, but we are not ready to blindly accept his pretensions. We may distinguish Hegel's various tactics and understand them, but we cannot remain silent for the truth must be told : between a 'flowing' but abstract dialectical account and a concrete but problematic one riddled with gaps and leaps, Hegel has opted and opted well. By this act he proves to be a good dialectician : not so much in that he defends the univocal primacy of the universal, but rather in that he puts forward that of the lived experience before which every thinker must bend. For, in the face of principles, reality is always right, even at the price of a certain lack of coherence.

Some people will say that one must be quite foolish to deny this obvious fact : the real existence of the sensible given. Yet, despite its imprudence, Hegel may be vindicated in the following terms.

Hegel, after Kant, brought into serious question the possibility of knowledge, say, **how can a subject know an object different from himself**[96]**?** On thinking it over it seems impossible, in so far as an unbridgeable gap separates the two entities.

Being aware of this problem, the philosophical tradition has proposed a number of solutions. One is **sceptical** : for man, as a limited being, knowledge remains out of reach. It is quite easy to show that this attitude is self-contradictory, since the very awareness of a problem of knowledge implies some knowledge already : if man really did not know anything, he would even not know that there is no human knowledge. Neither does the **dualistic** solution give satisfaction, for it accepts the fact of knowledge without being able to explain it : after having separated the two poles of knowledge (subject/object), it no longer manages to reunite them[97]. Yet, we want to understand.

Let us then consider the reductionistic way : in that case, the unbridgeable gap is not filled at all but is purely and simply denied, and the Other is reduced to a Same. Thus, the **subjective idealist** reduces any object to the subject's thinking, and the trick is played! As for the **objectivist,** he holds that thought is composed of chemical reactions or electromagnetic fields—hence the suppression of the subject as subject.

Hegel, for his part, is sometimes **idealistic**, such as when he asserts that Being is Thought. Sometimes he is both **idealistic and realistic**, such as when he suddenly reintroduces the sensible given he had previously set aside. Indeed, the first and third moments of dialectic may be regarded as idealistic (at these levels, there is no real other) ; the second one is realistic, inasmuch as it posits the moment of the difference between the developing determinations[98]. Of course a critic, who first of all asks for coherence, will reject such a hybrid solution. However, when considering the various available paths, a serious thinker will perhaps content himself with that practical though inelegant response, refusing to sacrifice the richness of concrete experience on behalf of the non-contradiction principle.

Now, the point is not to get out of trouble by a brilliant formula, but to answer a pressing question : how is human knowledge possible, if a gap separates the subject from his object? As for me, I do accept the ambiguous solution presenting Hegel as idealistic and realistic at the same time. If one adopts Vico's thesis, which asserts that man is able to know what he has created himself[99], then a certain knowledge becomes possible and the Kantian path, open to us ; that is : knowledge is to be considered as a self-constituting activity.

As for matter proper structured by knowledge, in principle it would remain unknowable—we thus encounter again the Kantian problem of the thing-in-itself. Rejecting the ghost of that thing-in-itself, Hegel wavers between abstract idealism—which considers the sensible given as not really existing—and the double way, which presents both spirit and matter as real. It seems to me that any *Logos* that is faithful to human experience, has to do justice to these two principles, reducible in part only.

Hegel is an idealist at the beginning of his central works, where he posits Being as Thought, as well as at their end, where the Idea and the Absolute are said to have overcome all otherness. However, in the 'middle' of these works, he appears to have an ambiguous attitude, advocating thus the **principle of the identity and difference of the subject and object.** In that sense, knowledge is understood as possible at the idealistic moment of identity, and as not possible at the realistic moment of difference. That is, knowledge is possible up to a certain point, demanding an unceasing double movement of identification and differentiation, to and fro between the subject and his object. Thus, matter proper cannot be moved but just restored by a non justified leap. And concrete dialectic comes to be the 'bringing together' of an abstract conceptual dialectic with a jerky, empirical reconstruction.

In this way, an **indirect path** to the knowledge of total human experience comes to light, the direct conception succeeding in grasping the absolute other—the sensible given—at the cost only of reducing it to the Same, the thought category.

Using these diverse strategies—while advocating ambiguity to the detriment of strict consistency—the concrete Hegel thus saves the world of phenomena without reducing it to an ephemeral shadow (as far as human reality is concerned). He also rescues philosophy itself from looking like a day-dream—to the extent, of course, that 'grasping together' the idealistic moment of identity and the realistic moment of difference proves possible. To that extent, Hegel somewhat uncovers to man the world of real, lived and thought experience : our specific reality.

CONCLUSIONS: NATURE, BOTH AS EXTERNAL AND AS INNER, PREVENTS HEGEL FROM REALIZING HIS PLAN

Hegel sees reality as the movement of spirit, positing itself as its other *qua* immediate and external nature. However, in so far as it concerns not formal but material nature, this is not the case and it cannot be so. If it were, we would not encounter a 'problem of matter and nature' but a flowing dialectical movement capable of moving any content, of reducing any given to its constitutive determinations, and that with no essential residue. The result would then be an ultimate 'absolute' philosophy.

However, man is both matter and spirit, and his experience is intelligible as well as sensible. Therefore, the other constitutes for him an obstacle not surmounted and, in principle, insurmountable. Hence an endless fight against that other, which both arouses him and limits his activity[100]. For man, then, there will always remain 'work' to be done; and the project of becoming 'Masters and possessors of nature' appears as a never-ending task, constantly dissolving otherness to find it again : like Sisyphus, the philosopher will never stop 'rolling' his system, which is unceasingly on the brink of breaking to bits[101].

Moreover, the **natural element** appears not only outside, in the non entirely explainable object, but also **within the subject**, himself inhabited by an opaque residue—man being not only universal thought but also private imagination, peculiar desire and emotion. Is it legitimate to neglect that dimension of human experience, when the aim is to give a comprehensive account of it? Do we not discern, behind his brilliant and bold argumentations, the individual Hegel with all his anguish, weaknesses and delusions? He who, by manipulating language, intends to convince the world that he has expounded the truth or, rather, that thanks to his work the truth has expounded itself. It seems that this pretension betrays a human being who is 'too human' ; who builds a palace and lives outside, as Kierkegaard observed ironically[102].

Consequently, if **external nature** univocally limits the possibility of developing dialectical movements, in principle **inner nature** seems to be just as a serious obstacle. Hegel asks the thinker 'to forget himself'[103]. By which he means : to overcome his **particular nature** in order to become **universal culture**. Of course, as a reliable researcher, the thinker has to tend to objectivity and impartiality. However, Hegel also knows that, as a 'child of his time', every thinker is characterized by his own personality which cannot fail to be reflected in his work—a fact that many critics have already noted.

In siding with the sceptics in their fight against the dogmatics, shall we leave the last word to matter? to nature? Of course not. For here also it is advisable to hold the golden mean and, in spite of the indicated difficulties, to assert that some explanation of human experience is possible—in this case with the help of the dialectical movement, which is immanent and necessary yet also external and contingent[104]. In other words, the **critical** point of view requires to see the thinker not only as a universal spirit grasping the world in thought, but also as a particular individual who imagines, desires and dreams.

Hegel's philosophy looks indeed like Hegel himself. Nevertheless, through it a certain world comes to light, a human world in which we may recognize ourselves, as least in part. In philosophy, Hegel has not overcome nature, the other of thought ; but neither has he been entirely overcome by it. Up to the present time, nature and culture evolve together, constantly intermingling and influencing each other.

Driven by its 'instinct', Reason seeks itself in its other[105]. It finds itself in it, loses itself, finds and loses itself again and again, in an unceasing movement that never succeed in 'dissolving' that other—despite Hegel's promises, manoeuvres and struggles, the sensible residue is still there! So seems to be human destiny : nature and matter constitute for man an obtacle and yet a necessary condition—like the sensible which, in Plato, makes the knowledge confused and yet awakens it.

PART FIVE

THE ANTINOMY OF LANGUAGE

Hegelian researches are often concerned with Hegel's thought, its nature and movement. However, we must not overlook the fact that this thought expresses itself through natural language, and must do so necessarily. From the outset, then, the **question of the exposition** *(Darstellung)* is raised : is natural language capable of adequately expressing dialectical thought[1]? It seems that it is not, in so far as natural language develops in a cultural environment that is stamped by formalism[2]. Nevertheless, Hegel has detected in that language the 'speculative spirit' at work[3]. There is no reason to wonder about it : if dialectic contains some truth, then this must appear in one way or another, in reality in general and in language in particular[4]. If that is the case, then, long before we deal with it, **dialectical activity** is already present, deconstructing and reconstructing the structures of being, thought and language. And under that condition alone are we able to bring out its movement, so as to form a dialectical mode of philosophizing.

The answer to the question of exposition is therefore twofold, both yes and no. **No,** since natural language is characterized by a discursive structure; **yes,** since speculative spirit is supposed to be already active within it. In that sense, dialectical thinking is possible for, although it expresses itself in a partly foreign medium, it somewhat finds itself there. If silent, dialectic would turn into a non dialectical 'pure thought', since to be dialectical is to appear to an other, to be exoteric, open to all. Or, in other words, a thinking that did not express itself would, at the very most, be a day-dream.

In this Part of the work, I treat of the fight dialectical thinking has against, and with, natural language in order to express itself and thus become real ; or, rather, I treat of the fight Hegel takes up against formalism, a fight where victories and defeats alternate. Eventually, I come to the conclusion that dialectic can neither be satisfied with natural language, which is too formal for it, nor dispense with it, given the absence of any serious alternative—a 'dialectical' situation, so to speak, insurmountable in spite of Hegel's endeavours.

Indeed, through various manipulations of language, Hegel strives to persuade us that the problems have been definitively resolved, and the 'other' entirely 'dissolved'. On this point, however, Hegel does not seem dialectical enough to understand that the 'other'—an obstacle to any dialectical movement—also constitutes a necessary condition of it[5]: without that stumbling-block, what would be the source of the 'negative' drive?

It is that inner tension, or self-contradiction, that comes to light in the elucidation of the problem of language in Hegelian dialectic, mainly in the form of 'The antinomy of language'[6].

1. THE EXPRESSIVIST DOCTRINE OF LANGUAGE

In my opinion, resorting to the expressivist doctrine of language may help to better understand the problematique in question.

Generally speaking, nowadays in the formal theories of language, by 'denotation' one understands the external relation existing between the sign and the denoted thing. Thus, it is said, some marks are used to convey various **pieces of information**, or still: the sign refers to an object existing in the world. That is, a ternary relation is at stake: with the help of **a sign,** such as a word, **I** learn something about **an object**. Language, as a system structured by signs, is considered to be the outcome of an agreement between different speakers, having decided to use that instrument of communication in accordance with well-defined rules. **This nominalistic point of view separates meaning from being,** the intelligible from the sensible.

Consequently, it does not acknowledge the universal as real : indeed, the universal cannot be pointed out with a finger[7]. According to this approach, in front of me stands a world of independent objects, a world that I can describe from the outside, by means of a 'subjective' language produced by my creative activity. As for the status of 'objectivity' conferred to language, it may be explained in so far only as that language is acknowledged and used by everyone in a certain group according to common rules[8].

Of course, Hegel understood things differently and, to the nominalism in vogue, let us in his name oppose the expressivist doctrine of language[9].

In Romantic Germany, Herder, one of the originators of the doctrine, rejected the modern objectivation of the human being, his inner division into independent elements—body and soul, feeling and Understanding—namely, the atomic individualism that takes away from nature and society[10]. Herder used to advocate a so-called **expressivist anthropology**, according to which man expresses his individuality in his action.

This doctrine can be understood as a certain return to Aristotelism, in that life is again conceived as a realization of some end. However, the discussion is then no longer about a 'highest form' to which one should draw as near as possible, but about an activity of self-realization that allows man to recognize himself in the product of his work. In other words, he is now expected to actualize his personal potentialities and, thereby, to grasp his own nature—historically, we are after the Copernican Revolution. In these circumstances **meaning and being become one**, the individual being regarded as expressing himself in his action and his life, which is now understood as a work of art[11]. Thus, Romanticism used to conceive language, and self-expression in general, as media essential to human being. Hence, among other things, **Herder's expressivist doctrine of language.**

According to this new point of view, a word is not only an external sign. It is also the expression of a certain human consciousness or reflection *(Besonnenheit)*, the language being mainly seized as fulfilling an **expressing function**. So that language cannot be separated from being, and it is equally through language that man realizes himself and learns to know himself : language has turned into a moment of a self-developing reality[12].

Whereas some Romanticists saw art, and especially poetry, as the **summit** of self-realization, others sought it in religion—for his part, Hegel proposes philosophical activity. In passing, we can notice that this expressivist approach, partly at least, explains the logical movement of Absolute Spirit, developing successively through art, religion and philosophy[13]. In effect, according to Hegel, philosophy alone—which exhibits its conceptual content in the form of the Concept—allows man to realize himself and reach self-consciousness. Art, of intuitive essence, and religion, perceiving mainly through representations, will not do : they are but preparatory stages[14].

So far, we have seen the explanatory power of the doctrine, regarding man coming back to himself, out of an alienating inner division. Now, the following question inevitably arises : is the human being truly capable of grasping himself 'from the outside' with objective and unbiased eyes, as the empirical method requires? Is this even desirable[15]? There is some doubt about this. As will be shown later, anyhow that 'coming back to oneself' demands a price : **the antinomy of language**.

Let us say that human language expresses a certain content, which may be called concept, thought, idea or spirit—in this Part, I argue mainly about **the proposition expressing Hegelian thought**. Therefore, the conclusion seems immediate : the unity of being with meaning, just recovered, disappears however in the exteriority of the proposition, which is itself composed of various separated elements[16]!

To clear a way towards the elucidation of this problem, first let us see how Hegel understands the nature of judgement.

2. THE HEGELIAN JUDGEMENT AS DIALECTICAL *SATZ* OR DIVIDED *URTEIL*

Some approach Hegel's work through a formal reading only, as if the work at issue were Aristotle's or Kant's. Yet, in composing his *Logic*, Hegel wrote not just a book on dialectic, but rather a book that is itself dialectical. That is why, in my view, things have to be understood differently. In particular, I ask what the Hegelian judgement is, a judgement here regarded as dialectical *Satz* or divided *Urteil.*[17].

2.1. Let us start with the judgement as dialectical *Satz* .

We are concerned with the judgement 'S is P' relating subject and predicate by means of the copula 'to be'[18].

First of all, we have to free ourselves from old habits and understand by 'subject' and 'predicate', not the usual determinations of Aristotelian substance and its attributes, but rather the **very categories of the *Logic***—such as Being and Nothing (which form "Being is Nothing") or Identity and Difference (which form "Identity is Different"). As for the copula, it is supposed to express the relation that develops between the terms of the judgement.

In Hegel, just as the concept is the essence of the thing, so **the logical judgement is the essence of the empirical judgement.** For instance, in the *Logic* the movement of Judgement opens with a judgement of the type "This rose is red"[19]. This judgement can be translated into Aristotelian language by 'This substance supports such an attribute' or, usually speaking, 'To this thing belongs such a property'. However, the Hegelian truth of that empirical judgement is: "The Individual is Universal" (Individual: that existing rose ; Universal: its quality of being red in colour)[20]—the discussion is really about a relation between logical categories.

This is the first characteristic of the Hegelian judgement as dialectical *Satz*; hence the second.

2.2. In the Hegelian judgement, **the predicate expresses the essence of the subject**[21].

In the empirical judgement 'This rose is red', 'red' is an accidental property of the rose, whose colour may also be white or pink. Whereas Hegelian logic is concerned with conceptual judgements containing **inner relations of necessity**, such as the judgements "The Idea is Actuality" and "The Absolute is Being", relating different categories of the *Logic* between them. In this connection, in the Preface to the *Phenomenology* Hegel mentions the 'counter-thrust' *(Gegenstoss)* that occurs in the usual mode of thinking, when it discovers that the predicate alone tells us what the thing is, and that at first the subject of the judgement is only an empty name or, at the very most, a non criticized immediate representation[22].

'Wondering', the non informed reader can learn from this that 'he has to understand things differently'. He used to think that his attention should be turned to the subject of the judgement, whereas the main content now lies on the side of the predicate : thus, says Hegel, in order to grasp the philosophical judgement adequately, **it is suitable to re-think it.**

According to Hegel, it is through this conflicting way that the novice begins to move towards the world of dialectical Ideas.

2.3. Hence a third characteristic : in so far as the predicate expresses the essence of the subject, **the Hegelian judgement has to be understood as identical**[23]. In truth, subject and predicate express the same content with different terms and, essentially, the copula signifies 'identical with'.

Leibniz has already proposed such a doctrine. Indeed, in his *Discourse on Metaphysics* he asserts that, as far as the monad (the individual thing) is concerned, the predicates are all contained in the concept of the thing[24]. Accordingly, in principle, the knowledge of the concept of the thing should allow the deduction of all its predicates. That transition, from ontological language, dealing with the individual thing, to logical language dealing with concept and predicate, is legitimate, given that Leibniz the idealist thinks that the monad **expresses** the whole world from his own point of view—here we can recognize a variant of the expressivist doctrine discussed in chapter 1.

However, while Leibniz still talks about analytical deduction, Hegel discerns there a dialectical movement : between the subject and predicate of the Hegelian judgement, a relation of dialectical identity or **reciprocal mediation** develops, and that through the copula which explicates its content.

The novice should focus his attention on neither the subject alone nor the predicate alone, but grasp the meaning of the judgement by a movement to and fro between them—hence the turning of the 'given' meaning into a produced one.

2.4. In my opinion, this line of argumentation throws some light on the problem created by the **conversion of the Hegelian judgement**.

This act becomes justified, as soon as the Hegelian judgement is conceived to be a reciprocal mediation unfolding between logical categories. Effectively, the relation of dialectical identity existing between the parts of the Hegelian judgement not only allows that conversion (of subject and predicate), but it even requires it. A certain immanent criterion of adequation may thus be brought out : if, in a given context, we do not succeed in converting significantly a dialectical judgement, that means we have to do with some defect. So that we should not be surprised by the fact that Hegel often ends his discussions with the expression *'und umgekehrt'* (and conversely).

In the *Phenomenology*, for example, the concept must correspond to the object, 'and conversely'. In the *Logic*, every determination is Something positing its presupposed Other, 'and conversely' every Other is also Something. In particular, if at the beginning of the movement of Judgement it is said that "The Individual is Universal" (i.e. the examined thing has such a property), it is also said that "The Universal is Individual", i.e. the thing as such is the totality of its properties, and each peculiar property is also an individual that exists in the world *qua* property.

On the other hand, in the ***Realphilosophie***, which displays the genesis of the empirical determination, that act of conversion becomes rare for, Hegel says, the Concept alone is dialectical[25]; namely, the logical movement is so and not the empirical reality. As for the novice, we saw that the suffered counter-thrust forces him 'to understand things differently', to understand that if the judgement reads 'S is P' it also reads 'P is S', the predicate giving expression to the subject.

The act of conversion thus brings to light the **movement of the Concept in its double direction**, explaining why Becoming is apprehended as a passage from Nothing to Being 'and conversely' ; that is, "The way up and down are one and the same" (Heraclitus). That dialectical vision of things requires the 'grasping together' of the universal particularizing itself and of the particular realizing itself, of the Concept or thought determining experience and of experience interiorizing itself as Concept or thought. Moreover, **these** different double movements are supposed to produce thought, at the subjective pole of human experience, and the 'given' external thing at its objective pole. Consequently, the Hegelian presentation of reality as Judgement, *Ur-teil* (division), becomes somewhat intelligible, for it means '**original division**' between a thing and its concept, between experience and the thinking of experience.

Of course, in the empirical judgement the law is different, the predicate there being an accident of the subject, and that in conformity with the adopted linear mode of thought: the substance is said 'to support' the attribute, but not the inverse. A conversion is thus practicable only in accordance with the strict rules established by Aristotle, and reconsidered by modern logic—when criticizing Hegel, formalistic thinking sometimes seems to overlook that difference[26]. Thus, according to Hegel, when inverting the above examples they become: "Actuality is Idea" and "Being is Absolute". Likewise, Hegel's famous aphorism may be interpreted according to its double movement (to and fro between 'rational' and 'actual'): "What is rational is actual, and what is actual is rational"—although in this case the example does not deal exclusively with logical categories, for the term 'reason' (which gives 'rational') belongs to the *Phenomenology* .[27]

2.5. Of course, the Hegelian judgement as dialectical *Satz* is not the proposition of formal logic

Generally speaking, formal logic deals with the rules of formation and transformation of various expressions, while abstraction is made from their content—whence its name.

In dialectic, a non formal logic, on the other hand terms such as the copula 'to be' or judgements such as 'The individual is Universal' develop their content without altering their form. In particular, the successive meanings of the copula, according to the movement of Judgement, are : being, inherence, subsumption and identity with no difference. Hegel distinguishes well between them—distinguishes, but without separating. That is, **instead of structural transformations we find there a development of content**.

Under these conditions, it seems that any attempt to formalize dialectic must necessarily lead to some impoverishment. Is it worth the trouble, solely in order to explicate the Hegelian logic somewhat? Indeed, if the movement of content proves here to be essential, how might a formal logic apprehend it[28]?

Hegel has warned us : the activity of the living and creative spirit is not reducible to a rigid system of univocal signs[29].

Let us suppose that a child and an adult consider some sentence. Although they both address themselves to the same passage, each of them makes a different reading of it, for the content, abstract for the child, appears to the adult as a rich exposition of human experience. Is not human experience like this : the explication of the present, out of deep reminiscences? At least Hegel thinks so, in my opinion[30]. And, as we saw, it is the same with philosophical judgements, which are really understood only when re-thought anew.

2.6. We proceed now to the judgement as divided ***Urteil*** .

This is concerned, first of all, with the empirical judgement, 'divided' in the sense that, from a semantic point of view, subject and predicate are regarded as independent. Its investigation shows very well that, in the *Logic*, **we have to do with a dialectical logic developing according to an increasing degree of truth, and not with a bivalent formal logic.**

Indeed, if we try to follow the movement of the judgement as divided *Urteil*, from positive judgement up to disjunctive judgement, then we can perceive that the movement does not develop from falsehood to truth, but from 'less true' judgements to 'truer' ones ; and that the degree of truth attained may be determined by the corresponding meaning of the copula, which progresses from 'being' to 'identity with no difference'[31].

For instance, the positive empirical judgement 'This rose is red' is characterized by a low degree of truth : it is only correct *(richtig)* when it corresponds to the external thing it describes. In effect, the rose has still other properties and, on the other hand, still other things in the world are red. That is why the relation between subject and predicate is said to be external, abstract, **accidental**—the beginning of the movement.

Later on, in the movement of the positive judgement, the meaning of the copula grows richer, from **simple being** (the rose is) to **inherence** (the property 'red' is inherent in the thing 'this rose'). The reflexive empirical judgements already express a higher truth, for the predicate here is a **necessary accident** of the subject, such as in the judgement 'Man is mortal' : although other beings are mortal too, that attribute is intimately attached to human nature. We have here a relation of **subsumption** holding between genus and species, indicated by the copula 'is'.

Next, in the disjunctive empirical judgement 'S is P_1 or P_2', the Hegelian judgement is supposed to be accomplished in the sense that, now, the genus thoroughly particularizes itself into its various species—for instance, any human being is necessarily either male or female. The corresponding meaning of the copula is then : identity with no difference. To be sure, the subject realizes itself through the mediation of its different predicates, but that relation lacks some 'and conversely'. Here the predicates, still natural, are 'found' and are not dialectically realized.

Consequently, the judgement passes on to the following movement.

TRANSITION TO THE ANTINOMY OF LANGUAGE: ALL IS *URTEIL*, DIVISION

The 'fulfilment' of the movement of Judgement as divided *Urteil*, in its double meaning of fulfilling (action) and being fulfilled (result), is reached with the judgement of the concept structured by the following form : 'A, as B, is adequate'[32]. This judgement expresses the adequacy (or non adequacy) of the judged thing A to its concept, while B constitutes the ground of the act of judging—a syllogism already appears here, that is, a logical relation connecting three terms with one another : the thing A, the ground B and the concept of the thing.

In order to understand this new movement, we should observe that when we utter the sound 'table', we are saying three different things : a word, a concept and an external object. In this connection, I wish to mention **the double subject of the Hegelian judgement**, its grammatical subject being grasped at one time as denoting an existing immediate individual, and at another time as expressing the self-particularization of the concept into its different determinations[33].

Above, on p.113, I presented the judgement, *Ur-teil*, as an 'original division'—the empirical thing is in question. That is, Hegel presents the partition of the thing *(die Sache)* into Being *(Sein)*, the immediate individual, and into concept or essence of the thing, its Ought-to-be *(Sollen)* ; from an empirical standpoint, meaning and being are really separated. In this perspective, therefore, judging is first of all comparing a thing with its concept and appraising the agreement or disagreement between the two.

In that sense, a realized thing is a thing adequate to its concept, that is a self-constituting spiritual (non natural) subject ; whereas the empirical, finite thing is characterized by the splitting, the inadequacy to its concept—hence its ineluctable destruction.

As an example of the apodeictic judgement of the concept, we can read : "The action, if of such and such a character, is just". This judgement asserts the adequation of the thing ("The action" as a being) to its concept ("action" as what ought to be), while the copula ("if of such and such a character") has become the ground of the act of judging—the thing is really understood both as an existing thing and as a concept to be realized. This value judgement states that the thing, formed in such a manner, corresponds to its concept, and conversely[34].

Since the judgement of the concept is itself an *Ur-teil*, that is, a division into external terms, the movement must proceed further.

To sum up, the discussion concerns a development **from correct empirical appraisal** ***(Richtigkeit)*** **to conceptual truth** ***(Wahrheit)***; or **from** the accidental and external relation existing at first between the parts of a judgement as divided (empirical) ***Urteil*****,** **towards** the necessary movement of a subject which, in the dialectical (logical) ***Satz***, realizes itself through the mediation of the different predicates it posited.

We are therefore obliged to observe that the predicative form of the judgement proves unable to express dialectical truth, whatever the mentioned level may be. For if 'All is *Ur-teil* ',division, the rational real is first of all a syllogism, namely, a coming back to mediated unity; hence, in the *Logic*, the transition to the movement of the Syllogism.

We began with the exposition of the nature of Hegel's judgement as dialectical *Satz*, and ended with the discovery of the inadequacy of Hegel's judgement as *Urteil*, i.e. as division between subject and predicate, thing and concept of the thing. In the following chapter I expound the 'antinomy' in question.

3. THE ANTINOMY OF LANGUAGE

In Hegel, then, the judgement as divided *Urteil* proves unfit to express the movement of the dialectical *Satz*.[35]

Let us examine more closely the point at issue.

Hegel often mentions the 'movement of the Concept' or the movement of 'Thought thinking itself'. However, as critical readers, we cannot overlook the mediator, the language without which thought would merely be that "night in which... all cows are black". Namely, as we saw it, dialectical thinking realizes itself through the mediation of non dialectical discursive language ; or still, Reason can reconstruct its original unity by means only of a product of Understanding, the discursive language[36]. In other words, the spiritual essence (as man) is living and creating activity through the sensible, spoken or written, phenomenon alone[37]. Mure concludes from this that, inevitably, the 'high' cultural activity depends on the 'low' natural language[38].

If the *Logic* shows us, more or less successfully, what that 'movement of the Concept' is, we have now to turn our attention to the linguistic dimension of reality.

As an educated being, man expresses himself, conveys his thought and accounts for his action. To this end, he makes use of an articulated language. Generally speaking, discourse is opposed to intuition in this that intuition is not directly communicable. At the level of intuition, indeed, man is unable either to realize whether he has really understood the matter at stake, or to learn and to teach, but he can only refer himself to some 'vision'—a non verifiable one.

Hence the following question : how can a delusion be distinguished from a true vision? Certainly not with another vision! The answer lies then in the discourse structured with concepts and words, allowing the comparison and estimation of the various sensible or representative intuitions. Seen in this way, discourse does not replace intuition but complements it. No wonder that Hegel seems to praise the creative power of the Understanding, "the most astonishing and mightiest power"[39]—a power without which our culture would consist only of a complex of orders and taboos sanctified by mere habit, violence and/or seduction. For, with discursive language, the possibility of criticism comes to light, and a basic antinomy too.

First, let us see in what terms Hegel presents this 'antinomy of language'.

In the Preface to the *Phenomenology*, Hegel talks about the immediate proposition, which he opposes to the speculative one[40]. By 'speculative proposition' *(spekulative Satz)* one may understand the judgement as dialectical *Satz* , the topic discussed in the previous chapter. The question is about the logical categorial judgement, wherein the parts stand in a relation of dialectical identity or reciprocal mediation between them, the subject realizing itself through the mediation of its different predicates. All this is well-known, and sends us back to the works of diverse researchers who endeavour to think the 'pure' Concept.

Yet, the thought judgement also has to be expressed in a medium alien to pure thinking, i.e. in discursive language : conceiving is not enough, for arguing is necessary as well. In this way, the conceptual inner comes to be 'thrown' into the spatio-temporal outer : essentially, we speak in time and write in space[41].

In this connection, Hegel treats of the immediate proposition *(der unmittelbare Satz)* 'S is P', which relates a subject to its predicate externally. By 'external relation' we must understand a 'non constitutive relation'. That is, it is supposed that the parts of the proposition make sense by themselves, independently of any relation to their other. Hence the reappearance of the judgement as divided *Urteil* : we have 'fallen' from dialectical concepts to empirical representations[42]. In other words, the discussion is now about **representative thinking**, which expects to find 'ready-made' determinations that it might use as inert things, for example in describing from outside some encountered 'given'[43]. Hegel points out that this thinking apprehends the subject of the proposition as a fixed point, to which the predicate is externally connected. Whereas, according to Hegel, the **conceptual thought** knows that truth is not a univocal value, an intended or described substance. It is a dynamical subject, namely the very movement of the speculative proposition forming the dialectical syllogism, which is still to be developed.

The antinomy lies there: to the intelligible essence of the speculative proposition is opposed the sensible appearance of the immediate proposition ; or, namely, the movement of the conceptual thought realizes itself by means of the rigid support of representative thinking.

Indeed, when I say **'S is P'** I say two different things : on the one hand, the question is about the self-realization of a living subject through the mediation of its predicates—what is called 'dialectic' ; whereas, on the other hand, the non informed reader understands that a certain attribute P is said of the substance S, the fixed and independent basis of the proposition—this is the usual standpoint of the empirical representation.

The antinomy opposing sensible and intelligible, representation and concept, is thus posited. But what is the point exactly? It is this : **the conceptual speculative proposition becomes real only through the mediation of the sensible immediate proposition**. That is, the divided judgement constitutes the way through which the movement of the Concept realizes itself. Therefore, the Concept has to split up in order to come back to itself, identical and different, dialectically united with its other and thus producing the syllogism—that is the truth of the development. As the interiorization of the linguistic and the real determinations of the surrounding cultural world, concrete thought is able to form itself at that cost alone. So that we can again perceive the **double nature of the Hegelian Concept** : it is both an activity of interiorization as conceptual thought, and an activity of exteriorization as representative thinking (and work).

However, the **double nature of the speculative proposition** is also to be noted : by 'speculative proposition' we can understand either the abstract **ideel** dialectical essence presented in the previous chapter, or the tension developing between the speculative form of the proposition and its immediate form—the double meaning is really unavoidable and comes to complicate the task of exposition.

Now, the truth may finally be expressed : **actually,** by 'dialectic' we must understand both the identity and difference of the opposites. That is, the tense speculative proposition comprehends within itself its own opposite, the immediate proposition that constitutes an obstacle to it... and yet proves necessary[44].

To put it another way, language is the other of thought, non conquered and unconquerable, as will soon be seen. In this sense I suggest to talk of the **identity and difference of thought and language**, which actualize themselves through the mediation of each other[45].

In view of these arguments, it could be added that language, as a 'necessary' moment, will itself be overcome. On paper, making such promises is easy. But, in fact, does the Concept actually succeed in 'overcoming' nature, does thought manage to assimilate matter?

In Part Four of this book, I showed that this does not occur : matter and nature persist as irreducible others[46]. In truth, the antinomy of language has no solution, if not the awareness of a basic issue—is **that** not the dialectical solution to a fundamental problem? Really, the tension that seemingly hinders the 'flowing' movement of dialectic proves vital. For in this way alone does the living dialectical thought find the obstacle that resists it and forces it to be 'tense', and thereby to become other : from this very tension springs the dialectical movement itself, in pursuit of signification and concretude[47].

My argumentation so far has been rather abstract. So I propose now to elucidate the point at issue with the help of a particular example, drawn from the *Logic*.

4. AN ILLUSTRATION: THE PROBLEM OF LANGUAGE AS EXPOSED IN THE DIALECTIC OF BEING, NOTHING AND BECOMING

In the *Logic*, in the second note to the dialectic of Being, Nothing and Becoming, Hegel presents the problem of language as it comes to light at that level[48]. I deal with this problem in the form of six main theses, each of which is explained and justified.

Thesis 1: The exposition of the dialectical movement in the medium of discursive language leads to an antinomy

At the beginning of the *Logic*, Hegel talks of Being as Nothing, for Being is the most general determination and therefore the most devoid of meaning. Conversely, he talks of Nothing as Being, as empty intuition or empty thinking : Nothing is, in the sense of privation, of potential of the development to come, and not in the sense of negation[49]. Thus he concludes that **"Being and Nothing are the same thing"**.

However, Hegel adds, if by 'the same thing' one understands the predicate of the above-mentioned proposition, whose subject is 'Being and Nothing', then the exposition of the matter proves inadequate : indeed, the speculative proposition is not the predicative one.

In truth, what is the problem? On the one hand, the proposition contains two different determinations, 'Being' and 'Nothing'; and, on the other, it states that these different determinations are identical[50].

Formalistic thinking does not encounter this sort of problem, in so far as it abstracts from experience and studies only forms and structures. Insensitive to the present difference, it focuses its attention on the affirmed identity.

In contrast, in dialectic the concrete linguistic experience is considered and explicated, hence the antinomy : from the standpoint of form we can discern an identical proposition of the type 'A and B are the same thing' but, from the standpoint of content, we distinguish two different determinations, 'Being' and 'Nothing'. A tension therefore develops between what that proposition says—**identity**—and what it really is—**differentiated** within itself.

Hegel holds that Reason *aufhebt* (transcends) Understanding. Nevertheless, to do so Reason needs a product of Understanding, the predicative proposition with which we are dealing : the transcending of Understanding is accomplished with the help of Understanding itself ; or still, the cancelling act of transcending is also an act of preserving.[51]

Thesis 2: The predicative proposition seems unfit to express dialectical truth

In the examined proposition a tension has come to light, opposing form and content, syntax and semantics : as *Ur-teil* or division of the Concept, the predicative proposition seems inappropriate to express dialectical movement. This proposition effectively affirms the identity of two determinations, which differ as to their proper meanings. Or it says that 'Being' and 'Nothing' are the same, while exhibiting them as different: any educated reader knows very well that these words are distinguished by their different, if not opposed, lexical values.

Formalistic thinking makes separations and considers only forms, so that it does not stumble over that antinomy. On the other hand, dialectical thought does not abstract but treats of totalities, thus apprehending together form and content, pronounced identity and actual difference. From this point of view, **difference forms an essential moment of logic, as well as identity**—hence the 'tense' relation between the opposites, constituting the source of movement and, Hegel says, the principle of all spiritual life[52]! Therefore, dialectical movement presupposes the identity and difference of the determinations 'Being' and 'Nothing', which are identical as indeterminate concepts but different as known representations *(Vorstellungen).*

Dialectically speaking, without the moment of difference we would not have to do with a living and creating spirit, but with the Parmenidean empty One. Discursive language, with its fixed terms, its articulated and relatively rigid syntax, is thus an essential moment of the self-realizing thought, essential and yet alienating—isn't this the basic structure of a dialectical paradox?

Hegel holds that the category develops through the mediation of its opposite, and is identical to and different from it. However, this creative tension expresses itself in the very medium of language, which represents the ethereal movement of the Concept empirically.

Dialectical thought thus accomplishes itself through the mediation of its other, here the language it oversteps and preserves at the same time : dialectical logic, said to be produced by the opposition between categories, is in fact produced by the opposition existing between 'pure' concepts and empirical representations[53]—the conflict opposing thought to matter, Reason to Understanding, inner to outer, proves constitutive of the dialectical movement.

In truth, if spirit had not appeared in history, then it would not have been what it is, a creative and living spirit. Likewise, thought is compelled to realize itself through discursive language, on which it depends and which it does not manage to overcome. At the 'apex' of the System we can again perceive the presence of the difference, in the form of empirical representations succeeding in time, without which we would not find a realized thinking but just a thought devoid of any concrete meaning. That is, the Hegelian Concept does not free itself from the **ladder** that has allowed its self-constitution : the negated and yet preserved representation[54].

We have discovered the essentiality of the moment of difference, and it must be exposed.

Thesis 3: Therefore, in dialectic we have to complement the positive proposition by the corresponding negative one, and to grasp the opposites together

The problem results from the fact that the examined proposition, while affirming the identity of the two determinations 'Being' and 'Nothing', actually contains them as different[55].

The proposition is thus required to express its animating tension : the moment of difference has to be posited. This time we may write **"Being and Nothing are not the same thing"**. This new outcome is not surprising since the movement of Judgement, exposed in the *Logic*, presented the positive judgement as developing into the corresponding negative judgement[56]. Even the *Phenomenology*, at the end (*Phän.* 542-543 (in English: 473)), acknowledges the ambivalence of the copula 'is' which, in the plural, means at once "are" *(sind)* and "are not" *(nicht sind)* .

Here, we encounter a known dialectical principle, according to which the positive is as much negative as it is positive, and conversely[57]. Consequently, the positive proposition "Being and Nothing are the same thing" has to be complemented with the negative proposition "Being and Nothing are not the same thing". In this way, the movement of completion allows the explicit exposition of the moment of inner difference, following that of the pronounced identity. As concerns the dialectical proposition, then, the identity and difference of its subject and its predicate may be mentioned—the antinomy has become manifest. More precisely, we can now discern a development, leading the original proposition "Being and Nothing are the same thing" to the explication of its material content, which is "Being and Nothing are not the same thing". We can say that the second proposition, bringing to light the conflict moving the first proposition, constitutes its 'truth'.

Hence the following formulation : **"Being and Nothing are identical and different"**. In this paradox lies the moment called by Hegel **"The union... only a restlessness between incompatible terms"** *(eine Vereinigung... nur als eine Unruhe zugleich Unverträglicher)* [58].

What does Hegel mean by that astonishing expression? I understand that the union is said to be "incompatible" in so far as its parts exclude one another. In formal logic, referring to the principle of non-contradiction, one would assert that 'A and B cannot at once be identical and non-identical in the same respect'. That union proves "restless" (not quiet), for the induced tension seeks some relief[59]: the third moment alone, a 'unity of opposites' realized as 'the same respect', is likely to bring the expected reconciliation.

In this way, through their dialectical identity, Being and Nothing therefore yield Becoming.

Thesis 4: In dialectic, every determination has to be understood with regard to its constitutive other

We can now better understand the determination of becoming, the outcome of the examined dialectical movement.

Becoming is not the abstract unity of Being and Nothing, but rather their 'restless', living unity ; hence the dynamic determinations of Coming to be and Passing away[60]. Therefore, in the double movement at stake the 'incompatible' opposites are supposed to find their solution. The achieved result, a new determination, implies that Being and Nothing are both identical as indeterminate determinations, and different in the usual meaning of the terms : 'full being' and 'empty non being', respectively[61]. This is why these opposites can truly be grasped with regard to this result only, that is, as moments of Becoming—here too one really understands at the end only. In Hegelian language we can say that Becoming is 'their truth' and that, as isolated, they are just abstract and fixed representations, for "The True is the whole".

From this discussion a **dialectical doctrine of signification** may be derived : every determination has to be grasped in relation to its other, its constituting term.

Thus, Being may first of all be explicated as the dialectical opposite of Nothing : Being is and is not Nothing, i.e. it is at once empty and full. Next, its other is also Becoming, which presents Being both as dynamic and as static (Being is and is not Becoming). Likewise, the Reason's concept of Infinite is the 'truth' of the concept of Finite, the two concepts having to be grasped one with regard to the other : Finite is Infinite in that it contains within itself the drive to pass beyond any limit, and Reason's Infinite—a circle of self-mediation—is Finite as determinate (it has a term). The concept of One has to be understood as the attraction of the 'Many' among themselves, and the concept of Many as a self-repulsion of the One, these opposites forming together the Quantity, One and Many at once, etc.

At every level of development, the categories draw their concrete meanings from their being placed in reciprocal relations ; whereas separated from their constitutive other, they are reduced to the usual representations.

It is the same for the whole of the *Logic*. For instance, the Doctrine of Being can be grasped adequately with regard to the Doctrine of

Essence only, from whose standpoint it develops ; and, in its turn, the Doctrine of Essence is intelligible only as a **reconstruction** of the determinations of the Doctrine of Being *qua* Appearance, Existence, Phenomenon, etc. By their reciprocal mediation, Being and Essence are supposed to produce the Doctrine of the Concept, which is the total context that explains them both.

The same holds for the *Encyclopaedia*, where the logical Idea is understood as an abstraction of concrete reality, and reality as a determination of the logical Idea, the creative thought. The totality is here comprehended both as human and as divine spirit, dividing itself into logical thinking and cultural product, and then, when accomplished, returning to its original unity. But **the converse is also true** : to us, particular individuals, with regard to our determining reflection, the totality, 'context of all contexts', itself appears as a certain determination, if any[62].

The dialectical doctrine of signification thus leads to the following paradox : the totality explains the part constituting it and, conversely, it is explained by the very part, its other. Thus, in so far as Hegel's work may be considered as forming a system, a certain activity of reciprocal mediation manifests itself at all its levels[63].

Thesis 5: Moreover, to the positive expression 'unity' should be added the negative one 'inseparability'

Hegel thinks that terms such as 'unity' or 'identity' are inadequate in principle. Indeed, they may be understood abstractly, as relations determined by external comparison—as formalistic thinking does when it affirms that a relation of 'identity' exists between two connected concepts. A presupposition is then introduced : the indifference of the concepts with respect to that identity relation. That is why, in this case, we find a Kantian synthesis of independent terms, and not a unification of dynamic moments—as is demanded by the dialectical approach.

'Unity' (or 'identity'), Hegel adds, is a representation signifying abstract sameness between static determinations. In that way, the moment of non unity is overlooked, even though it is essential to the progression of the movement. Consequently, instead of talking of 'unity', a concrete exposition of the matter should express itself in terms of unseparateness or inseparability *(Ungetrenntheit, Untrennbarkeit)*, expressions that have the advantage of making the present tension explicit.

Thus, in Hegel the 'unity of the opposites' is sometimes stated negatively as the 'separation of the inseparable'[64]. But this time the other side tips the scale : the positive moment is no longer expressed. As a remedy, some paradoxical expressions could be proposed, such as the 'separation of the unified' or the 'unity of the separated'. However, these far-fetched solutions might render the dialectical discourse very complicated, and with no important benefit. In my opinion, in order to introduce the novice to a paradoxical mode of thinking, a non natural language, itself paradoxical, will not do: I understand that, to this end, an intelligible ladder must be made available.

As we shall see later, Hegel's tortuous style is thus partly accounted for. That obstacle does not originate in the 'German spirit' alone, but also in the thing itself : so to speak, the paradoxical dialectical thought has some difficulty in explaining itself analytically[65]. Moreover, because of the inner tension that constitutes the driving force of the movement, any solution that would abolish that tension would thereby abolish the dialectic itself...[66]

Thesis 6: Hence the presence of an inexpressible dimension in the dialectical proposition

At the beginning of the *Logic*, Hegel recognizes that the difference holding between Being and Nothing is inexpressible, for that difference, existing between empty, undetermined determinations, is itself empty and undetermined. So that this difference is but intended, *'nur gemeint'*.

Indeed, the articulated language is just able to pronounce the articulated itself, which relates subject and predicate in a meaningful proposition. At the outset, however, no thing is, neither determination—hence the ineffable *(das Unsagbare)* . Hegel explains this inability with the help of an analogy : no thing can be seen in either pure light or in pure darkness. The determinable, and then the expressible, implies a union of shade and light, or of Nothing and Being, to go back to our subject-matter. In a certain sense, therefore, Becoming proves to be the true beginning of the *Logic*.

So far we have talked of a beginning that is and is not, ineffable as an 'indeterminate determination' devoid of any inner differentiation. However, there are good grounds for suspecting that the same problem may arise at other beginnings of the dialectic and the System.

Thus, in the *Logic*, on the threshold of the Doctrine of Essence, Hegel observes that the exposition of the movements of reflection is not easy, for it opens with pure identity lacking any determination. If pure Being and pure Essence cannot be known, how is it possible to know what the pure Concept, their dialectical unity, is[67]? As for the *Phenomenology* which opens with Sense-Certainty, it shows a similar ineffability. Here the question concerns the sensible Certitude, the 'this, here and now', particular and empty according to Hegel (as pure darkness) ; there, in the *Logic*, the question was about pure thought (as pure light)[68]. With regard to the different dialectical transitions, the same problem may arise : Hegel claims that every mediation produces a 'new immediate', a simple unity that is difficult to explain, if not with the help of vague metaphors.

All of this obstructed Hegel's plan, whose intention was to exhibit truth and reality through a language that could unite the subjective intention of the author with the objective signification of the proposition, *Meinen* with *Bedeutung*. Such a plan seems impracticable in that the predicative proposition, in the form 'S is P', remains inadequate despite Hegel's various manoeuvres to attempt the impossible : to express the uniting movement of spirit through a static and divided medium, composed of audible or visible marks—the words rooted in a people's linguistic tradition.

Nonetheless, does not that ineluctable failure bear witness to a certain success, the inner tension being a *sine qua non* of the dialectical movement?

Conclusions

The above study leads to the following paradoxical conclusions.

On the one hand, the purpose of dialectic is to give èxpression to the movements of being and thought, while entirely rejecting any mystical vision of things or any 'immediate intuition'—it is in this activity that the power specific to Hegelian Reason is supposed to reside. On the other hand, however, Hegel has at his disposal only a discursive language, that is unsuited to the realization of his ends : it reduces the thinking of the absolute to *"nur ein gemeinter Gedanke"* [69], to a thought 'just aimed at'. In fact, the predicative proposition does not express what it ought to say, the *das Gemeinte* (intended) of the speaker differs from his *das Gesagte* (achieved) and, in that splitting *(Ur-teil)*, *Sollen* and *Sein* oppose again.

The isolated proposition thus refers to its other, which is supposed to provide the required concrete signification. However, that other is itself a partial moment in need of completion. In this way, in striving to achieve significance, from 'other' to contexts more and more comprehensive, one is brought to pass through the whole system, and then sent back to the starting-point for, relative to the part, the totality itself is only a part starting on a new cycle.

Thus the Idea, which contains all the categories within itself, is itself a determined category : the one which, as the other of Being and the Concept, closes the *Logic*. Philosophy itself, the 'summit' of the *Encyclopaedia*, is just the 'other' of human experience that is partly interiorized and explicated as real philosophy.

Yet, if spirit "is not a bone", does the System consist in that collection of books (as material things) published here and now? Actually, the philosophy that claims to grasp and expose the whole reality, is likewise but a moment of itself, 'abstract thinking alone' or 'printed book alone'—as a reality comprehending this time its own self-consciousness within itself[70].

Hegel sometimes complains about the formalists who "are supposed to mean something different from what they say". In our turn, we can reply to him that neither does he manage to convey his thinking in a satisfactory way. However, from the dialectician we are entitled to expect an acute consciousness of the problem...

Reality is judgement, *Ur-teil*, Hegel says. Indeed, the one pure thought has to express itself but, thereby, it becomes divided within itself. Consequently, the following dilemma seems to arise : either an inescapable division or a silent unity. Hegel has chosen and rightly chosen, even if in our natural language the actual signification sometimes does not correspond to the speaker's intention. In principle, then, ambiguity reigns, at least according to the dialectical doctrine of signification that requires, for every determination, a completion through the determining context. However, because that context is itself underdetermined (it is but partial), the intelligibility of the discourse progresses only through the mediation of the complementary terms clarifying one another, yet without attaining entire transparency. Even in the linguistic realm the 'other' subsists, not surmounted and insurmountable, bringing about the presence of an **inexpressible** dimension Hegel wished to eliminate, but in vain.

Do we have to desperate, to admit defeat and surrender ourselves to dark, violent or mystical forces? I do not think so, for a certain communication is available : in part at least, natural language succeeds in expressing the dialectical movements—we are not obliged to argue by 'all or nothing', since a third way takes shape[71].

5. HEGELIAN DISCOURSE AS THE DOUBLE MOVEMENT OF THE DIALECTICAL SYLLOGISM, ACCOMPANIED BY EXPLANATIONS GIVEN IN COMMON LANGUAGE

By 'Hegelian discourse' I understand a certain philosophical content expressed by Hegel in the medium of discursive language. This means, in other words, the unification of the inner Concept with the outer Judgement, thus constituting the dialectical Syllogism.

At the beginning of the Subjective Logic, Hegel exposes the movement of self-division of the Concept into the Table of judgements of Kantian logic. However, as far as the nature of the dialectical syllogism is concerned, his doctrine of the Syllogism is not very informative : rather, it seems to consist in a logical reconstruction of the method of mathematical and empirical sciences[72]. Obviously, such a 'methodology of sciences' cannot satisfy, for what is required here is the fittest way to expose the dialectical movement. Therefore, I shall attempt to open up my own path, relying on various notes provided by Hegel here and there throughout his works[73].

Let us recall the linguistic experience in question : the presentation of the dialectical movement in discursive language, which brings about an opposition between the dynamic, living and semantic content of the Hegelian proposition and its static, divided and syntactical form. Consequently, in order to express the present inner difference, we first have to complete the positive proposition with the corresponding negative one, and then to grasp the opposites together. This act is said to produce a new proposition, the 'truth' of the original one that has thus been explicated.

Basing myself on that Hegelian doctrine of language, I propose to build a pattern of the dialectical syllogism, developing from the positive to the negative proposition and leading to the 'unity of the opposites'.

For example, we can develop the following syllogism:

Positive proposition: Being and Nothing are the same thing;
Negative proposition: Being and Nothing are not the same thing;
Unification of the opposite propositions: Being and Nothing are and are not the same thing; they become.

Or, following another example:

The True is the whole (it is totality);
The True is not the whole (it is moving);
The True is and is not the whole; it is concrete and living totality, that is, a totality which becomes.

It is also possible to build a similar syllogism with the proposition "The Absolute is a result". Of course, these examples belong to the domain of the Doctrine of Being, but examples belonging to the Doctrine of Essence may also be formulated. In that case, instead of talking about **transition** from one 'independent' term to its other, we can say that the opposed moments **are reflecting** in each other[74]. Given the reciprocal mediation in presence, now the converse proposition itself will play the part of the complement. Indeed, the moment of the 'and conversely' must also appear, this moment expressing the activity of the determining double movement.

This time we can write:

The Identity reflects itself in Difference (Difference exists only between Identical terms, *viz* terms in relation of reciprocal mediation);
The Difference reflects itself in Identity (as a relation, the Self-Identical contains the moment of Difference within itself)[75];
In the process of reciprocal reflection a Contradiction arises, seeking for a solution; and the opposites *'zugrunde gehen'*, *viz* pass away, going back to their Ground[76].

Or, according to the following opposites:

The Inner exteriorizes itself as Being (or Phenomenon);
The Outer interiorizes itself as Essence (or thought);
That double movement constitutes Actual Reality.

As for the realm of the Doctrine of the Concept, the 'truth' of the movement, we find there a **development** *(Entwicklung)* of the determination which, in its other, is said to remain 'close to itself'. To sum up, we started from the change of so-called independent terms, then we passed on to the double movement of moments reflecting in each other. Now, we are at the level of the Concept, mediating itself through its three essential determinations and thus forming a circle. In particular, the Doctrine of the Syllogism seems to be structured according to the following syllogism, where the middle term *(Mitte)* is exposed as realizing itself *qua* unification of the Universal, Particular and Individual:

> The Individual develops as Universal through the mediation of the Particular (the Individual realizes itself);
> The Particular develops as Universal through the mediation of the Individual (the Particular is realized by the individual);
> The Individual develops as Particular through the mediation of the Universal (the Individual particularizes itself, he acts)[77].

From this it results that, in dialectic, the adequate exposition of the conceptual content requires the constituting of the diverse arguments into a **syllogistic form** : truth cannot be expressed in an isolated proposition, as we already know.

We have now discovered the three different forms of the Hegelian syllogism : the form of the Doctrine of Being, that of the Doctrine of Essence and that of the Doctrine of the Concept ; the question is to know which of the three is the most fitting. Hegel seems to expect us to opt for the syllogism of the Doctrine of the Concept, the 'truth', as it were. Nevertheless, from a critical standpoint, to the harmonious circle of a flawless system I suggest to prefer the **double movement** of moments reflecting in each other[78].

Indeed, in my opinion the concrete truth lies half-way, at the level of the Doctrine of Essence, non external as the Doctrine of Being, and not beyond real human experience as the Doctrine of the Concept[79]. We thus again meet the dialectical ***Satz***, the 'truth' of the divided ***Urteil***, this time as a syllogism or double movement of mediation leading to a certain unity of opposites.

Does this discussion imply that the problem of language has been solved? Certainly not. In principle the problem is unsolvable, since dialectical thought must necessarily express itself through an inadequate discursive language. Even the dialectical syllogism, our subject-matter, develops through this unsatisfying medium[80]. In consequence, if the way out is to bypass the obstacle and to provide the required ladder, then another road has to be opened : Hegel proposes an 'analytical' comment on the nature of the dialectical movement.

Now, the pattern of the Hegelian discourse may be specified further. Having opted for the movement peculiar to Essence, I suggest to refer to the **double movement of the dialectical syllogism**, developing through the reciprocal mediation of the opposites. As for each of these opposites, it may itself constitute a dialectical syllogism, as the above syllogism of the Doctrine of the Concept shows explicitly. That supra-syllogism, a double movement structuring chapters and paragraphs, is presented with the help of explanations set forth in the current language : I mean the various **Prefaces, Introductions and Notes**, found frequently in Hegel's central works.

Accordingly, Hegelian discourse is structured as follows. First, there is the twofold movement of a suprasyllogism ; i.e. the discussion develops through diverse chapters and paragraphs that enlighten one another—in Sartrean language, we may mention the 'enriching movement to and fro' of a 'quasi-synoptical' argumentation, both 'progressive and regressive'[81]. Moreover, in every paragraph we can discern a syllogistical movement relating various opposed and complementary arguments, bringing through mediation to a positive result. To end with, Prefaces, Introductions and Notes furnish the required elucidation in a common language.

Of course, this pattern must be understood not in a rigid but in a 'plastic' manner. That is, it all depends on the issue concerned and on the surrounding context, since in dialectic we have to follow 'the thing itself' *(die Sache selbst)*, without complying with any strict order given in advance. The same holds with respect to the structure of the Hegelian discourse, which varies with the requisits of the thing *(Sache)*. Instead of enforcing distorting symmetries by architectonical instincts, is it not better to let spiritual life speak freely in its ever surprising and creative activity[82]?

Indeed, even a furtive look at Hegel's dialectical works shows that, in general, the argumentation develops according to a certain movement of reciprocal mediation between complementary opposites, each development itself constituting a moment of another more comprehensive development, until the whole work has been traversed. Within this we also find several Prefaces or Introductions and many Notes breaking the rhythm of the discourse.

However, it is in the *Logic*—the dialectical work *par excellence*—that this structure is realized at its best. There we can see 'dialectical' general divisions, themselves dividing into syllogistic triads, down to the different paragraphs that oppose and unite one another, producing thus their 'truth'[83].

The *Encyclopaedia* itself attempts to reproduce that inner organization. However, its ***Realphilosophie*** does not seem fitting to that task, if not at the cost of the presentation of a reality somewhat falsified ; what Hegel, as a bad dialectician at this point, does not hesitate to do[84]. As for the *Phenomenology*, its problematic composition is well-known[85]. Yet Consciousness, Self-consciousness, Reason and Spirit are understood adequately only when grasped in their reciprocal relations of mediation[86]. We may also perceive in that work a lack of inner Notes : the *Phenomenology* does not explain itself sufficiently, but it is explained by the *Logic* and by the *Encyclopaedia*, which do explain themselves. In that sense, it can be said that the *Phenomenology* still belongs to Hegel's early works.

In that way, **dialectical exposition and predicative explanation are complementing each other**. A work written in a predicative language only would appear formalistic, and a work written in a thoroughly dialectical language would become unintelligible, since it would lack the indispensable 'ladder' to get in. Sometimes, Hegel labels these different explanations (Prefaces, Introductions and Notes) as just 'historical' or 'preliminary'[87]. Admittedly, however, without these unscientific ladders, 'science' itself could neither develop nor be understood and open to all : in that sense, 'science' comprehends within itself the non-scientific that seems to hinder it, but without which, in fact, it would not work—a typically dialectical paradox.

Thus we must not be too impressed by the abuse Hegel gives against these 'unscientific' Prefaces, Introductions and Notes. Even the 'Compedia' (as the *Encyclopaedia* and the *Principles of the Philosophy of Right)*, which only sum up the original dialectical movements, find their elucidation in public lectures, delivered in the usual language[88].

6. CONSEQUENCES CONCERNING THE NATURE AND POSSIBILITY OF A DIALECTICAL APPROACH

In our investigation we have seen that some Hegelian propositions, ordinary as regards their structure, such as "Being and Nothing are the same thing" or "The True is the whole", prove paradoxical as soon as the linguistic experience in question is examined. Hence the following conclusions, concerning the nature and possibility of a dialectical approach[89].

6.1. The logical category realizes itself through the mediation of the empirical representation

When Hegel talks of the movement of the Concept, he presents it as being constituted by the category that posits its presupposed opposite and, through reciprocal mediation, produces a new category still developing, for the lack is still active within it : the Hegelian category is 'erotic'.

This quite formal exposition of the thing makes believe that the *Logic* moves in the high spheres of 'pure thought', delivered from the sensible, spatio-temporal stuff. Yet, critical studies of dialectic, from Trendelenburg to the present, show that the so-called pure thought is conditioned by sensible intuition, with no hope of eliminating it. In other words, sensible intuition, which is set aside at the beginning of the *Phenomenology*, is nevertheless a necessary constitutive moment of dialectic. That is, if the concept asserts itself as the 'truth' of representation, conversely that very representation functions as an embodiment of the category : the pure category realizes itself through the mediation of the empirical representation, its other that examplifies it[90]. Does not Hegel himself affirm that, in the dialectical movement, **the second moment is the moment of the finite and the difference**? As, for instance, the Real moment succeeding the Formal moment[91], Quantity succeeding Quality, or as the 'second' categories of being-there, external reflection, matter, appearance, causality, particularity, judgement and objectivity—to quote a few of them only.

The above inquiry concerning the problem of language seems to confirm this point. On the one hand, we saw that the predicative proposition, as an object of representative thinking, is unfit to express dialectical truth ; and, on the other hand, that this insufficient support however proves necessary, since dialectical thought needs embodiment—this is the paradox which constitutes the problem of exposition. So, in the production of the dialectical movement the predicative proposition plays an essential part, the meaning of terms building itself out of the differentiated structure : in the milieu of natural language, semantic unity and syntactic division form the driving tension that advances the process. More precisely, the so-called 'inadequate' predicative proposition is the very element that introduces the constitutive moment of the difference.

Consequently, the following double meaning must be conferred to the word 'dialectical' or 'speculative' : first, the **abstract meaning** corresponding to the movement of category or judgement that is supposed to unfold in the medium of pure thought; and, second, the **concrete meaning** corresponding to a movement going to and fro between pure category and empirical representation, between dialectical judgement and predicative proposition—this complex movement producing the speculative proposition. To put it another way, the **concrete dialectical** is the living unity of the abstract dialectical and the non dialectical, the latter meaning a 'representative' or 'predicative' component, in any case a 'sensible' one[92].

It must not be forgotten that dialectical thinking is said 'to comprehend' human experience in all its richness and diversity. And, in my opinion, this claim seems legitimate if, in one form or another, **the sensible is present in the intelligible**. For this very sensible constitutes the concrete content of the category, as the phenomenological exposition of the thing has well shown, from the exclusive opposition between subject and object up to their dialectical union.

Yet, the *Logic* is not the *Phenomenology*. That is, the bare sensible given, which is still encountered in the phenomenological way, is said to have been ***'aufgehoben'*** in the *Logic*, an assertion I understand as follows : it is through the empirical representation that the **negated** sensible intuition is **preserved**, as we can read in the *Encyclopaedia*, par.451-464. Indeed, there Hegel presents this representation as the synthesis of the sensible and the intelligible components.

In that sense, the real is truly a syllogism where representative thinking plays the part of the middle term, sensible intuition and conceptual thought being the extremes : that 'psychological' movement, exposed in the *Encyclopaedia* par.445-468, shows how the sensible intuition interiorizes itself as empirical representation and, through a complex process, gives pure thinking. As for language, here described as a system structured by auditive and visible signs and utilizing images and symbols, it is a 'representative' moment of that 'psychological' exposition[93]. Mure's assertion is thus verified again: thought realizes itself with the help of a lower moment conditioning it ; it is the same with empirical representation, a moment conditioning the thought category.

Hence the following conclusion: let the question be about phenomenological, logical or real movement, in dialectic the sensible dimension always fulfills an essential constitutive function, ensuring that at the summit we will not find the empty Parmenidean One but a real content, grasped and expressed so that philosophy may be asserted to be the "comprehension of the present and the real".

If it is true that an *a priori* dimension actively dwells in human experience, then representation, as a synthesis of the sensible and the intelligible, comes to be assigned an essential function : it is the royal way to the concept, interiorizing and preparing the sensible given so that it may be, at least in part, 'digested' by conceptual activity—in this sense, thinking is truly like eating[94]. From that a new ambiguity emerges : in the Aristotelian sense of the word, the best is the summit (the extreme), the concept or 'pure' thought, and **also** the middle, the representation as a concrete synthesis of *a priori* and empirical dimensions. All things considered, the partly opaque, empirical representation proves necessary, not only 'in order to make dialectic plausible', but also in order that dialectic should be[95]!

Consequently, dialectic may be expressed as a double movement from category to representation and vice versa, or as the movement of the speculative proposition that is, actually, also a predicative proposition. The apprehension, together, of the identity and difference of the opposites—sensible and intelligible, syntactic and semantic—does produce that 'tense unity', an animating power that creates the social, cultural and spiritual life. In this sense, the synthetic Hegelian proposition is also analytic[96].

6.2. There is no real alternative to the unsatisfactory, discursive language

The **formalized, artificial language** is obviously inappropriate, and that for a number of reasons.

Generally speaking, in dialectic the aim is to study not the structure of language, but human reality in all its manifold historical, social and cultural dimensions ; so that natural language alone, as real (as really spoken), full of distinctions and rich of inexhaustive contents, is able to serve as a starting-point and a medium for the development of a concrete approach[97].

In this context, how could we construct an axiomatic system? What would be its premises, its transformations rules? Could we decide on them according to our arbitrary will, our needs of the moment? Now, our purpose is to explicate human experience, and not just to create another analytical tool ; that is why arbitrariness is entirely rejected here[98].

More specifically, we saw that the Hegelian judgement does not oppose true and false, but develops from less true to more true, according to a growing degree of self-realization. Instead of the structural transformations of formal logic, we have here a shift in meaning, produced in one and the same sentence grasped differently in different places. In any case, since the meaning of a term depends on its context, strictly speaking the principle of identity cannot be respected. Hence the 'plastic', fluid identity of living terms, unceasingly becoming richer and richer as the movement progresses. As for the dialectical contradiction, of course it is not the formal contradiction[99].

For all these reasons the artificial language, although effective in its own realm, proves inoperative here.

To the abstract character of formalized language, it could be opposed the concreteness of **poetic language**. But this way is not satisfying either, in so far as science means intelligibility and communicability. Now, poetic language always seems to be the peculiar representative and coloured language of an original creator. Although the poet may unveil to us the 'nature of things', the essence of the world, the trouble is that he does it in his own way, making very free use of linguistic signs in order to produce new images and unusual associations[100]. Certainly, much can be gained by looking into the poet's work, by paying attention to his sensual language, but that is not the question. In science, another way of research and expression, exoterism and opening to all is indispensable, for it allows the required communication between the different individuals who exchange pieces of information in a publicly acknowledged, conceptual language.

Great is the work of art, yet we must not confuse the various domains. If dialectic intends to be a science in any sense whatsoever, it has to reject that peculiar subjectivity constituted by pure poetic language, despite its attractive richness and depth. For the thinker cannot confine himself to the spontaneous but non critical, creative intuition, however interesting it may be. Entirely turned towards the universal, in principle he has to talk a non poetic universal language, the sensual image being, at best, able to constitute a moment of the representation in its way up to the level of the concept—which is the specific element of philosophy[101].

Lacking any real alternative, the dialectician must therefore recognize **the predicative proposition as an unsatisfactory and yet necessary moment**. He cannot do without that self-diremption *(Urteil)* of the conceptual thinking, for the movement of the Concept has to accomplish and express itself : thus, the articulated language functions both as a *sine qua non* and as a non transparent medium[102].

If it did not express itself, dialectic would become mystical, abstract, undialectical, that "night in which... all cows are black". If it does express itself, an insurmountable hiatus separates the subjective intention of the speaker from the objective reached signification, *Meinen* from *Bedeutung*. It remains an ought-to-be, expressing and communicating, which is not realized in being, here the still opaque isolated proposition. Accordingly, to acquire an adequate understanding of a single proposition leads—strictly speaking—to go through the whole system, step by step, if not through total human experience[103].

"The True is the whole", Hegel said in an isolated proposition. We saw that the **one** unified thought becomes actual through the mediation alone of its own moment, the discursive language. The difference is therefore present even at the apex of the System, for every attempt to understand and explain again introduces a gap : the intended unity of sense and being is not achieved[104]. So, as 'reflective', the dialectical approach provides **a sense not entirely real** (the abstract category is 'only thought') and **a real not entirely significant** (the opaque experience is 'still given'). An ineffable residue persists, *nur gemeint* , and, eventually, in human experience *Logos* and *Chaos* continue to co-exist...[105]

6.3. Double meaning and ambiguity are characteristic of the dialectical approach

In Hegelian logic, every determination is said to posit its presupposed opposite. For instance, being is nothing as lacking any determination, the infinite of the Understanding is finite as ever-finding an other beyond it, the universal is particular with regard to other universals. More specifically, it is important to grasp the determinations of identity and difference in their reciprocal mediation : dialectical identity, as self-identity, contains the difference within itself, and dialectical difference is identical in the sense that it is self-mediating. Thus **the dialectical determination is characterized by a double meaning** : when isolated, it has the immediate meaning of the current representation ; and when related to its other, it acquires its concrete meaning as a logical category.

The same holds for the Hegelian proposition. In effect, the positive proposition 'contains' within itself the corresponding negative proposition it has to posit[106]. Therefore it can be said, likewise, that being is not nothing (it is that which comes to be and passes away), that the infinite of the Understanding is not finite (it is endless, in-finite) and that the universal is not particular (it is individual) ; moreover, as everyone knows, identity and difference are not the same. In short, the Hegelian proposition has a double meaning, either representative or conceptual, either formal or dialectical[107].

Let us recall the point in question : according to the dialectical doctrine of signification, every term acquires its meaning only when related to **its other**, constitutive for it. It is thus itself and an other, self-identical and self-different at once[108]. However, if the other of the identical is the different, the other of the part is totality. Hence the following generalized doctrine : in dialectic, a determinate term acquires an adequate meaning only when put in **its proper context**. That meaning then changes, depending on the position of the term in the considered text, in its linguistic surroundings. Thus, being can be understood as the other of nothing (being is), of becoming (it is static), of being-there (it is not determined) and of essence (it is external, at the surface).

Likewise, at the other extremity of the *Logic*, the Concept comprehends different opposites constituting it, such as judgement (the Concept is inner unity), syllogism (it is only a germ, pregnant with all the development to come), objectivity (it is a subjective principle) and Idea (it is still abstract thought).

We see that, at every level, a shift in position brings about some shift in meaning. As the movement advances, the meaning grows progressively richer in new determinations—strictly speaking, the principle of identity cannot really be respected.

Ambiguity seems to be everywhere and dialectic, indeterminate to the utmost. Nevertheless, there is a remedy : **in every case, the close context allows a reduction of the equivocation**. As for the double meaning, it proves fundamental : in principle, every term is 'pregnant' with its presupposed other that it posits. To me, that seems the price to be paid to enable a progressive and enriching dialectical movement.

Formalistic thinking, on the other hand, demands univocity and strict obedience to the principles of identity, non-contradiction and excluded middle. It regards any ambiguity as a problem to be solved through analysis of the given into its elementary components, first isolated and then classified. Indeed, the formalistic mode of thinking assumes that there are immediately intelligible simple elements or significant atoms, and that the complex thing is obtained by combining some of these simple elements[109].

Dialectic also deals with the simple but, in opposition to the formalistic approach, it sees in it the product of a large abstraction fraught with presuppositions, that cannot be understood independently of the process of abstraction itself. In its isolation, the atom is just an empty name and, when related to its different presuppositions, it shows itself in all the wealth of its many determinations : dialectically speaking, the so-called 'simple' is not so simple as it appears... That is, seen from this standpoint, the simple and the complex mediate each other and thus constitute the real thing.

Formalistic thinking builds various combinations of simple elements and, in so doing, is engaged in a mechanical reconstruction of reality[110]. There lies the problem: with respect to the dynamic character of human experience, the simple atom is but the abstract lifeless One. **Must not the living being be distinguished from the mechanical aggregate?**

If so, *vis-à-vis* the rigid deduction of 'the same' in tautological systems, stands the dialectical development of 'the new'[111]. Then the univocity of the formalistic thinking gives place to the ambiguity of an ever-creative, living dialectical thinking. It looks as if the admirable precision of the formalistic approach was demanding its due : an abstraction and a reduction of the real to a combination of atoms, points and numbers. In contrast, the concrete character of the dialectical approach, which claims to explain human reality, goes together with a basic polysemy that is quite obstructive when the communication of ideas is concerned. An adequate expression of the content is not enough : intelligibility is required too.

The dilemma seems to be as follows : **either precision and abstraction, or concreteness and double meaning.**

In science, however, one has 'to fasten thought by some definitions', to set determining limits to the fluidity of ideas. Consequently, to avoid sinking into vain sophistry, one must recognize the essential function of the dividing Understanding, that "most astonishing and mightiest power" according to Hegel. For even in dialectic, sensible intuition and empirical representation are constitutive moments not to be purely and simply set aside : a dialectical approach, which wants to be regarded as scientific, has also **to preserve** the moment of the independent determination, and not just **to cancel** it[112]. Dialectic itself requires it, since with no pressing tension there is no movement either! It is in this very meaning, however, that the non satisfactory discursive language proves indispensable.

6.4. Consequently, the philosopher has to develop the 'speculative spirit' actively present in the language of his people

Having at his disposal only the discursive language, which is inadequate in principle, Hegel finds himself compelled to create his own language on that problematic foundation. Torturing the natural means of expression, he intends to arouse certain associations and to give thus an impression of movement and fluidity. To this end, I understand that he makes use, *inter alia*, of the following devices.

Firstly, referring to the **roots** of different German words, Hegel enriches their lexical meaning. Thus we can read: Ab-grund, Be-dingung, Bei-spiel, Da-sein, Er-innerung, Not-wendigkeit, Un-ruhe, Ur-sache, Ur-teil, ver-söhnen, wahr-nehmen.

Secondly, he makes the most of the **family ties** that appear to hold between words like Gedächtnis-Gedanke, Monstration-Demonstration, Meinen-mein, Sein-seinige, setzen-Gesetz-Satz, Wirklich-wirken, Zeigen-Zeugen, Zufall-fallen. He also produces **semantic affinities**, taking advantage of the ease with which the German language generates new words by means of various prefixes. In this way he connects : schliessen, ausschliessen, beschliessen, entschliessen, umschliessen, zusammenschliessen; or setzen, entgegensetzen, fortsetzen, gegensetzen, gegenübersetzen, herabsetzen, voraussetzen ; and stellen, darstellen, herstellen, hinausstellen, verstellen, vorstellen, wiederherstellen. We also find : an sich, in sich, für sich, which give Ansichsein, Insichsein, Fürsichsein, and das Diese, das Etwas, das Andere (which gives: Anderssein, Sein-für-anderes), where we see diverse particles turned into substantives ; 'sein' gives also 'das Seiende' by nominalization of a verb.

Thirdly, Hegel recalls the **metaphorical value** of philosophical terms, whose original meaning was sensible, such as: begreifen (to seize with the hands or in thought), Mitte (spatial middle or middle term), Moment (mechanical or dialectical term), Reflection (as movement of light or of thought), Sinn (sense as organ or as signification)—we thus encounter again the intelligible 'high' level explaining the sensible 'low' level, and explained by it in return[113].

Fourthly, he discovers an original double meaning in some German words, and sees in it **'the speculative spirit of language'** at work[114]. The corresponding typical example is provided by the term **'aufheben'**, where he keeps the meanings of 'to come to en end' and 'to preserve'. But, of course, some others relatively appropriate may be mentioned, such as : auflösen (to dissolve and to resolve), Bestimmung (determination and destiny), Darstellung (embodiment and exposition), das Endliche (the finite in the sense of the limited in its being and in time), Sache (spatio-temporal thing and matter in question), zugrunde gehen (to progress towards the basis or to go under)[115].

This viewpoint, referring to the 'speculative spirit' of language and to its fluidity *(Flüssigkeit)*, accounts for Hegel's interest in **etymological studies**. There is no use disparaging these studies which, in philosophy, are out of fashion (must truth keep up with the fashions?), for one can discern in them a certain awareness of linguistic issues, against which any investigation may come up sooner or later.

In particular, Hegel turns his attention to the Greek origin of terms such as 'logic' (the science of the Logos), 'method' (from 'met-hodos', a way towards) and to the double meaning in Greek of 'Logos' (as Reason and word). In the expression 'Dasein' he perceives the sense of 'being in some place', and in the Latin word 'aliud' he discerns the double meaning referring to Something and an Other. But sometimes he is mistaken, as when he fancies that there is some kinship between 'das Ding' and 'das Denken'[116].

Hegel therefore believes that, despite some linguistic resistance here and there, **the philosopher, for the most part, needs no peculiar terminology**. Relying on the cultural patrimony that animates him, his essential task consists in drawing from his people's language the appropriate words fraught with diverse significations, in using the different semantic affinities he has found out and in creating some others, all according to contextual necessity. Thus, with these more or less suitable procedures, Hegel takes from natural language various representations *(Vorstellungen)* , so as to turn them into living concrete concepts.

Can we conclude that Hegel was successful in building up a **'philosophical language'**, the philosopher's stone of contemporary thinking? While some doubt the very possibility of achieving such a result, others see in that *'philosophische Kunstsprache'* the main source of the renowned abstruseness of Hegel's work.

Whether it is cunning or legitimate manoeuvre, a faithful reading of Hegel requires special attention to these different manipulations by which the philosopher seeks the right expression, one that may express the dialectical idea in a non-dialectical language—as we have seen, here lies the origin of the difficulty... and of the creative tension.

When understood in this way, the diverse unavoidable strategies may be admitted more willingly : they animate the thinker's fight against the sensible and spatio-temporal 'other'. That is why the **reader** ought to study the text, to re-read it methodically and to discern in it the different levels of signification : etymological, original and contextual significations, double meanings, semantic affinities and occasional associations. As for the **philosopher**, of him are required a good knowledge of the language of his people, continuous research centred on the various sources and meanings of the terms used, and a 'fluid' use of the results in order to achieve, as far as possible, an adequate exposition of the dialectical movement, which is complicated in any case.

If so, no wonder that **Hegel talks in praise of the German language**. Indeed, it is well known that, following Wolf, he recommended philosophizing in German and created a new philosophical dictionary—a step that renders the reading of Hegel still more difficult, especially to us, readers foreign to his language and culture, and who come a long time after[117]. If it can be said that Kant prefered terms of Latin origin, Hegel, on the contrary, expresses himself mainly in the German language. However, even on this point things have to be regarded in a 'plastic' way, for it is a matter of context[118].

For instance, in the *Logic* 'Essenz' turns into 'Wesen', which Hegel brings together with 'gewesen'[119]; but as regards 'reflection', the ambiguous Latin term appears to be just what is needed. Likewise, for his long list of logical categories, Hegel sometimes draws from Latin sources and writes : Qualität, Quantität (Quantum), Attraktion, Repulsion, Kontinuierlich, diskret, extensive, intensive, Form, Materie, Existenz, Absolute, etc., using the Kantian terms when it suits him to do so. Nevertheless, he can say equally well : Darstellung or Exposition, Einzelheit or Individualität, zweideutig or doppelsinning[120].

In sum, dialectical thought requires the philosopher to express himself in a living language, so that he may be able to raise the representation *(Vorstellung)* up to the level of the concept. We must not be mistaken, for we have not to do with some nationalistic spirit, but rather with a necessity proceeding from the 'thing itself': to express oneself as adequately as possible in a non adequate language.

A living thinking, grasping a living reality, can embody itself in a living language only, the language of the people among which the philosopher is working, and whose culture and creative activity he expresses. Instead of the abstract dream of a universal language, what is required here is an authentic insertion in an original culture : is not philosophy called by Hegel "the age grasped by thought"? Therefore, in contradistinction to the Leibnizian plan of an artificial, non living *'Characteristica Universalis'*, Hegel advocates letting the moment of particularity express itself in the universal medium of scientific language. Whence the 'concrete universal' of a system of thought where a people might recognize his own specificity[121].

For example, of the contemporary philosopher it is required to study the language of his people, its sources and history, to bring out the 'speculative spirit' that dwells within it, so as to form a 'tensed philosophical language' full of significations and nuances, and able to express human reality in general and the 'spirit' of the people in particular. "Philosophy moves essentially in the element of universality, which includes within itself the particular", says Hegel, and there is no reason why the philosopher should express himself in another language, natural or not.

PART SIX

DIALECTICAL EXPLANATION

Much has been written about the 'dialectical method', its nature, whether it is a method in the usual sense of the word, and whether it can be regarded as valid. In this final Part, I deal with these questions in terms of 'dialectical explanation', a modern version of that topic.

Moreover, in the Hegelian philosophy a distinction must be drawn between the 'method of the System' and the 'method of the researcher' : no need to say it again, despite his own statements, Hegel is not the System! Hegel wrote at lenght about the **method of the System** in his *Science of Logic*, summarizing his point of view at the end of the book[1]. However, with regard to the **method of the researcher** he seems less loquacious, so that I have to sort things out myself.

The exposition of the movement of the speculative proposition may be an answer to this problem, a suggestion that leads me to the following remark: if the speculative proposition was really moving 'by itself', Hegel would be right, and the two meanings attributed to the 'dialectical method' could be identified with each other. But Hegel knows very well that the speculative proposition starts moving only when considered by the researcher (or by a reader).

In other words, the 'method of the System' realizes itself through the mediation of the 'method of the researcher'. Consequently, I do not examine again the 'method of the System', a movement of objective thought, nor the speculative proposition, a movement of language, since these questions have already been discussed[2]. Here I am concerned with the constitutive role of the researcher who, *qua* researcher, is the moving principle of the philosophical proposition, thus bringing about the production of a **dialectical explanation**.

As for modern science, first of all, it makes use of **deductive** and **experimental proofs**. By the former it understands a procedure of deduction founded upon a precedent truth and carried out according to accepted rules—the grounding truth being constructed out of a consistent and complete set of independent definitions, axioms and/or postulates. The latter refers to a conclusive experiment, that can be reproduced in principle, at any moment and in any place by anyone, to arrive at the same result within a certain margin of error calculable in advance.

If the **formal** and the **empirical** were the only dimensions of reality, then the matter could rest there. However, apart from the world of formal sign and sensible given, there is still the world of **meaning**, of **thought**[3]. It is true that modern logic constructs various highly articulated systems, and that experimental sciences offer a detailed and precise image of material reality. Yet, human reality appears as especially problematic: man thinks. Certainly, between matter and thought diverse intermediate links may be inserted, such as various organic, psychic, social and cultural entities[4]. But that is not the point. For my part, *grosso modo* I suggest distinguishing between social, cultural and historical reality—as human reality with which I am concerned—and physico-chemical, biological reality—as natural reality investigated by empirico-mathematical sciences[5].

With regard to the human world, the one-sided empiricist can, *inter alia*, employ two different strategies : either he can deny the existence of a relatively autonomous non material world, purely and simply, talking of it in terms of 'fiction', of 'merely subjective' phenomena or even of 'metaphysical divagations' ; or he can resort to ontical reduction, claiming that the movements of thinking are understandable as forms of electromagnetic or chemical processes.

At this point the divergence becomes clear : in principle, dialectic is opposed to any form of reduction. In my view, it considers that in the world there are different ontological realms, each one being characterized by its own mode of explanation—and we must respect the difference. In consequence, dealing specifically with human reality, dialectic endeavours to grasp the cultural object in its living and creative singularity, without reducing it. For, in this domain, **to explain is not to cancel**[6].

In this Part of the book, I raise the following crucial question : is the often proposed plan for an approach proper to the world of meaning and thinking finally workable? Or rather, when reflecting on the cultural object impregnated with sense, creator of sense, may dialectic *qua* **'organon'** allow us to build scientific demonstrations[7]?

In chapter 1 I show that, in term of **justification**, the answer is affirmative for, in that realm, no apodeictic proof is available. Later I expose the forming of dialectical justification through a double movement, which develops *a priori* and reconstructs *a posteriori* the creative activity of thought and culture. From that will emerge the **dialectical, self-explaining explanation**, which solves in part the thorny problem of the beginning. Finally, with the aim of making my thesis more concrete, I examine in some detail the dialectic of Something and an Other, as it is presented in the *Logic*. Hence I reach certain conclusions concerning the nature and limits of the dialectical explanation: dialectical development is not organic genesis!

1. DIALECTIC HAS NO APODEICTIC PROOFS, BUT DISCURSIVE JUSTIFICATIONS ONLY

In this chapter, I characterize the dialectical justification in a negative manner: it is not the deductive proof used in Mathematics[8]. Later on, I shall present it positively too.

In a broad sense of the term, classical philosophy may be called 'empirical', in that it thinks the manifold determinations of human experience. No one would consider taking it for an experimental thought, that investigates various phenomena in a laboratory or describes and measures natural objects observed from the outside. Yet, the **mathematical method** periodically attracts the philosopher, who is impressed by the elegance of its demonstrations and the precision of the achieved results: how could the philosopher resist the weight of its decisive arguments?

Given the well-known advantages attributed to that method, if it were really possible to philosophize in a mathematical way, philosophy, as a separate and autonomous branch of knowledge, would have disappeared a long time ago. Such has not been the case, however. Maybe that this results from the fact that philosophy differs from mathematics, not only with regard to its object but also with regard to its form.

With regard to the relevant **object**, I suggest the following argument: mathematics deals with discrete or continuous quantities, or with notions of order, whereas philosophy, according to tradition, deals with 'Ideas', the 'essence' of the world—to put it in modern language, with the meaning of things or events[9]. With regard to the **form**, i.e. the method, following Kant's and Hegel's arguments, I mean to show that the mathematical deduction cannot meet the concerned requirement: in general, it proves unfit for the development of a non formal way of thinking and, in particular, for the dialectical approach, that is our subject-matter[10].

In the course of the discussion I refer to formal and empirical sciences, in the classical meaning of the terms, in so far as both are founded on deductive thought. Besides, it is worth noting that, without hesitation, Hegel seems to rank mathematics among the formal sciences...

1.1. Dialectic cannot base itself on exhaustive, first definitions

The problem is as follows[11]:

First of all, how could we define what is a definition? Is it possible to avoid that circle? Moreover if, in principle, to define is to set a term, to delimit, in fact, in Aristotle from the outset it is asked to give the essential characteristics of the thing to be defined—in dialectic, that requisite will not do.

In this connection, it is advisable to recall the problem of the beginning : at the start we do not know what the object of our inquiry is, and the aim is to grasp it through thinking. Under these conditions how might we, from the outset, furnish a list of relevant characteristics? We must open either with a stipulating definition or with a vague description relying on the usual representation of the thing. As for the 'intuitive' definition there is no ground to trust it, every individual having his own peculiar intuition: **we thus always begin with the non true**, the arbitrary, uncertain or merely probable.

Consequently, dialectic will not base itself on pre-established definitions for, here, all real knowledge concerning the treated object is regarded as a purpose and not as a starting-point. Hence the question posed by Hegel at the beginning of the *Logic* : "With what must the Science begin?".

At the start, the dialectical thought addresses itself to what appears to be an immediate given. But, in the course of reflection on the thing, that 'immediate given' reveals itself, partly at least, as a 'mediated product' determined by human thought. Aristotle himself notices that dialectical syllogism opens with subjective opinion.

As for the dialectical examination of that first opinion, it means its **criticism**, namely the bringing to light of its different presuppositions and the development of its fruitful content. Therefore, only **at the end** of the examination will it be possible to propose a set of enlightening concepts, the required essential 'definition'—if that term is still to be preserved. That is, in Aristotelian language one can speak of a movement **from nominal to real definition**, the founded knowledge being assigned to replace the fluctuating representation.

The dialectical beginning is thus abstract, requiring some explanation ; and Hegel himself often opens his dialectical studies by anticipating the development to come. However, he asks us to regard these 'additions' explicitly as the non-scientific, only preliminary part of his work—for truth unveils itself within 'the science' alone.

In concrete terms, Hegel remarks that, in the *Logic*, every new level, every new category, is a new definition of the thing in question ; and that the exhaustive, 'total and absolute' definition is given in the context of the Idea, the last category of the logical development[12]: in so far as "The True is the whole", from that synoptical viewpoint alone may every term become truly comprehensible.

For instance, if a novice asks 'What is the Idea?', in the hope of receiving a satisfactory answer, the dilemma of the dialectician will then be either to give an **answer in one single proposition,** deforming more than informing, in the following terms : 'The Idea is a twofold movement of subjectivation and objectivation' ; or **to repeat the whole *Logic***, the latter playing the part of the required definition.

At the beginning, then, the novice cannot obtain satisfaction for the 'short way' is too vague, and the 'long way' too complex.

So far, I have based myself on Kantian arguments, but Hegel adds the following[13].

In his opinion, mathematics inquires about discrete (arithmetical) or spatial (geometrical) magnitudes. Since it is concerned with simple-structured objects, an abstract definition may account for them. As for empirical sciences, a definition of matter in spatio-temporal movement can be given there. Whereas dialectic deals with **cultural** (Hegel says 'spiritual') **things** and, in that realm, the wanted distinction between the different characteristics—to form a good definition—causes many troubles. For instance, how can we define such terms as 'concept' or 'thought'? Aware of the problem, the formalist assumes that the thing is 'intuitively known' or indefinable[14]. But in dialectic, understood as a concrete thinking, the **entire explanation** of the thing is said to provide the required definition.

From this it results, according to Hegel, that **the pre-eminence of the exact—empirical or formal—sciences is due to the abstract nature of their fields of inquiry**. In this context, however strange it may be, the empirical sensible thing is grasped as abstract, since abstraction is made from its cultural dimension. Indeed, a table is not only a wooden implement, but also a thing that acquires sense and function in a determined human environment. When seen from the angle of an other culture, that thing will not appear as a table but, for example, as an altar for worship, a bench to sit on, a stool to climb up, wood for burning, and the like.

Formal and empirical sciences thus rest on abstract, stipulated definitions, or on definitions of seemingly simple elements such as points, numbers or atoms ; while dialectic is concerned with the complex entity in the process of developing, that is, with living, cultural things. How could they be apprehended by means of 'clear and distinct' definitions laid down at the outset?

Human reality, the object of dialectical thinking, shows itself to be diverse and rich in content, with an ever-surprising creativity. Therefore, at the beginning of the examination it is not possible to conceive the essential characteristics that are required to obtain a satisfactory definition[15].

1.2. Dialectic has no evident nor arbitrary axioms

By 'axiom' I mean here an *a priori* or arbitrary principle, or one admitted as true in an immediate way. Obviously, in dialectic there is no place for such an axiom.

1.2.1. Dialectical truth is not expressible in one isolated proposition

When separated from its determining context, any proposition is here understood as abstract and 'lacking in truth'. In this sense, it implies some presuppositions that have to be brought out. And the whole 'proposition and its context' alone forms an intelligible unity.

This is why the axiom, laid down at the beginning, is regarded as poor in content and as requiring completion through its dialectical development. Only after such a development can the preliminary axiom become a grounded truth ; but then, as an immediately true proposition, it has vanished.

1.2.2. Anyhow, at this point **'immediate' means 'not true'**.

In the dialectical examination, as we saw, the immediate reveals itself as mediated, at least in part, that is, as the product of an act of positing, of a movement of genesis. Then, how could we base any knowledge on it, when it itself is in need of being based, justified? For, as Hegel noticed, the 'familiar' is not 'cognitively understood', and 'intuition' proves to be a mumbling knowledge.

1.2.3 As for the **arbitrary axiom**, it is definitely excluded from dialectic. Dialectic, so to speak, thinks that reality can be explicated by itself (by reality) only, namely by non arbitrary real propositions[16]. When Hegel asks "With what must the Science begin?", he is inquiring about its **necessary** starting-point, on the basis of which he will build the scientific edifice[17]. The *Phenomenology* opens with Sense-Certainty, and the *Logic* with the category of pure Being. Here we have not to do with arbitrariness, but with a beginning at the most abstract level of experience, at the lowest degree of truth according to Hegel : sensible intuition on the one side, empty thought on the other.

In general, empiricists and formalists suppose their subject-matter and the corresponding method of work to be determined from the outset, and build on the basis of past results : starting afresh every time is out of question, they say realistically[18]. However, that is exactly what dialectic claims to do: the re-thinking of all concepts, judgements, proofs and grounds—in that sense, Heidegger appears to be a good pupil of Hegel[19].

If there is no beginning, neither with immediate evidence nor with arbitrariness—it may be asked—how is it possible to start at all? Yet, there is a way out: to consider that reflection itself, to re-think the concept of the beginning.

1.3. Consequently, the deductive method proves unfit for dialectical argumentations

Considered in their deductive aspect, the empirical and formal sciences are based on either generally admitted, or intuitive or stipulated definitions, and on either evident or arbitrary axioms. Now, in dialectic these procedures are regarded as unacceptable, so that the deductive method must really be rejected.

In addition, Hegel makes the following remarks[20].

1.3.1. **The movement of the mathematical demonstration is an activity external to the achieved result**

Passing judgement on the pre-Fregean approach of his time, Hegel says that, in mathematics, the demonstration does not follow from the theorem in question, but advances in the form of orders to be carried out, beginning at a certain point and progressing according to certain procedures, although other procedures are available too. The way remains external to the achieved result, unintelligible in that sense, or based on an 'illuminating intuition' that is quite unexplainable.

1.3.2. Conversely, **the mathematical result is presented as true, independently of the procedure used for its justification**

Indeed, the demonstrated theorem is here recognized as such once and for all, without thinking it necessary to reconstruct the demonstration every time : from here on, with regard to its truth and its meaning, it is considered as autonomous.

Inasmuch as, in mathematics, abstract objects are concerned, that method seems suitable. However, for non abstract thinking, any abstraction turns the true into non true, meaning into empty words. As we saw it, dialectical truth is not expressible in an isolated proposition, and the attained result must not be separated from the justifying process leading to it: truth and the way to truth form here a concrete whole to be grasped. Or, in Kant's language, **the 'deduction' must follow from the concept to be proved** and, in Hegel's language, way and result mediate reciprocally, each conferring meaning to the other[21].

For Hegel, then, the character of evidence attributed to mathematical results comes from the abstract nature of the examined area : there, the movement progresses 'at the surface', without really getting inside the thing itself. Referred to the *Logic*, the matter is about the quantitative categories of the Doctrine of Being, which exposes the world of phenomena as deprived of its essence[22].

Instead of a conceiving *(Begreifen)*, Hegel finds there a mechanical calculus combining magnitudes or relations, **indifferent** to the corresponding qualities. As concerns concrete human reality, however, the principle proves to be quite different: the human domain requires a movement of thought that respects the observed quality, while, in its turn, the quantitative determination becomes secondary[23].

As a justificative process, the deductive method of mathematics therefore has to be ruled out of the dialectical way of thinking, despite the evidence and exactness of the achieved results. For Hegel, these advantages are obtained **to the detriment of the content, emptied of any immanent movement and any concreteness**: constituting an external means, mathematical deduction renders the object itself external. Hence the dichotomous approach, opposing *a priori* and empirical dimensions, and leading to the following alternative : either formal science or empirical science, no room being left for an available third way[24].

If this is the case, a *Logos* (an account) of human reality becomes possible, provided only that thought be given an intuitive dimension and human experience, an *a priori* dimension ; that is, provided that thought and being, subject and object, universal and particular, are reconcilied—a result achieved, in part at least, by the dialectical approach constituted as an **organon,** or as a form that develops a content[25].

Conclusions

From this study it follows that, in dialectic, **there are no apodeictic proofs**.

The **linear** model of justification implies an 'Archimedean' starting-point and a way of progressing univocally ; whereas here the beginning is made not with the true but, say, with the non true, the 'immediate given' which, on thinking it over, shows itself as mediated, that is, as the result of a previous development. Consequently, in the course of its progressive march, dialectic has to correct itself unceasingly, **the proof being a moment of the result** and the 'true definition', reached at the end only. In that way a **circular** mode of justification is shaped, so that any demand for strict consistency appears out of place.

Indeed, strict obedience to the principle of identity would prevent any development, namely any possibility of exposing the movements of human reality in their wealth of sense and figures. For it seems to me that consistency involves a rigid faithfulness to a univocal starting-point, and a total abstention from any deviation whatsoever.

In dialectic, on the other hand, the cultural object is respected in its multi-dimensionality and manifold diversity as far as possible. When experience shows that things are not as expected, in order to attain 'the thing itself', the concrete dialectician therefore does not hesitate to stray from the beaten track, to contemplate with new eyes. In this resides the power of the dialectical approach : its ability to be 'plastic', that is, to follow the movements of the thing without imposing on it, *a priori*, any distorting fixed model.

Uncertain about its oncoming movements, dialectic requires the leading strand of human experience, such as the history of culture in the *Phenomenology* and the history of philosophy in the *Logic*—really, it cannot be satisfied with the 'familiar' or the 'generally admitted'.

The dialectical approach can therefore offer **only discursive justifications that are nevertheless concrete** in so far as, in opposition to Kant, it is dealing with a universal concept that is supposed to contain within itself the corresponding particular[26].

It does not define but exposes, in the two meanings of the German term *'darstellen'* : in every enquiry, it 'embodies' the latent content by unveiling the relevant presuppositions and it 'presents' the different developments in a more or less coherent way. With this procedure, the dialectician somewhat 'illuminates' the thing in need of explanation, or 'elucidates' its signification : **proving** *('démontrer')* **is showing** *('montrer')* [27].

Finally, it may be observe that, as provided with 'plastic' justifications, dialectic is characterized neither by the evidence of its arguments nor by their exactness or univocity. Therefore, it seems that a thinker, in search of certainty, is likely to find little satisfaction in that 'quite flexible' approach : here everything turns to its opposite, and no indubitable, decisive truth can be obtained.

For example, the following paradox may be pointed out : on the one hand, human experience is said to be subjected to dialectical criticism ; and, on the other hand, in the dialectical movement experience itself serves as a leading strand...

The explanation of this paradox, if any, lies in the activity of mediation unfolding between cultural fact and theory : in a progressive movement, dialectic is supposed to illuminate and correct the examined fact which, in return, enriches the mediating theory. As for the formed circle, to me it does not seem 'vicious' but rather indispensable, 'virtuous', so to speak[28].

For how could we inquire what thought is, if we were not thinking beings? And how could we grasp significations, if we did not know in advance what to look for[29]? Plato and Aristotle themselves already noticed that a beginning is never made out of nothing, but out of an implicit and latent foreknowledge which, in the process of learning, becomes explicit and manifest: that is, **to explain dialectically is to develop**.

This assertion leads us to the following.

2. DIALECTIC AS THE *A POSTERIORI* MOVEMENT OF SYSTEMATIC RECONSTRUCTION

First, let us examine the general context of the problem : here we are concerned with a new apprehension of the relation holding between truth and falsity.

Some scholars complain about the absence, in the Hegelian system, of a univocal criterion for truth. Consequently, a mixing of true and false would reign in it, preventing any serious inquiry[30]. For my part, I think that things have to be understood in a different way.

The scientific nature of the domain lies in the form of an **organic system**, conferred on the set of philosophical propositions that found one another, that is, in their hierarchical organization around a common aim : pure Knowledge, logical Idea or absolute Spirit. According to the positions they occupy in the system, concepts and judgements acquire in it corresponding significations—hence their degrees of truth and concreteness[31]. Thus the *Logic* presents, in an ascending onto-logical order, the categories of Being, Essence and the Concept, with the Idea at its summit.

The strict formalist, for his part, being in search of clear-cut criterion, is inclined to ask: 'Is such a proposition true or false?' His question implies an external and excluding relation between the two logical values 'true' and 'false' ; no wonder, since the dividing Understanding does not recognize the relation to an other as constitutive for any determination[32].

In contradistinction to that, for dialectical Reason, the non-true is the isolated, the partial and the one-sided[33]. As such the isolated term contradicts itself, claiming to independence, whereas it can acquire an adequate meaning only when brought in relation to its other. In this way, its **inner contradiction** produces a drive to develop towards a higher level of truth.

It must be added, however, that the achieved result is supposed to contain the traversed path as a constitutive moment: we do not find here some 'scaffolding' to be removed 'at the end'. Even the expression "the false is a moment of the true" proves inadequate, for as a moment the false already contains some degree of truth ; the error rested only in the pretension of autonomy of the so-called isolated[34].

Hence the following crucial consequence: the opposites isolated/related, partial/total, one-sided/synoptic and abstract/concrete, seem more appropriate to the evaluation of dialectical concepts and judgements than the classical pair false/true.

This new apprehension of the thing forms the basis of the Hegelian doctrine of systematic reconstruction, that is the point at issue here. In the study of this issue I distinguish (without separating) between the epistemological and the ontological standpoints.

2.1. From an epistemological standpoint, considering that truth unveils itself progressively, Hegel grasps the acquired pieces of knowledge as moments that together constitute the totality of cognition, expressed in his own philosophy.

2.1.1. Let us start with the **diachronic aspect** of the thing.

Hegel apprehends the great systems of the past as different moments in the development of truth. Namely, he understands the history of philosophy as an arrangement of the different truths scattered in time and space, which thus form a genesis leading to the 'summit', occupied by his own system. Working *post-factum* (after the event), Hegel appraises the different discovered principles and organizes them into a 'total system', on the basis of reciprocal relations of founding and elucidating.

Therefore, the **refutation** *(Widerlegung)* of a past system is not made from an external viewpoint that judges without trying to understand. Now **refuting is developing**, complementing what is lacking by a **negative** act of criticism, which is also **positive** since it is producing : in dialectic, to negate A comes to affirm non-A, a higher, 'truer' concept[35]. What is thus negated is the one-sidedness of the different systems which, little by little, turns into a comprehensive view. That is, in the history of philosophy, which does not amount to a sequence of contingent stories, every system is supposed to contribute to the common work of unveiling and explaining[36].

As a typical example, let us consider Spinoza's system[37]. For Hegel, Spinozism is valid in that it reveals the standpoint of the absolute. However, it also proves non valid in so far as it sees itself as the 'summit'. The explicit plan of the *Ethics* is to take us from *imaginatio* to *scientia intuitiva* through *ratio*. Unfortunately, Hegel says, the still abstract exposition of that philosophy, which is interesting in itself, sets limits to its bearing: here, absolute, attributes and modes do not develop inner relations of constitution and mediation. That is why the true refutation of Spinoza consists in the **negation** of its abstract exposition, in the **preservation** of the absolute's standpoint, and in the **development** of its content shaped that time into an 'organic totality'. Estimating dialectically is also ***aufheben***.[38].

It is in this connection that Hegel mentions its method of dialectical **reconstruction** *(Wiederherstellung)*.

The idea is not new. Indeed, in the dialectical movement Hegel regards the obtained **new immediate** as a 'reconstruction' of the original one, developed through the mediation of **its other**[39]. That moving 'unity of opposites' is supposed to produce the 'true' system, that which is to manifest itself 'at the end only'. Thus, in his philosophy Hegel opposes and unites, *inter alia*, the thoughts of Parmenides and Heraclitus, the Epicureans and the Stoics, Bacon and Böhme, Spinoza and Leibniz, Fichte and Schelling, etc.

Therefore, Hegel adds, true criticism 'comprehends' the examined object and does not cancel it in a one-sided manner, as does external criticism, which is blind (it does not grasp the concrete whole) and impotent (it does not succeed in assimilating the other but only in abolishing it).

In short, in the Hegelian dialectic **to refute is to develop, to reconstruct or *aufheben*,** according to the following way :

a) The **necessity** of the examined moment should be grasped, together with its **positive** side.

b) The unveiled **lack** (its negative side) requires its raising up to a more concrete level.

c) The 'immanent' development of the examined point of view—with the help of the leading strand of experience—unites within itself negative and positive sides, blame and praise, and constitutes thus its *Aufhebung* in the System[40].

Hegel believes that, in principle, all is already accomplished 'in itself', and that just the moment of self-knowing is lacking, a moment developed in his own philosophy. That is, surprising though it may be, while a philosopher who saw the systems of the past as constitutive moments of the 'truth', Hegel does not seem to have forecast that his own system could be transcended : with regard to himself, Hegel the critic remained still naive...[41]

Among contemporaneous works, the work of G.R.G. Mure seems to provide a fruitful example of a critical reconstruction of Hegel's thought. Mure first wrote *An Introduction to Hegel*, in which he **exposed** Hegel's philosophy with the help of Aristotelian, Kantian and some other discussions. Then he published *A Study of Hegel's Logic,* a **critical** work that brought to light different problems to be discussed. He thus showed that it was possible to join exposition and criticism together and to form a refutation that is also a reconstruction[42].

However, despite the real fecundity of that **inner criticism**, requested by Hegel and realized by Mure, I understand that, in all reliable research, the complementary dimension of **external criticism** must not be overlooked : dialectic itself recognizes the external standpoint as a constitutive moment[43]. Consequently, it seems that a good criticism ought to unite the immanent development of the considered system with the enrichment of the view through outer, and even opposing, objections.

It is according to that very approach, both inner and outer, that I address the various issues raised by the Hegelian dialectic[44]. Hence the following conclusion : despite the undeniable progress accomplished, Hegel, like Spinoza before him, is able neither to grasp the whole—from his peculiar historical and cultural position—nor to overcome the other, for the inert and opaque matter sets univocal limits to the domain of explanation[45]. Thus, the comprehensive and coherent system remains an unrealized and unrealizable ideal.

2.1.2. Now, as concerns the **synchronic aspect** of the Hegelian reconstruction, it should be noticed that Hegel *'aufhebt'* (comprehends) in his system the results of the different arts and sciences of his time—or at least he claims to do so.

Even from that new angle a large work of synthesis can be perceived, namely an apprehension of the various cultural products as constitutive moments of the whole truth. The *Realphilosophie* (the philosophy of nature and of spirit in the *Encyclopaedia)* is supposed to present that reconstruction in the form of an exposition of the concrete categories of human experience. However, this seems easier said than done : the movement here appears quite artificial...

2.2. It remains to consider **the ontological point of view**.

According to Hegel, even **world history** shows itself as the product of an activity of *Aufhebung*. Effectively, we can discern in it diverse social phenomena at first opposing each other, then melting into each other, and finally engendering new structures.

Hegel thinks that the achieved result preserves the positive content characterizing the opponents of yesterday, and cancels their mutual exclusive relation.

Accordingly, if the 'cunning of reason' really guides the practical, blind movement of history, on the contrary in philosophy, "its own time apprehended in thought", we are supposed to discover a lucid reconstruction of the past, advancing together with history itself. This allows Hegel to believe that the 'last' philosophy, his own of course, manifests itself as soon as 'all' the historical patterns have been unveiled, thus reproducing in a total and harmonious system the movements of reality said to express the life of spirit; hence the pretension of absoluteness that sometimes renders Hegel's work ridiculous, however prodigious it may be from another standpoint.

The way of the Hegelian reconstruction thus constitutes an inner criticism of the examined thing, affirming and denying, exposing and developing into a system, reflecting and correcting according to the requisites of the thing itself. Whence also its univocal limits:

a) Since the cultural object must equally be grasped from the outside, a phenomenological (or yet another way of) **description** has to complement its teleological explanation.

b) A '**total and last**' point of view is presupposed, which seems hardly understandable, since every man is a "child of his time", including Hegel himself.

c) At the very most, world history constitutes a system in a weak meaning of the term only, the **external** dimension of matter proving important in that field[46]; hence a dimension of **contingency** and exteriority, brought into dialectical explanation by the long human experience playing the part of a leading strand.

d) Hegel assumes a more or less **linear** development of human culture, from the 'less true' to the 'more true', from abstract to concrete. However, current ethnological research shows that things are not so simple : Eurocentrism is not the whole truth. Human history can be divided into specifically distinct cultures, so that to a differentiation between ascending degrees must be added a qualitative differentiation. Maybe, after all, that human reality is richer and more complex than Hegel supposed it to be[47].

Of course, dialectic does not consist in the *a posteriori* reconstruction alone. If that were the case, we would obtain a historical and synthesizing genesis, and not a dialectical movement. Actually, dialectic also progresses in an *a priori* way, a dimension constituting 'the logical' *par excellence.*

We now proceed to that new dimension.

3. DIALECTIC AS AN *A PRIORI* MOVEMENT OF SINKING AND RESTRAINING

In dialectic, Hegel says, the researcher has 'to consider the thing itself'. It is now time to elucidate the meaning of that expression.

Again, I open with a preliminary discussion, referring this time to the nature of **determining reflection**.

According to the *Logic*, the external reflection, a necessary but insufficient moment, does not grasp itself as positing the determinations it meets ; it sees them as 'given'[48]. Dialectic, on the contrary, considers reflection as determining. In order to understand that paradox of a 'reflection' that is also a 'determination', let us go back a moment to Kantian philosophy.

In his *Critique of the Judging Faculty*, Kant separates the **determining** judgement from the **reflecting** judgement[49]. His first *Critique*, for its part, assigns to Understanding (which "legislates on nature") a constitutive power that acts through categories and transcendental principles. As for Reason, it only possesses a regulative, non constitutive, function. When Reason thinks about experience, Kant says, it posits an Idea which in fact does not take part in the constitution of the examined object, but guides Understanding in its progressive search.

From this dichotomous point of view—opposing constitutive to regulative elements, determining to reflecting power—a gap appears, separating the **knowing** Understanding from the Reason that only **thinks**. If however, as Hegel supposes, Understanding is a constitutive moment of Reason, the excluding relation then turns into an inner relation; **determination becomes reflecting** and **reflection becomes determining**[50]. Namely, on the one hand human thought, in the product of its conceptual activity, now recognizes itself as acting ; and, on the other hand, thought is assigned a non formal, intuitive dimension : with Hegel, logic becomes an organon explicitly![51]

Let us see how this 'determining reflection' proceeds.

The discussion is about the movement of the Concept. Hitherto, we have considered negation or self-contradiction as the driving principle of the movement. The matter now is **to consider the act of consideration itself as moving the dialectic.**

External reflection 'surveys' all content it examines in a 'detached' manner, Hegel says[52]. And, instead of getting involved in the thing and surrendering to it, it is preoccupied with itself; that is, it refuses to sink *(sich zu versenken)* into it, to remain in it and 'to forget itself'[53]. Its prejudices are stronger. Therefore, it does not bend over the **peculiar object** facing it, but turns away towards something else. It is not ready 'to deny itself', to deny its rigid attitude, for it does not know that the negative is positive as well. For fear of losing itself, it obstinately persists in its external behaviour, not moved and not moving—a dead point!

So to speak, addressing it Hegel advocates : sink into the thing itself, forget yourself, forget your prejudices ; and, through the activity of your thinking you will discover the 'movement of the Concept' unfolding from the positive to the negative, then to a new positive. In this way you will not lose yourself ; on the contrary, immersed in the **particular** determined content, your **general** apprehension will constitute a **concrete unity**.

The matter does not stop at that point, however. For if we ought not 'to survey' the examined content 'from the top', equally we ought not to sink into it as far as to direct the movement arbitrarily. Consequently, from that standpoint a certain '**restraint**' *(Enthaltsamkeit)* is required: to avoid interfering in the movement of the thing itself. Here Hegel addresses the contingent representation *(Vorstellung)*, and asks it for an attention respecting the examined object and **its inner necessity**. Since this time the matter is about particular apprehension, the latter has to unite with the universality of the conceptual content—in this way too, a concrete unity is to be produced.

The requisite is twofold. Of **formal thought** it is required to linger in the thing, so that its universal activity should move the cultural object and fill it with some content. Of **representative thought** it is required not to divert the movement of the thing according to its own interests : on the contrary, in moving with the examined universal content it brings about a conceptual self-development.

As a third way, dialectic thus consists in **sinking into** the thing itself in order to move it, as well as in **restraining from** a deviating mixing. Hence 'the attention centred on the Concept' that shows itself as both active and passive—as always in dialectic, ambiguity is inescapable.

From this, the following conclusion can be drawn.

In lingering in the thing, the dialectician **moves** it and, in renouncing to its own particular 'I', he himself **is moved**. Accordingly, it is necessary 'to consider the thing itself' and not to aim at something else. Moreover, the examined content has to be respected and not distorted. That is, one ought to be **close** to the thing in order to observe it in its peculiarity, but **not too close** in order not to hinder its necessary movement[54].

These opposing requisites may be called the 'dialectical identity of sinking and restraining', an activity said to yield the movement of the Concept and, likewise, the movement of the thinker himself as a thinker. At this point it seems relevant to mention the 'three subjects' of the Hegelian proposition : **the considering subject, the considered object and the Concept (of) consideration, which are moving together**[55].

In view of the importance of this ternary movement, let us consider it further.

The dialectician examines an 'immediate given' and discovers that it is mediated ; for, from his own point of view the so-called 'given' is in fact produced. In that development, then, from an immediate term to a mediated one, the movement of the Concept reveals itself as the movement of **the thing itself: out of an object of sensation, it has become an object of thought.**[56]

Furthermore, the movement of the Concept implies the activity of the thinker, supposed to give up his rigid position (as formalistic thinking) and his arbitrary operation (as representative thinking). This very sinking and restraining reflection is said to constitute the concrete unity of the self-animating object-subject. In that **dialectical unity of the thinker, the thought term and the act of thought itself** producing the Hegelian Concept—the 'true' subject—every moment realizes itself through the mediation of its other. In this circular flow, the mover itself is moved[57].

On that occasion Hegel talks of 'the cunning' of knowledge[58] which, on the one side, as reflection seems to abstain from any active interference ; and which, on the other side, does just the opposite. That is, as an object of the act of sinking, the examined content becomes dissolved *(aufgelöst)* and is thus transformed into a moment of the dialectical movement. In this way, Hegel says, reflection, which looks like keeping itself from any activity, actually does determine and move.

Is not this 'determining reflection' what we have just dealt with? Indeed, it is! In a concrete language, the discussion is about an active and living thought—'spontaneous' Kant would say—moving itself and recognizing itself in the moved conceptual content.

With regard to the rigid external thinking and the arbitrary representation, Hegel says, dialectical thought alone is able to realize the examined content and, thereby, to realize itself. In Hegel, then, the mystery of the 'thought thinking itself' is resolved in this way : the 'divine' (absolute) thought accomplishes itself through the mediation of human thought, which recognizes itself in the considered 'given' for, at least in part, the latter proves to be the product of human 'determining reflection'[59].

We have thus studied the dialectical explanation as an *a priori* movement of sinking and restraining, and as an *a posteriori* movement of systematic reconstruction. That bidimensionality can also be accounted for in the following manner.

The formal criticism shows that the *a priori* movement of dialectic meets with leaps for, from an immanent standpoint, the transition from the mediated determination to the new immediate is impracticable[60]. It therefore becomes necessary to resort to experience. This allows us to find out in the history of our culture the required third term, a synthesis of the original immediate with the mediated, which *a priori* reflection alone is not able to produce[61]: the dialectical movement, continuous and discrete at once, truly develops *a priori* through 'restrained sinking' into the examined content, and *a posteriori* through systematic reconstruction according to the leading strand of human experience. Dialectic is really unable to make any predictions whatever, if not of mere illusions...

In short, **dialectical justification is both exposition and explanation**—an expression that reminds us of Kant's discursive justification. Furthermore, in Hegel's system the onto-logical basis is supposed to base itself: the problem of the **infinite regression**, displayed by a ground itself in need of grounding, is solved by the **circle** of the realized self-grounding.

Indeed, if grounding is **regressing** to a first truth, and if explaining is **progressing** towards an elucidation of the thing, in dialectic self-grounding also proves to be self-explanation, for here *explicans* and *explicandum* are exchanging roles—as the following chapter intends to show.

4. THE DIALECTICAL EXPLANATION EXPLAINS ITSELF AS A DOUBLE MOVEMENT BOTH *A PRIORI* AND *A POSTERIORI*

In this chapter, I examine the nature of dialectical explanation in the context of **the problem of predication**. This problem arises when an attempt is made to think the relation holding between a certain determination *(die Bestimmung)* and the corresponding determined term *(das Bestimmte)*.[62]

The Platonic philosophy distinguishes between **the world of transient phenomena** made of shadows and images, and **the world of eternal Ideas**, which constitutes the essence and truth of the sensible things. The 'presence' of the Idea in the sensible thing accounts for its property and, from that point of view, the **determined** thing 'takes part' in the **determining** Idea. For instance, in the ethical realm a good act is one that 'takes part' in the Idea of Good.

However, the problem becomes complicated as soon as one tries to understand the Idea itself. Concerning its own determination, it may be asked : **'What is the Idea of Good?'**. At this point, two ways are available. Either an **infinite progress**, every determination being explained by another, which is in turn explained by a new determination, and so forth ; or a **circular** reasoning considering the Idea as self-determining. The passage from the expression of *methexis* (participation) to that of *mimesis* (imitation) does not alter things at all.[63]

Plato is acquainted with this problem, and rejects these solutions, both seeming invalid to him. In the *Republic* he puts forwards the doctrine of the **intuition of Ideas**, signifying that what the mind's eye directly apprehend—the truth—cannot be expressed in discursive language[64]. That doctrine, about the impossibility of stating what really is, leads to a mystical vision of things : the relevant explanation does not explain itself.

Furthermore in the *Parmenides*, a late dialogue, Plato discovers another paradox in this connection[65]: if we attempt to think the Parmenidean One as One, in the full meaning of the word, then we can tell nothing more about it, for any additional qualification would turn it into Many. Even Being cannot be attributed to it, for if we were to say that 'the One is', then it would no longer be one but two : One and Being. In this case, the patent contradiction would allow us to deduce no matter what; for instance, to pass from the tautological 'One is One' to the self-contradictory 'One is Being', from mystical silence to vain chattering. Talking significantly about the Parmenidean One thus proves impossible. From this we can understand that Plato did not manage to solve the problem of predication.

According to the *Republic*, the determining essence remains inexpressible, and the *Parmenides* only offers empty or self-contradictory propositions. As for the *Sophistes* it shows some progress, since it proposes a doctrine of **the community of Ideas**[66], according to which the different determinations do explain one another. Unfortunately that doctrine, which is interesting in itself, was not developed by Plato—Aristotle was the one who carried on with his plan.

The problem of predication comes out in **Aristotle**'s theory of science. At the outset, the Aristotelian science lays down some axioms, common presuppositions, hypotheses and definitions. The role of the definition is to indicate the essence of the examined object, its genus and specific difference.

Now, the diverse essential determinations really explain one another, for 'genus' and 'species' have turned to relative concepts 'in community', but still not in 'mutual community' : the genus prevails! Therefore, the following decisive question is to be asked: **is the defining genus itself definable?**

According to the Aristotelian method, the definition of that genus needs a higher, more comprehensive genus, in relation to which the former appears as a subordinate species. When climbing that 'natural' ladder, one is inevitably led to the highest genus which, for its part, is not definable. Indeed, there is no more defining genus above it; and, in addition, Aristotle explicitly affirms that it "is not possible ... that the genus, taken apart from its species, should (may) be assigned to its differences".[67]

In other words, Aristotle rejects any solution that might bring us either to an infinite progression or to a circle. He learned very well from the teaching of Plato's *Parmenides*: : as fundamental concepts, Being and the One are indefinable. In order to escape this dilemma, Aristotle, like the young Plato before him, resorted to the strategy of immediate knowledge : rational intuition should provide the required highest genus.[68]

From this discussion it follows that Ancient philosophy, although rich in explaining principles, does not succeed in explaining, in a detailed and consistent manner, these very principles of explanation—**the rational *Logos* has to rely on the given intuition**. Consequently, we can really understand neither the determination (Idea, genus and species), nor the determined (sensible phenomena, first substance), and nor the relation between them.

If that is the case, any discourse about the real may give rise to the following trilemma : either a founding on rational intuition (which, in Aristotle, is at the origin of the basic principles), or a self-contradiction (the One turning into Many, according to Plato's *Parmenides)*, or still a founding on sensible intuition (with which the empiricist starts).

However surprising it may be, Hegel seems to take all of these different ways. Thus, his system exposes a reciprocal mediation between the intelligible and the sensible elements : as we saw above, dialectic is at once *a priori* and *a posteriori*, advancing through the activity of pure (rational) thought, according to the leading strand of human (sensible) experience—hence the circle and paradox, and also the self-contradiction, of a pure thinking that is also empirical.

However, where the formalist makes separations, the dialectician, who does not fear antonymy, only distinguishes. And to the extent that, for us, in human reality every entity acquires its meaning only when related to its other, we have to agree with this assertion : when dialectically interpreted, human experience is truly intelligible, sensible and self-contradictory at one and the same time.[69]

In the problem of predication, then, Hegel adopts a new approach, expressible in terms of Greek philosophy.

The Hegelian dialectic considers the **Ideas** as active, living and moving. Now, these Ideas relate to each other and also to themselves through the mediation of their other. In consequence, in order to understand the world of Ideas it is now advisable to address the world of phenomena : the Idea of Good is good through good actions, and conversely. Namely, the two movements of 'presence' and 'participation' complement one another : Good and good actions account for one another ; or, paradoxically, the phenomenon is also 'present' in the Idea and the Idea also 'takes part' in the phenomenon.

It is this very bidirectional doctrine that shows how dialectical explanation may explain itself; the discussion is really about **the self-explanation of a thought turned towards experience**, a doctrine already detectable in Kant, in so far as abstraction is made of the transcendental Ego and of the Thing-in-itself[70].

Here is the place to mention, so I understand, **Hegel's inverted world** *(die verkehrte Welt)* , the 'specular' relation or the double movement of the speculative proposition. In that sense, the Other is my reflection only, the *Spiegelbild* revealing the Other as a Same and, reciprocally, the Same as an Other—the optical metaphor being logically translatable, I think, in terms of reciprocal mediation[71].

In opposition to Aristotle, we can say that, likewise, the genus may well be determined by its own **differentia**, as a genus self-particularizing into its different species and realizing itself through their mediation. Therefore, instead of an 'absolute' beginning with **dogmatic** intuition or of a **sceptical** infinite progression, we find here the **critical circle of a double movement** that is both regressive and progressive, from fact to theory and vice versa.

The linear model of explanation rejects the vicious circle but, at the outset, runs into difficulties : must it start with intuition, axiom or fact? Whereas the circular model of explanation founds itself through the mediation of the explained things—the beginning being really understood at the end only, after the different latent determinations in play have been unveiled[72].

5. AN ILLUSTRATION OF THE CIRCULAR WAY OF EXPLANATION: THE DIALECTIC OF SOMETHING AND AN OTHER

Thus far, I have highlighted the relevant principles. All that remaining quite abstract, I now propose to see concretely what the matter is about.

The Hegelian logic is said to possess a certain explanatory power. In addition, as a mode of explanation it presents itself as self-explaining. Accordingly, the purpose here is to show that, partly at least, Hegel's *Logic* explains things while explaining itself, according to the suggested pattern of the double movement[73].

5.1. The macrodialectical movement

In his *Logic*, Hegel exposes the movements of human thought. To achieve that task, he divides the book as follows : first, an **objective logic** dealing with the produced object ; then a **subjective logic** dealing with thought producing its world[74]. Therefore, in view of that unavoidable partition due to the exposition of philosophy in discursive language, the logical work, abstract as such, becomes still more abstract—in any study of the text, that difficulty must be taken into account[75].

Moreover, to that binary division coming from the *Phenomenology*, an original ternary division is added: the *Logic* is also divided into the **Doctrines of Being, Essence and the Concept**, and it closes with the presentation of the Idea—from the outset a tension arises, as a result of the above-mentioned twofold inner division. If "The True is the whole", as Hegel asserts, truth then comes to light at the level of **the Idea**, the last and total category (a paradoxical one as a totality being also a moment), as the Concept effectively realized or the still subjective thinking having grown into a thought world.

The *Logic*, as a categorial explanation of the world, is self-explanatory to the extent that, first, objectivity and subjectivity enlighten one another : according to the well-known Kantian pattern, subjectivity (here the Hegelian Concept) is such only through the mediation of objectivity which it constitutes and, reciprocally, objectivity (the thought world) is such only through the mediation of the presupposed subjectivity.

Furthermore, Being, Essence and Concept form an explanatory syllogism in so far as, on the one hand, there is a double movement from Being to Essence and return, the movement of Being being exposed from the standpoint of Essence, and Essence, in its turn, reconstructing the categories of Being in the form of Appearance, Existence, Phenomenon, etc.[76]; and in so far as, on the other hand, that double movement constitutes the Concept, as the subjective Concept positing objectivity and returning to itself, realized as the absolute Idea[77]. In the developing circle of mediations, the different moments explaining one another are supposed to shape a whole, *qua* a self-explanatory system that confers to everyone of its moments signification and concreteness.

From this point of view, while explaining the basic structure of the real, the *Logic* at the same time seems to explain itself—if at least only.

At the **macrodialectical** level in question, the *a priori* movement of sinking and restraining does not appear. On the other hand, the *a posteriori* movement of systematic reconstruction does come to light.

The **history of philosophy** being presupposed, in the doctrine of Being the discussion seems to be about the reconstruction of Presocratic and mathematical categories. In the doctrine of Essence, it is about the reconstruction of Platonic and Aristotelian categories and of Kantian, Spinozistic and modern physical thought. As for the doctrine of the Concept, we can discern in it the reconstruction of formal traditional logic and of the philosophy of Hegel's time, such as the thoughts of Fichte and Schelling.

If the *a posteriori* systematic reconstruction looks acceptable, from the standpoint of the *a priori* movement we are to encounter inescapable leaps, the most notorious being the transition from objective to subjective logic. A certain response to that issue can be found in the logical development according to the leading strand of human experience and to the leading Idea of an organic system.[78]

That is, beginning not with no thing but with a rich tradition of philosophical researches, and advancing towards the constitution of his System, Hegel can arrange the various materials at his disposal at least externally, from partial to total, from the isolated to the related, from abstract to concrete. For, as Hegel observes, "What is first in science has had to show itself first, too, historically"[79]. Witness the opening with the presocratic categories of Being, Nothing and Becoming, and the closing with the Hegelian categories of the Concept and the absolute Idea.

All of this does not yet form a dialectical work, but rather a collection of categories arranged according to general and external guiding principles[80]. For the *a priori* movement of sinking and restraining is still lacking in it, that is, the dialectical *par excellence*. This emerges only at the level of the **microdialectic**, of the detailed movement of individual categories.

I therefore proceed with the examination of a specific example, the important **dialectic of Something and an Other**[81].

5.2. The microdialectical movement

It is well-known that the dialectic of Something *(Etwas)* and an Other *(ein Anderes)* constitutes a criticism of the Kantian *'Etwas'*, a term by which Kant sometimes denotes the Thing-in-itself. In Kant, this Thing is said to constitute the true real, albeit unknowable—only thinkable...[82]

With the help of his dialectic, Hegel, by an ironic reversal, wants to show that what Kant regards as the highest, the essence *qua Etwas*, in fact consists in an abstract elementary category, since every determined being is something : this table, I, the leaf or the mud... For instance, in sharp contrast to Kant's position, Hegel asserts that, for us, **to be is to be determined, thus knowable** ; so that the unknowable (that is, the non determined) is not, or is only the dead One and the empty Nothingness.

Consequently, as a *Gedankending* determined in its apparent indetermination, the Thing-in-itself may be known : it is but the empty abstraction of a view that isolates the thing from its being-for-another. Thus, Hegel notes, **to be determined is to be related to an other.**

5.2.1. The context of the examined movement

If the true is really "the whole", from a didactic point of view it seems suitable to start the exposition of the movement from the very end. Anyhow, in that circular and paradoxical approach, the beginning is truly understood only in retrospect, the determining context having to be taken into account[83]. Hence the following study.

The category of Something is to be found in the Objective logic. Its exposition thus deals with thought directed to the world of objects surrounding us[84]: in everyday life we have always to do with 'something', that is with no matter what.

Moreover, this Something is neither Essence nor Concept, but simply an external elementary category of the doctrine of Being, with neither depth nor ground ; so that to understand the world as an aggregate of things is to understand it quite superficially.

Hegel holds that this is the point of view of atomistic thinking, which examines the various things separately and gives them a name. From this viewpoint, reality appears to consist of an indefinite quantity of points, numbers and atoms that combine into various aggregates. The endlessness of that universe, which captivates the child so much, is boring for Hegel, who perceives there just a world of mechanical repetition, of the diverse but not of the different for, after all, it is always the same story[85]. Hegel believes that human events alone can really be of interest to human beings...

In the doctrine of Being itself we meet, surrounding Something, the categories of Determinate Being and Finite : Something is finite since it is faced with an Other. As we saw it, **the absolutely isolated is devoid of any reality** for, to us human beings, to be isolated is possible only against an other—in this paradoxical approach, isolation itself is grasped as a relation.

Hence, it is in the context of that atomistic vision of reality that we have to think the dialectic of Something and an Other.

In the reconstruction of that logical movement, I do not intend to reproduce the Hegelian argument faithfully, but I translate it into a language intelligible to us, twentieth-century readers : we have seen that **predicative** exposition in a modern language and **dialectical** explanation must complement one another.[86]

5.2.2. Thinking 'Something'

The purpose is to think the category of Something, to think something. To this end, the following preliminary remarks must be made:

Firstly, I did not say 'to think **about** something', about that which comes into our mind, but 'to think the category of Something', that is, something in general as an object of thought. Here the object is a particular category, a spiritual or cultural entity as a determination of thought—the intelligible ***Denkbestimmung***—and not an entity belonging to the empirical world. Thus 'thought is thinking itself', so to speak, through the mediation of the sensible things divested of their material attributes[87]. In short, thought constitutes the *Sache* in question, the 'thing itself' and not this or that empirical thing.

Secondly, the purpose is to think, and not to remember or imagine ; it is to engage in 'pure' thinking and not in empirical thinking[88].

Yet, what thinking is can only be understood at the end, after having passed through the whole logic... That circle proves quite inescapable.

Now, Hegel says ironically, in order to learn we first have to jump into the water, rather than to learn to swim at first... That is, for all that we are not at a loss. We shall use a guide, a **leading strand** : Hegel himself, a known commentator[89] or our own experience and culture—a leading strand but not a despot, for the aim is to think according to our own personality, our own creative power.

A final point must still be specified: once having thought Something, it will be necessary to start again and think the Other ; indeed, the explanatory double movement requires a double beginning.[90]

My question is thus the following : when I am thinking some determined thing, in general what am I thinking?[91]

To help in my task I can appeal to Kant who, at the beginning of his *First Critique*, asked a similar question concerning the judgement '7+5=12', and concluded that, when we think the sum '7+5' we do not think the number 12 [92]. The judgement '7+5=12' is then synthetic (it extends our knowledge of reality) and *a priori* (it is not in need of empirical experience but of an *a priori* intuition : counting with our fingers or using representive points on a sheet of paper).[93]

Based on this example, I again ask : what am I thinking when I am referring to 'something'? In this discussion it will be supposed that man thinks, that we think and that this view is meaningful.

Hegel's answer may, in the first place, be expressed as follows: when I am thinking some determined thing in general, I am also thinking the other of that thing. Or: **to be something is to be related to an other**.

How does Hegel achieve this result? First, he already knows the whole, and has classified the different categories he drew from the history of philosophy; he then reconstituted the traversed path genetically—this constitutes the external, *a posteriori* moment of systematic re-construction.[94]

In addition, dialectically speaking, the discussion is also about immanent thought; that is, **to think some determined thing is to think an other!** As in Kant's transcendental deduction, the derivation (of that other) has to be made out of the examined concept, that is internally: the *a posteriori* reconstruction must thus be completed by the *a priori* dimension of sinking and restraining.[95]

In other words, it is within the very category of Something that its opposite has also to come to light, and not just in Hegel's knowledge or in the dictionary, however sophisticated they may be—for here **critical** thinking is at work, whereas the 'familiar' is not as familiar as it seems.

This discussion may appear quite abstract or abstruse, so that I propose the following concrete example.[96]

This table, for instance, when I perceive it as something, or as a determined being, I perceive it in the room, standing on the floor, in front of me while I am leaning on it. I am unable to grasp that table in a total void! Expressed in a Kantian language, it becomes: **the necessary condition of perceiving a table *qua* table** is its location in a fitting milieu. In this case the milieu, as its other, is its physical surrounding or some privileged object such as a chair, a man leaning, etc.

Likewise, **any thought concerning something proves contextual** ; I am really unable to think in the void, outside the human context.[97]

'Are we here concerned with a discovery or with a creation?', it may be asked. 'With both', replies the dialectician[98]. For, on the one hand, here Hegel tells us what he means by 'thinking' and, if we want to cooperate, we have to follow him—in that sense he **creates** and imposes his point of view on us.

On the other hand, he succeeds in convincing us to the extent only that he is talking about us, that **we recognize ourselves** in his discourse : indeed, Hegel is a man, like me. To that extent, then, through his particular way of thinking he helps me to know myself better, **to discover** my own being ; for the thinking individual I discern in myself also constitutes the universal thought, present in all men in so far as they are men who have grown up and learned in one and the same cultural milieu.

If, for an African living in his native village, Hegel may appear as an absolute 'other', here, on the contrary, the argumentations of the *Logic* address themselves to readers belonging to our Western culture.

In that context, then, **following Hegel** is at once following him on its particular way and following ourselves on our universal way. Moreover, fortunately we do not follow him very well, and in place of his thinking we also put our own. In other words, **we tend to interpret** from our own point of view, which is universal as well as particular[99]. What allows us to be, at the same time, ourselves and an other, of Hegel's period and of our own.

Thus, following Hegel is also following an other than himself, ourselves as an other and the same—we see that the categories of the *Logic* (here : the Same and an Other) apply to the reading of the *Logic* itself! In this way, we can manage to avoid both extremes : either becoming the reflection, the parrot of the master, or its violent and destructive opponent.

To sum up, to be something is to be in an inner constitutive relation to an other ; that is, **the Other is the determining source of Something**[100]. Something entirely isolated would not be something at all but nothing, an empty Being or an abstract Nothingness.

From a Hegelian point of view, I suggest illustrating my thesis with the following example.

As far as the atomic world is concerned, this issue may be raised: let the atom be separated from an other atom by a total void; it is then not a real thing. So that the scientist is brought to find some relation between these atoms, such as the field of universal gravitation—in this context, I understand that this very field confers to the atom its essential determination[101]. If so (am I right?), even for the physicist there is no question of conceiving the atom outside of a universal magnetic field. As for the inter-atomic void itself, it is **necessarily** (rationally) filled with magnetic, electric fields and the like.

Therefore, from a Hegelian point of view again, it seems that, considering the fact that the scientist talks about something and not about nothing, as science progresses the atom and the void gradually fill up with such a multiplicity of determinations that the distinction between atom and void may become problematic.[102]

Now, at a human level the same applies to the isolated individual, whether Robinson Crusoe on his island or the city dweller lost among the skyscrapers: 'everyman for himself' means in fact 'nothing for anybody', for man, as a political animal, really exists only when related to an other, to its other—the friend, beloved or fellow worker.

Dialectically speaking, then, Something posits an Other which it presupposes, and relates itself to it. This act constitutes the *a priori* movement of sinking and restraining. **Sinking,** because an inner consideration is concerned ; freeing myself from my usual prejudices I have to sink into the thing itself, not beside it, and to animate it with my living and creating thought. **Restraining** too in order **to keep my distance** and perceive the examined object in its right perspective, grasping it in its proper context without interfering ; that is, following that individual thing in its necessary movement. In this way the paradox of a thing, which is at once moved (by my reflection) and self-moving (through its inner contradiction), is again brought out.[103]

5.2.3. Thinking the Other

Something is, then, only when related to an Other. Isolated, it is nothing and only in the presence of the other does it become something. Even in common language, to be something ('someone' in English) is to be recognized, and to become something ('someone') requires the help of an other.

In the exposition of the examined dialectic, a **double beginning** was mentioned. Therefore, basing ourselves on the just presented analysis we have to think the Other. This new consideration gives : **the Other is**

Something too! Or, in more precise terms, to think an other is, in general, also to think a determined thing, the Other of Something.

For instance, again if the other of the table is the chair, apprehending the chair is still apprehending something. That is, expressed in a Hegelian language, Something relates to an Other and, in its turn, the Other is Something. But that Other, in order to be Something, must also find its own determination in an Other, such as the chair as against the table, or as against the man who is sitting on it, or the room in which it stands, etc. For us, then, a new movement develops.

5.2.4. The infinite progression of Something and an Other

In concrete terms, the **inner contradiction** is the driving force of dialectical development[104]. Let us see how that inner tension develops in the case at issue.

We saw above that the Other is the determining source of Something: Something, which believes to be such **in itself,** in fact is such **for another** only. A difference thus arises between what **it claims** to be, something in itself, and what **it is** in fact, something for another and nothing in itself!

The reader may be inclined to wonder about the use of anthropomorphic language in the examination of a logical topic. If so, I beg him to turn his attention to the following.

First, for Hegel Something is truly a subject, the first form of a subject as negation of negation or self-relation, as we shall see later. Next, this examination brings us to produce a movement of thought, and a living thought is always a thought of a human being, here ours which constitutes that dialectic. Naturally, that movement is not for the table nor the chair, but **for us** only as **subjects** who are examining the category[105], sinking and restraining, moving and moved at the same time—indeed, we are already in possession of some truth...[106]

In order to develop that dialectic, we first have to let us lead passively ; then, on a second reading we have to reconstruct the examined movement, this time both passively and actively. In the context of this approach, **to understand is really to produce from within**[107]; criticizing from the outside would only bring to see nothing, not even the specific problems at stake. As for the above-mentioned "jump into the water", that is to be done in so far only as one somewhat recognizes oneself in the Hegelian point of view—indeed, a certain culture is always presupposed.

In itself (as isolated) Something is truly nothing, we have said, for it is only **for another** (in relation to an other); this constitutes its inner contradiction. Therefore, simultaneously it is (as something) and it is not (its being lying outside) ; hence the tension and movement : Something becomes.

To put it another way, Something as Finite has a **limit** *(Grenze)*, the determination that constitutes it[108]. Within its limit something **is**—*qua* this determination—**and is not** since its own being is beyond it. And, conversely, the limit **is**, as that which separates something from an other, **and it is not**, for it does not succeed in holding back the moving Something.[109]

This situation, a most dialectical one, brings us to a **necessary** movement[110]: Something, *qua* Finite, contains a **Sollen** (an ought-to-be) which pushes it forward to realize itself and attain its being, there at the other, beyond its own limit which, at once, separates from and unites with that other.

Now, at the outset, Something as one-sided does not recognize itself in its Other, grasped by it as an Other only (in fact **we** are speaking here, as subjects sunk in the category). In this **excluding** relation, the moment of the destructive contradiction comes to light[111]. The other being for it just an other and not something, that something does not find satisfaction: isolated, it remains abstract, empty, unrealized.

However, in fact, that other is also something that is in search of its being. In its turn it passes beyond its own limit, in a second other where again it does not find the expected realization.

In the course of this process, the **relation of otherness** is never overcome : it passes away in something that does not recognize itself in the other, and again comes to be as a new other—unceasingly.

In this way, that which, at the beginning, was just an exclusive relation has now become an **endless progression** of a developing something that cannot reach its true being[112]. From one other to the next, it is continually in search of itself without being able to find itself.

There lies the contradiction : let it be Something or an Other, it contains within itself its own opposite. Consequently, not only is it an other for an other but also **an other for itself**—as a negative relation within Something which, at that stage, cannot find inner peace or reconciliation. Launched into its frantic pursuit, it bears within itself the seed of its own alienation: alien from itself, from the other, in the state of splitting it finds its own destiny, its **Bestimmung.**[113]

This so expected self-realization, it could find it if it were also something **for itself**, and not just for an other. However, given that it does not recognize itself in its other, from one **Sollen** to the next it is thus endlessly induced to re-create the failure from which it flees—a crazy running in search of sense, encountering only emptiness and monotonous repetition.[114]

5.2.5. The completion: a reconciliation in the circle of the true infinite

To be a subject means to be for itself. However, that level can be reached only if the negative self-relation—being another to itself—is altered into a positive one. Such a transformation requires flexibility and self-recognition in the other.

Now, **man thinks**[115].

Therefore for us (for the discussion is about **our** thinking as creative of a human world) the **conversion** proves possible, even necessary.

To this end, it is sufficient to grasp Something in its determining context (the Other) and that context itself as a determining Something (as the Other of Something). In this manner, the annihilition of Something is itself annihilated, forming a productive 'negation of negation'.[116]

It is that very **experience** and **reasoning** that allow to rise from the alienating exclusion[117], a first abstract negation, up to the realizing unification throught the path of the negation of negation, the second living and concrete negation.

In other words Something—having experienced and suffered, and now reasoning—finally understands that the Other, the source of its inner splitting, may also become a source of life : in truth, to be related to an other is to be related to oneself, and reciprocally.

The other is still myself[118]. As soon as I grasp this truth, I also grasp that the other does not determine me univocally. Rather, I must say that **I determine myself through the mediation of the other, and vice versa.**

At this point, the destructive inner contradiction is resolved *(aufgehoben)*, a certain peace is attained and the one-sidedness is overcome. 'For us', now, Something turns out to be for itself and for an other, something and other at once—purely and simply a subject[119]. It is then itself and different from itself ; a new contradiction arises, this time a creative one.

Something no longer seeks its being beyond, but in its self-identity, that is, in its identity with the Other. The dialectical unity thus achieved gives the **true infinite**, a circle of the determination recognizing itself in the other and vice versa : there is no longer a flight forwards, but a **coming back to itself** which is mediation and self-realizing reflection. At present, in the Other, Something feels 'at home', *bei sich*, reconciled with itself and with its surroundings.

Beforehand, we had to deal with two separate worlds, that of the finite Something grasped as real, and that of the expected infinite looking as ideel (as conceived only)—**this was** the dualistic point of view, criticized by Hegel. But now, according to the image of the **inverted world**[120], Hegel asserts that the finite (as a moment to be overcome) is ideel only, and that the true real consists precisely in that reflecting and mediating circular infinite, an in-itself that has become for-itself through the mediation of the for-an-other—**this is** the Hegelian idealism.

As for me, grasping these two 'inverted' thoughts (dualism and Hegelianism) together, I propose to talk of **the finitude and infinitude of the finite and infinite terms**—a really paradoxical and illuminating expression, in my opinion at least[121]. Because a movement and not a formula is concerned, Hegel asks to avoid using such formal expressions as "the unity of the finite and the infinite", which is too rigid in his eyes[122].

Now, '**an** Other' of Something, a multiple and recurrent one, has become '**its** Other' as its individual opposite!

Although the contradiction has truly found a certain solution, this does not mean that it has disappeared, however. On the contrary, from the **destructive contradiction** of Something ever seeking itself in the Other and ever failing—that is, from the dichotomous thinking—we have passed to the **creative contradiction,** as a unification of the differentiated terms, which is also a differentiation of the united terms.

Thus, we see that the renewed reconciliation preserves the 'overcome' *(aufgehoben)* inner contradiction, this time as a source of life and not of death, of communication and not of emptiness and solitude ; that is, as a product of the reflecting and determining dialectical thought.

6. FINAL CONCLUSIONS

6.1. Dialectical explanation develops as a **double movement** between the concept in question and its opposite, **each one** being at once an *a posteriori* movement of systematic reconstruction and an *a priori* movement of sinking and restraining. In this way, a **syllogistic movement** forms itself as the movement of the speculative proposition or the specular relation of the inverted world. That is, the explanation itself comes to be explained, for the dialectical circle, when turned to human reality, avoids the one-sidedness of the deductive-nomological explanation.[123]

Indeed, in the D-N pattern of explanation, the beginning is made with a supposed true starting-point, such as a theory now in favour or an admitted general proposition. In the dialectical explanation, on the other hand, *explicans* and *explicandum* reverse their roles : from 'A' 'non-A' is deduced, then from 'non-A' 'A' is deduced, the two opposites, through their reciprocal mediation, producing B, a new concept explaining the just-mentioned double movement.

Hence, this mode of explanation explains itself through its own logical **genesis** in a movement that is both progressive and regressive. It is **progressive** in that new categories are posited, and **regressive** in that there is also a return to the original ground. In Hegelian language, we can say that the last, from the standpoint of **discovery**, is the first from the standpoint of **justification**—B is really the 'truth' of A and non-A.

Obviously, this dialectical mode of explanation constitutes a philosophical **justification** only and not a scientific proof, since apodeicticity is not attained. That is, in the course of the growing elucidation, examined concepts and principles are more clearly determined and specified, in the context of the surrounding culture.

However, it must be pointed out that the thinker is 'situated', as a "child of his time", being the product both of a particular and a universal experience and that, furthermore, he explains a movement of which he is an **observer** as well as a **participator**. So that the dialectical explanation is impregnated with this double dimension, showing itself as particular and universal, that is as subjective and objective at once ; in other words, it is objective up to a certain point—various alternatives are always available.[124]

Within the framework of a given culture, the dialectical explanation may be regarded as satisfactory, although 'provisional' and 'on the way', since the theoretical progress of knowledge and the practical progress of history demands a permanent revision of the current ideas, 'taken for granted'. Thus understood, in principle dialectic proves to be **flexible** and **critical**: here fundamentals, thoughts and cultural things are said to be moved continuously—while rigidity, dogmatism and petrification are swept aside, as are therefore also anarchism, scepticism and nihilism, their opposites.[125]

In short, that mode of explanation proves at once **historical** and **systematic**, diachronic and synchronic, within the circle of an elucidation comprising also its own **real genesis**: "the path to science is itself already science", says Hegel[126]. To put it another way, we have to do with a development **from less true to more true**, and not with a leap from falsehood to truth. In principle, every proposed truth is corrigible, including this truth concerning the very nature of explanation, that is of dialectic itself! [127]

At this crucial point, I propose to turn to Kant: "Reason must not, therefore, in its transcendental (read here: 'dialectical') endeavours, hasten forward with sanguine expectations, as though the path which it has traversed led directly to the goal, and as though the accepted premisses could be so securely relied upon that there can be no need of constantly returning to them and of considering whether we may not perhaps, in the course of the inferences, discover defects which have been overlooked in the principles, and which render it necessary either to determine these principles more fully or to change them entirely"[128]—when seriously conducted, dialectic is **self-correcting**.

Thus, the explanation develops in a given context, such as the 'spirit of the times'. However, the context itself cannot be accounted for in a still larger context. Indeed, it is no use going back to the Aristotelian solution of the 'highest genus' furnished by rational intuition. For here, the circle of argumentation brings the ground back to the grounded. In other terms, the explicative totality (or larger context) is in return explained by its own part, and a new dialectic develops, Part being the other of Totality.

If, as Hegel asserts, "the True is the whole", that is, if, in the course of the **regressive explanation**, the part is elucidated through totality, conversely, in the course of the **progressive explanation,** that very totality itself is elucidated through its own genesis, that is, through the part self-moving towards the totality. Therefore part and totality (the larger context) advance together, explicating one another and thus producing the dialectical justification[129]. For instance, in the *Logic* Reflection and Idea mediate reciprocally, the Idea determining Reflection as a moment, Reflection bringing division in the Idea.

Nevertheless, a **residue** is left for, actually, the totality is not entirely determined and the part is not entirely grounded, so that the development of knowledge continues[130]. At this stage, the **synthetic** viewpoint of the totality may be distinguished from the **analytical** viewpoint of the part[131], or again, the standpoint of the **thought** of experience from that of **experience** itself—does not general opinion hold that the whole is as thought only[132]?

That is why, depending on the adopted angle, any determination may be considered either as **ideel**, as a moment to be overcome, or as **real**, as a being affirming its own existence against an other. In this way, the finite proves at once ideel and real ; and, dialectically speaking, the human thing is cultural as well as empirical.

In brief, in the dialectical explanation **double meaning** and **ambiguity** are ineluctable, the explainer elucidating ambiguously a reality which is itself ambiguous.[133]

6.2. In other words, the dialectical explanation **explicates the** common **abstract meaning** of an isolated determination (in itself, what is Something?), **brings out its presuppositions** (Something is in relation to an Other), **and constructs the concrete meaning** of that determination when linked to its living context (Something and an Other acquire their respective meanings against the background of the surrounding culture). In this sense, the approach in question is truly **heuristic**.

A similar development is also made **out of the opposite determination:** 'In itself, what is an Other?', it may be asked, 'what are its presuppositions, and what is an Other in-itself-for-itself?'[134]

At this point, it is advisable to pay attention to the following : what is now in question is not the opposition existing between the examined determinations, but **the opposition within each of them!** Effectively, the task is to show, in both directions, how the category 'contains' its own opposite within itself or, conversely, how it posits its presupposed opposite (Something is an Other to itself, the Other is also Something).[135]

It is this very **inner contradiction** that compels Something at once 'to cancel' its opposite—in order to become its own ground—and 'to preserve' it—merely in order to be[136]. In another language, we find here an *Aufhebung* (or comprehension), in the two meanings of the term. In this way a reciprocal mediation develops, producing a 'solution' *(Auflösung)* which, in its turn, is to expose its own contradiction.

We saw that inner contradiction is a basic characteristic of dialectic. This is why, as Hegel says with regard to the finite things, "The hour of their birth is the hour of their death"[137]. Namely, the same immanent contradiction that constitutes the moving drive of life and creativity, proves after all to be also a source of decrepitude and death.

In Hegel's view, the infinite alone is the true life, the whole truth that subsists thanks to the transient activity of its different elements, the various individuals in search of their realization. Yet, in so doing, those give rise to the life of spirit or culture, while themselves dying as sacrificed—Hegel's 'cunning of reason' has its own reasons...[138]

That inner contradiction, which constructs and which destroys, appears in the Hegelian logic in the following forms.

6.2.1. In the movement of the categories, with respect to each other the opposites stand at once in a relation of exclusion—at the moment of analytical Understanding—and in a relation of attraction and repulsion—

at the moment of dialectical Reason. From this, it results that those opposites are different from one another (according to the formal or empirical viewpoint), and identical with one another (in the meaning of dialectical identity). In this way **the identity and difference of the opposites** arises, producing a **first form** of dialectical contradiction.[139]

Thus, Something and an Other differ as separated by a limit. Yet they are identical (or necessarily related) in that Something is for-another and determined by its Other, and that the Other is as such for Something : in each term, there is a contradiction between its claim to **autonomy** and the fact of its **dependence**. In this sense, **the outer is inside**—i.e. the immanent determination comes from an other—and **the inner is outside**, in that both opposites have their being (their original being) out of them. In this connection, Hegel sometimes talks of **the reflection in itself which is a reflection in another and reciprocally**.

In Hegel, then, the inner contradiction shows itself in that **second form** too[140]. For instance, the dialectics of Being and Nothing, Finite and Infinite, Real Ground and Formal Ground, may be explained by the first form of contradiction (on the identity and difference of the opposites) ; and the dialectics of Something and an Other, the End and the Means, by the second form of contradiction (on the outer which is inside and the inner, outside).

In addition, a third form of contradiction may be proposed : **the whole is a moment** with regard to another moment—it may be said that it posits itself as an other to itself—and **the moment is a whole** : as a microcosm, it contains its own opposite within itself[141]. Explicitly, it is through such an inner contradiction that Hegel asks to grasp the 'reflected' categories of the doctrine of Essence, such as the categories of Identity and Difference or of Form and Matter—to which may be added the later categories of Universal, Particular and the Individual.

Of course, actually, things are most complex and rich in nuance, and various explanatory principles are to be used, depending on the case and the moment. Undoubtedly, my suggestions are only of indicative value, with a view to elucidating somewhat the dialectical elucidation.[142]

6.2.2. The inner contradiction also shows itself in the linguistic dimension, that is fundamentally constitutive for the dialectical movement : for Hegel, in natural discursive language, in principle, what is effectively said, ***das Gesagte***, differs from what is intended, ***das Gemeinte***. That is, differentiated syntactic **form** and united semantic **content** oppose each other[143]. Above, mention was made of a contradiction between **claim** and **fact** ; here, the contradiction arises between **intention** and **fact**.[144]

Namely, in the predicative proposition as speculatively understood, *Sein* (being) and *Sollen* (ought-to-be) do not square with one another. Therefore, the isolated proposition fails to express what it is supposed to mean, so that to obtain a concrete signification requires the addition of the implied complementary proposition : its opposite or its converse.

From this it follows that, in dialectic, the terms are adequately understood only when grasped **in their appropriate context**, their other, in their determining linguistic environment. From a linguistic point of view, it is this inherent contradiction in the Hegelian proposition that explains the (dialectical) necessity of a self-development towards a wider comprehension, towards a more adequate context : what moves the isolated proposition is here also what cancels it as an immediate, 'intuitive' or axiomatic truth.

6.2.3. As far as the cultural world is concerned, from a theoretical point of view it may be said that sometimes **the practice**, i.e. what actually occurs, does not square with **the proposed theory** : the deeds does not correspond with the words. It is this essential **hiatus**, Hegel says, which causes things to progress in the historical world.

Thus, a philosophy that promises more than it gives in fact, comes to be forced to develop itself, even to become different. Its actual content not realizing what the philosophical plan had announced, basically it contradicts itself.

Likewise, it happens that a recognized ideology does not tell what it achieves in reality, but contents itself with idealization, and no matter whether it calls itself 'realistic' or 'idealistic'—as in business, one cannot trust the wrapping. Here also the obtained results differ from the declared projects. So, no wonder that the sons rise in revolt, on behalf of the values proclaimed by their fathers, intending only to actualize the principles inculcated into them since childhood.

Unavoidable from a dialectical point of view, this fundamental contradiction incites a rebellion. Yet—what a surprise!—the attempt to realize those very principles alters them to some extent : **practical reality is not theoretical reality**. At the end of the process, the 'finally satisfied' sons having become fathers, a new contradiction comes to light, and the cultural history continues...[145]

From a general point of view, mention can be made of a hiatus that separates being (practice) from knowledge of being (theory) or, in dialectical language, of **the identity and the difference of being and knowledge**. On the one hand, this basic tension enables the dialectical explanation to develop and, on the other hand, it also sets a limit to its extent.

Torn within itself, dialectic however lives on that tearing...[146]

6.3. I understand that the cultural thing can become intelligible in so far only as dialectic is itself so. The onus then rests with me to bring out some difficulties that taint the dialectical explanation.[147]

6.3.1. We must note that in the *Logic* some non systematic **mixing of the explanatory categories** is to be found. Although Hegel claims to advance from the part to the whole, from abstract to concrete, he is obliged from the outset to use indiscriminately diverse categories belonging to the whole *Logic*—in that sense also, intelligibility only comes out in the end.

From the very beginning, the categories of Essence and the Concept are employed. For instance, in the dialectic of Something and an Other, we can find the following 'higher' determinations : in-itself and posited, Reflection and mediation, Identity and Difference, Syllogism and middle term. As for the doctrine of the Concept and its 'highest' categories, in its turn it makes use of the 'lowest' categories of Being, such as Other, Determination, Finite and Infinite, For-itself, etc.

Whereas, explicitly, Hegel asks us not to elucidate the 'high' by the 'low'—and does so himself—he seems less preoccupied with the reverse movement, since the abstract low categories have to be criticized concretely (from a 'high' standpoint, of course).[148]

However that may be, this mixing of categories constrains us **to read over and over** that work where, in principle, as with Kant, **acquaintance must be prior to employment**. Yet, in the *Logic* the movement of reciprocal explanation proves to be constitutive so that, accordingly, dialectical intelligibility is truly **circular** without being vicious ; or, more precisely, it is **spiral**.[149]

From a similar point of view, W. Becker points to another issue : as regards its meaning, **the explained category differs from the used one**.[150]

Again, that criticism seems to me relevant and yet it is not destructive, for the dialectical movement necessarily forms itself at that cost! In effect, if dialectical signification is truly contextual, then the dialectically explained category cannot be identical to the same category when predicatively used : the very notion of use does not leave the category indifferent, so to speak[151]. In other words, the use of a term implying an external approach, the concerned term itself becomes external, thus recovering its current meaning. In this way the logical category, said to be an end 'in-itself and for-itself', falls back to the rank of representation *(Vorstellung)*, merely a 'mean for an other'.

In so far as that 'imperfection' proves inescapable, it does not render dialectic impracticable. In my opinion, rather the opposite is the case, given the fact that empirical and formal thinking constitute necessary moments, for instance as the 'familiar' representation from which a beginning is made.

Consequently, it is also through that **tension** or **inner contradiction** that the dialectical movement constructs itself, here as a producer of sense. And to show that dialectic contradicts itself comes not to point to its defects, but rather to indicate one of its main characteristics...

6.3.2. Grasping 'dialectically' the logical movement consists, then, in 'grasping together' the representative (empirical or formal) meaning and the dialectical meaning of the examined categories. Is that grasping possible? The answer I propose is doubly double.

Firstly, dialectic consists in that very mode of grasping, the moment of representative thinking being a constitutive one. It is to the extent only that we are able 'to grasp together' these opposites that dialectic becomes really practicable.

For, what in fact is the dialectical approach?

In my view, it is the apprehension of the movement of the Concept 'in its immanence and necessity'—which constitutes the moment of the **abstract dialectic** excluding its other—together with the apprehension of its representative moment, thus giving the **concrete dialectic** that comprehends its other within itself. This is a most complex situation, where inner contradiction and driving tension are the most evident![152]

Secondly, Hegel himself defines **speculative thought** as "the comprehension of the unity of opposites". That is, in agreement with W. Becker I want to say that, from a theoretical standpoint, that act of grasping (the opposites as a unity) proves possible and not possible at once ; and, at variance with him, from that I conclude that this act (this comprehension) is possible in a certain sense.[153]

Speculative thought is **not possible** in so far as the following question is raised directly : what are we thinking when we grasp 'together' Being and Nothing, Finite and Infinite? The answer is not obvious and, as with Kant, it may be said that when we are thinking the two opposites-- 5 plus 7 in Kant, Being/Nothing or Finite/Infinite in Hegel—we do not think their positive union as a new category—the number 12 for Kant, Becoming and For-itself for Hegel.

Speculative thought is made **possible**, however, as soon as, *a posteriori* , the leading strand of the history of philosophy is used. Therefore, the raised question (what is this expected 'third term'?) will only be answerable *post-factum* , the error lying in the previous abstract procedure—being *a priori* alone. That is, dialectic is in fact not just *a priori* development, since we saw that it is also *a posteriori* reconstruction.

One may thus talk of a speculative thought, **indirectly possible**. If Hegel is right, then, now (at 'the end' of history), when we are thinking Being and Nothing dynamically we are also thinking Becoming ; and when we are thinking Finite and Infinite dynamically, we are also thinking the true subject as For-itself or as the Concept—and conversely, of course.[154]

6.3.3. If, in dialectic, the inner contradiction functions as a principle, how can we avoid falling into the most savage **sophistry**? Indeed, if the basic laws of formal logic are not strictly respected, it seems that one may take all liberties and **say no matter what**—Kant has correctly forewarned against such a danger.

However, if that danger threatens any dialectician, a safeguard exists. On the one hand, as D. Marconi showed, the deductive power of dialectic is not sufficient to render the laws of formal logic univocally applicable[155]. Those laws truly show that, under certain conditions, from a contradiction any statement can be derived. Yet, from a dialectical proposition **nothing at all can be deduced analytically**—if not that proposition itself—for dialectical development always needs the leading strand of human experience.

Consequently, the prudent dialectician will not think 'in emptiness', but really **according to an appropriate leading strand** : that of his own experience, of common experience, of the scientific results of the time or of the *Logic* itself—his gaze fastened on the guiding Idea of the organic system. Looking at both sides at once, the prudent dialectician endeavours to read in the text of life as well as in the text of scientific knowledge. In so far as he is a cultured man, he may fruitfully compare claims with facts, what is aimed at with what is really produced, and come to a conclusion... In this way, perhaps, he can succeed in remaining strongly bound to earth while pointing his eyes towards heaven—ambiguity is always reproduced, so as to get the last word. Concrete from that point of view he nevertheless remains abstract, since he appraises every determination according to the criterion of logical thought.

Now, reality is not logical only...

I therefore suggest the following final conclusion:

It is globally, as a whole, that **dialectic contradicts itself**—through a 'dialectical' contradiction—and it is that inner contradiction that animates it.[156]

If this were not the case, if the contradictory relation existed only between the moments while the whole remained coherent, then we would not have to do with dialectic but with an organic pattern.

Now **dialectical development is not organic** (or biological) **genesis**.[157]

There lies the basic difference: in dialectic, the contradiction dwells in the whole, and that whole can live out of its death only, like the

Phoenix that unceasingly rises born again out of its ashes. The true is really that "Bacchanalian revel in which no member is not drunk" but, besides, it seems to me that now the true itself is drunk...

The onus rests with the dialectician, who thus not only has to look after the contradictions of the others, formalists or empiricists, but also after his own. Under these circumstances only can he perhaps understand why progress is sometimes accompanied by some regression.

Indeed, the power of the negative is active even at the summit of the system, or of culture. In Hegel's view, if any negative entity contains within itself a positive dimension—a **determined negation** is concerned here[158]—conversely any positive entity, even and especially a dialectical one, contains within itself the negative.

Every element finds its life and death in the bosom of the totality; and every partial totality is itself an element of a larger totality. As for the 'totality of all the totalities', it is not an 'absolute absolute' but only a 'relative absolute' for, for us human beings (and the discussion is about us, our world and our vision of the world), our eyes or reflection, turned to that totality, produce again the unavoidable gulf which re-creates both life and death.

The aim is not to gaze at Being from the outside, as a 'detached' thinker, but to express the movement of the Concept, a movement wherein we are at once both observer and participant. For it is always from a particular and universal point of view that we conceive Being, accordingly particular as well as universal. In so far as the real point of view (from below) and the ideel point of view (from above) mediate reciprocally, they constitute the dialectical explanation which, *qua organon*, both criticizes and develops—of course, within the above-mentioned limits.

A last word: time and again, the study of the dialectical explanation has shown that performing a logical elucidation of the cultural thing cannot suffice. If dialectic truly wishes to contribute to the understanding of human reality, then it has to tackle the problem straight out, for the real cannot be just thought and thinking: it is also work, fight and socio-economic relations, the 'Other' of thought.

Here, a new horizon is opening out.

NOTES

NOTES TO THE INTRODUCTION

1. Of course, I mean Hegel's *Wissenschaft der Logik*, F.Meiner Verlag, Hamburg ed.1975 (2 vol.); or *Science of Logic*, transl.Johnston & Struthers, G.Allen, London 1929 (2vol.). Later on I mention also Hegel's *Phänomenologie des Geistes*, F.Meiner Verlag, Hamburg ed.1952, or *Phenomenology of Spirit*, transl.A.V.Miller, Oxford Univ.Press, 1977.

2. Or any other treatise intended to make things clear 'from the outside'. As for a dialectical comment on dialectic, its comprehension already implies some knowledge of the dialectic...

3. Thus A.Trendelenburg's *Logische Untersuchungen*, Hirzel, Leipzig 1970, and W.Becker's *Hegels Begriff der Dialektik und das Prinzip des Idealismus*, Kohlhammer, Stuttgart 1969, helped me a lot in my understanding of the problematique proper to the Hegelian dialectic.

4. Obviously that proposition is self-contradictory, since it says in one proposition, that truth cannot be said in one proposition...

5. This means that formal thinking constitutes a moment of dialectical thinking, necessary but not sufficient as such.

6. In the *Logik*, vol.1 p.51 (in English: I, 79=vol.1 p.79), Hegel talks about the beginning of Science.

7. Since "the True is the whole"... Hegel regards any idea of an introduction to speculative philosophy as external, as of an 'historical' value only. We meet thus again with the necessary, but insufficient, formal dimension specific to dialectical movement.

8. The *Phenomenology*, as its name shows it ('Phenomeno-logy'), consists precisely in an exposition of the development of spirit from appearance to truth.

9. In other words, the paradox lies in the fact that the attainment of the whole requires beginning with the part, whereas the part itself is really understood in the context of the whole only.

10. In this connection, in the *Phänomenologie* p.27 (in English:16=p.16), Hegel talks of the individual's 'inorganic nature'.

11. As noticed by Hegel in *Phän.*. p.52 (39). A reciprocal mediation really develops between the part (the read proposition) and the whole (the clarifying context).

12. This issue is the central subject-matter of M.Clark's *Logic and System*, Martinus Nijhoff, The Hague 1971.

13. See *Phän.* p.59 (45) where Hegel says that the individual "must ... forget himself"; see also p.11 (3).

14. We can observe that, even if the author is still alive, it may be said that for him 'reading himself again' is 'inventing anew'; for he is a self-developing being, ever-creating some new meaning.

15. Mockingly, Kierkegaard claims: "I cannot even forget myself when I am asleep". See *The Journals of Kierkegaard 1834-1854* (Collins, London 1967) p.51 (1836).

16. See J-P.Sartre, *Critique de la raison dialectique* (Gallimard, Paris 1960), in particular p.142. Especially, Sartre underlines the fact that man is always 'situated', i.e. always acting and thinking from a particular historical and cultural position.

17. Nietzsche is known especially to have grasped man as an 'evaluating' being.

18. In particular, my understanding of Hegel's thought depends on my scientific past and my experience of the kibbutz.

19. If this was not the case, according to Hegel we would have to do with a torn 'Unhappy consciousness', a subjective 'Beautiful soul' or with an objective 'abstraction', to speak Kierkegaard's language.

20. As a borderline case, this thesis may even apply to Hegel himself, in so far as his thinking is alive and ever-renewing itself. On this point, see also Note 14 above.

21. A moment which is formed by the examining subject, and also by the specificity of the examined object, of the content in question.

22. On this point see Nietzsche, *Thus spoke Zarathustra* (Penguin Books, London ed.1969), at the end of the first part, p.103.

23. Anyhow, we always read with our own eyes, which are partly coloured by our social surroundings and our personal history.

24. Moreover, Hegel has said many things, he is not always coherent and, in every case, the convenient quotation may be selected.

25. On the *a priori* movement of 'sinking and restraining', see *Phän.* pp.11 (3), 48-58 (36-44), and below Part Six, chap.3.

26. I find it regrettable that similar conclusions have been drawn by the two researchers quoted in the Note 3 above, A.Trendelenburg and W.Becker.

27. On this point see L.Wittgenstein, *Tractatus logico-philosophicus* (transl. C.K.Ogden, Routledge & Kegan, London 1981), prop.6.54.

28. In this sense, it can be talked about the 'identity', within the dialectic, of the dialectical and the non dialectical, or of the intelligible and the non intelligible—it is in this particular way that I understand the presence, within, of the external. Or, in other words, 'the external is both external and internal'. This proposition gives expression to the basic ambiguity characterizing the Hegelian dialectic.

29. He who asserts, in the *Concluding Unscientific Postscript to the philosophical Fragments,* that "Subjectivity is the truth".

30. I understand that the basic ambiguity comes from the fact that, in Hegel, every logical category contains its opposite within itself. Thus, for ex., from a dialectical point of view, necessity is accidental (all real development implies essential alternatives that could also have been actualized), the accidental is necessary (as a constitutive moment of reality), the universal is particular (with regard to other universals) and the particular is universal (as a determination of the Concept).

31. I am concerned with the domain studied by dialectic, that is with human (social, historical and cultural) reality. As for natural (physical, chemical and biological) reality, it constitutes the object studied by the empirico-mathematical sciences.

32. Concrete dialectic may thus be grasped as the unity of abstract dialectic (from an inner point of view) and formal thinking (from an external point of view).

NOTES TO PART ONE

1. See Plato's *Parmenides* 127e and comments in Taylor A.E., *Plato, The Man and His work*, Methuen, London 1960 (ed.1966).

2. For instance, two tables are said to be identical as 'tables in general'. They differ as 'this table which I perceive here and now'. In this case, it may be said that the matter is about two particular things realizing the same universal.

3. See *Parmenides*, second Part.

4. Antisthenes seems to deny the very possibility of predication. He thinks that one can say 'A man is a man' and 'The good is good', but not 'The man is good.' This example is to be found in Plato's *Sophistes*, 251b-c.

5. Hegel perceives a contradiction between the content of that proposition taken as identical, and its form which exhibits 'A' and 'B' as differing in themselves. For this problem of language, see Hegel's *Phänomenologie des Geistes* , Felix Meiner Verlag, Hamburg ed.1952, pp.49-54 (or *Phenomenology of Spirit*, transl.A.V.Miller, Oxford Univ.Press 1977, 36-40=pp.36-40), and *Wissenschaft der Logik*, F.Meiner Verl., Hamburg ed. 1975 (2vol.), vol.1 pp.75-76; or *Science of Logic*, transl.Johnston & Struthers, G.Allen, London 1929 (2vol.): I, 102-3 (= vol.1 pp.102-3).

6. I distinguish between historical, social and cultural 'human reality' and physico-chemical and biological 'natural reality'. I see dialectic as a logic of human reality. As for natural reality *qua* natural, I think that dialectic is not very instructive on that matter.

7. Of course, these propositions imply a certain interpretation as regards the structure of the dialectical movement. This interpretation is exposed in my doctoral Dissertation: *La dialectique de Hegel en tant que logique philosophique: exposition et critique de ses principales caractéristiques*, Hebrew Univ. of Jerusalem, 1983 (in Hebrew, summary in French).

8. Hegel himself presents his logic as a metaphysics or ontology; see *Logik* , vol.1 pp.23-47 (I,53-75). We may thus talk about a movement of categories or about a movement of reality. As for *'erinnern'*, Hegel plays upon the morphology of the word, hence its double meaning: to recall and to interiorize *(er-innern)*.

9. For the *'Reflexionsbestimmungen'*, see also H.G.Gadamer, *Hegels Dialektik* (J.C.B.Mohr, Tübingen 1971) pp.53-56.

10. Kant talks only about 'concepts of reflection' *(Reflexionsbegriffe)* , with the help of which we compare the various representations. See Kant's *Kritik der reinen Vernunft* (F.Meiner Verl., Hamburg ed. 1976) 316B et seq.

11. It is well-known that Hegel is opposed to any philosophy of the representation *(Vorstellung)*.

12. When the Doctrine of the Concept is concerned, he talks of 'manifestation' or 'development'. See *Logik*, vol.2 pp.170, 240 and 269 (II,174,235 and 262-3); see also Hegel's *Enzyklopädie der philosophischen Wissenschaften 1830* (F.Meiner Verl., Hamburg ed.1969) par.142 note, 161,240 and 244.

13. The matter is really about a logical development into three different levels. But Hegel thinks that this development 'reflects' the movement of history.

14. In dialectical logic, the movement is convertible directly: 'and reciprocally' may always be added.

15. M.Clark, in his *Logic and System* (Martinus Nijhoff, The Hague 1971, pp.70 and 90) explicates that which follows from Mure's discussion in his *A Study of Hegel's Logic* (Clarendon Press, Oxford ed.1967) pp.79-91.

16. If the expression 'internal relation between the independent determinations' is more precise, the expression 'relation of the independent terms' seems more fit to the movements of the Hegelian dialectic. By 'inner relation' a constitutive relation is meant; in the opposite case, we have to talk about 'external relation'.

17. We find there some Kantian terms *(Verstand, Vernunft)*, interpreted in a Hegelian way. See e.g. *Phän.* p.29 (18-9), *Logic* vol.1 p.6 (I, 36) and *Enz.* par.79-82.

18. It must be observed that Hegel's philosophy, known as idealistic, is also realistic. On this point see *Logik,* vol.2 p.444 (II, 429), and my *Dialectique de Hegel* 1°Part chap.4.

19. Until now we talked about determination as reflective. With the 'determining reflection' we obtain an example of the double movement constituting dialectic; on this point see J.P.Sartre, *Critique de la raison dialectique* (Gallimard, Paris 1960) pp.41, 94 and 238.

20. Within the limits of its validity, the dialectic of the Part and the Totality shows that every particular determination, every part, must be grasped in its proper context, namely in relation to the totality, since "The True is the whole".

21. And conversely, of course. Thus, the Hegelian infinite consists in the self-reflection constituting the finite in its essence, passive matter is a determination posited by the active form, and the universal has to particularize itself in order to become real.

22. Hegel grasps the category as a subject. He is therefore allowed to talk of 'its pretension', 'its illusion', etc. By these expressions, one has to understand the movement of the thinkers' thought which considers the category, isolates it and moves it. On this movement of 'sinking and restraining', see *Phän.* p.48 (35-6) and my *Dialectique de Hegel,* 2° Part. chap.3 par.2.

23. Human thought being considered as active and determining, we may also talk, according to the context, of the 'identification of the different terms' and the 'differentiation of the identical terms'.

24. In dialectic, using a formal example is not very adequate. However, I make use of it with a view to maintaining the dialogue with the non dialectical thinkers. Later on, in chapter 4, I shall employ an empirical example, also 'not very adequate'.

25. Thus Leibniz thinks that in a garden you cannot find two leaves which are strictly the same. Besides, it must be noticed that the Hegelian principle of the difference of the identical terms does not exclude the Leibnizian principle of the identity of the indistinguishable: even according to Leibniz the two signs 'A' and 'A', of the identity 'A=A', are different. As for the relation of self-identity, it signifies that I am able to identify myself as 'the same subject', self-identical over the various changes in time and in space happening to me. In this case, we also have an 'identity of the different terms'. On the problematique caused by simple identity, such as 'a=a' or 'a=b', see Frege G., 'Sens et dénotation' in *Ecrits logiques et philosophiques* (trad.Imbert C., Ed.du Seuil, Paris 1971) pp.102-126.

26. The plan of treating of totalities is not considered as arbitrary but as necessary, for here man is regarded as living a total experience. It is true that, empirically, the whole is never given, and Kant is right on this point. But if the element has to be grasped in its context, then we cannot dispense with the study of totalities—in his *Critique de la raison dialectique* Sartre has really understood the point. The price paid for is the formation of paradoxes; hence the need for a 'logic of ambiguity'. We shall see again and again that dialectic may be understood in this way.

27. See W.Becker, *Hegels Begriff der Dialektik und das Prinzip des Idealismus* (Kohlhammer, Stuttgart 1969) pp.31-36.

28. Whitehead and Russell, in their *Principia Mathematica* (Cambridge Univ.Press, ed. 1964) pp.41 et seq., talk also of "systematic ambiguity".

29. This ambiguity takes its origin in the fact that, in dialectic, the result contains the traversed path within itself: the Hegelian *Aufhebung* which cancels, preserves too. On dialectic as analytic and synthetic, see Sartre's *Critique de la raison dialectique*, pp.94 and 115.

30. Likewise, Spinoza distinguishes the imagined quantity from the thought quantity, the infinite in its kind from the absolutely infinite; see *Ethics*, 1° Part, prop.XV scolie and def.VI expl.

31. We can say that the genus self-differentiates into particular species, and that the species identify themselves as expressing the same genus.

32. If the judgement relates two concepts, Hegel, for his part, calls 'syllogism' a movement which dialectically relates three judgements to each other. The problem of the *quaternio terminorum* has been examined by A.Trendelenburg in his *Logische Untersuchungen* (Hirzel, Leipzig 1870), p.107.

33. See *Logic*, vol.1 pp.193-194 (I, 213-5).

34. For the opposition Likeness/Unlikeness *(Gleichheit/Ungleichheit)* see *Logik*, vol.2 p.35 (II,46) ; for the opposition Positive/Negative see p.41 (II, 51).

35. In the *Logik*, vol.1 p.59 (I, 86), Hegel mentions the principle of the "identity of identity and non-identity". There, in a too harmonious manner he forgets the moment of the difference (of identity and non-identity).

36. The Hegelian contradiction may thus be expressed in two different manners, equivalent from the viewpoint of the dialectical movement: either as identity of the different terms and difference of the identical terms when grasped together, or as the identity and difference of the identified and differentiated terms. On the Hegelian contradiction see also F.Grégoire, 'Hegel et l'universelle contradiction' in *Revue philosophique de Louvain*, **44** (1946), 54-66, and Note 38 below.

37. One-sidedness is also to be avoided here. If in the movement of the *Aufhebung* the moment of negation alone is perceived, then a 'well-rounded' and perfect result might be obtained, but the necessity of the external moment will not be understood. Now, dialectic being a "logique vivante de l'action" (see Sartre's *Critique*, p.133), the movement cannot cease. If, on the contrary, the moment of preservation alone is perceived, then an empirical and/or formal aggregate will be obtained, which can in no way be seen as dialectical. It seems to me that things become clearer, partly at least, when the present ambiguity is properly grasped. See also Note 29 above.

38. According to Hegel, the finite contains within itself the following contradiction: as real it exists for itself, isolated and independent, and it exists for another as ideel, containing within itself the impulse to further development towards infinite.

39. Of course, we are not dealing with empirical judgements like 'it rains' or 'it does not rain'. I recall that dialectic treats of historical, social and cultural reality, and not of natural reality *qua* natural. I see the idea of a 'Dialectic of nature' as not valid. On that point see Sartre's *Critique*, pp.123-128 and Note 6 above.

40. A similar expression may be found in *Logik*, vol.2 p. 49 (II, 59).

41. 'Repulsion' and 'Attraction' are themselves categories of the Hegelian logic; see *Logic*, vol.1 p.160 (I, 183). As for the relation 'dependent/independent', cf. N.Rotenstreich, *From Substance to Subject* (Martinus Nijhoff, The Hague 1974), pp.90-92.

42. It must be pointed out that, in dialectic, language and thought are considered as constitutive moments of reality.

43. See Notes 21 and 38 above. Dialectic can also be defined as a movement of self-actualization. In that meaning, every determination contains its opposite 'potentially'.

44. On this point see G.R.G.Mure, *An Introduction to Hegel* (Clarendon Press, Oxford ed.1966) p.124, and *A Study of Hegel's Logic*, pp.351-54. See also *Logik* vol.2 pp.256-57 and 298-99 (II, 249-50 and 291). From there we can learn that Hegel distinguishes between *'widersprechend'* (concerning a dialectical contradiction) and *'kontradiktorisch'* (concerning a formal contradiction).

45. If the contradictory opposites divide the universe into two sets excluding one another, the contrary opposites, for their part, let room for a middle term within the range of a defined genus.

46. To be sure, the Hegelian synthesis is itself problematic. In that Part of the work I do not treat this question. I can however add that it implies the Kantian synthesis.

47. In other words, context and figure clarify one another (see Note 20 above). There lies the power of the dialectical explanation—and its problematique.

48. Effectively, it is as a double movement that dialectic becomes intelligible, in part at least.

49. The dialectical contradiction being defined as the identity and difference of the identified and differentiated terms, we can equally say that the 'thinkable' dimension refers to the moment of identity, the 'non thinkable' dimension to the moment of difference. Indeed, absolute difference is unintelligible: we are not able to think a world supposed to be entirely 'other'.

50. A dichotomous standpoint will perhaps conclude that contradiction is not thinkable at all. In contradistinction to this, the dialectician does not reason in terms of 'all or nothing', but only according to diverse degrees of Being, Truth, etc. Moreover, the question is also what is understood by 'thinking': on the whole, my book provides an indirect answer to that question.

51. The Hegelian dialectic implies as well the history of human culture in general and the various sciences of the time—in these also it finds a leading strand. For dialectic as positing *a priori* and reconstructing *a posteriori*, see below Part Six, chap.2-3.

52. For this point see e.g. Fleischmann E., *La science universelle ou la logique de Hegel*, Plon, Paris 1968; Meulen J.v.D., *Hegel; die gebrochene Mitte*, F.Meiner Verl., Hamburg 1958; G.R.G.Mure, *An Introduction to Hegel* and Y.Yovel, *Kant and the Renewal of Metaphysics* , Bialik Inst., Jerusalem 1973 (in Hebrew).

53. Becker W., in his *Hegels Begriff*, pp.63-64, points to this ambiguity in terms of 'expressible/not expressible'. I see the works of ancient and modern philosophy as thinkable to the extent that, as childs of our time, we are necessarily thinking in the context of these works.

54. Or still: dialectical necessity contains within itself its own opposite, the accidental (see *Logic* vol.2 pp.171-84 or II 174-186). In concrete language, this means that, for human reality, there is no fatality to develop just so and not otherwise: Parmenides and Heraclitus could have not existed or have been other... For the leap to the 'new immediate', see the criticisms of Becker W., *Hegels Begriff* , pp.62-65, and Meulen J.v.D., *Hegel*, pp.45-53; for the presence of the empirical in 'pure thought', see Trendelenburg A. , *Logische Untersuchungen*, e.g. pp.42,47,48,70 and 75.

55. Either their likeness or unlikeness. I prefer to talk in terms of identity and difference for, being inner (constitutive), these relations develop a dialectic. Whereas 'likeness' and 'unlikeness' constitute external relations, that is, relations posited by a comparing third, the Understanding.

56. Hegel seems to adopt Leibniz' theory concerning a continuous development from sensation to thought. On this point, see at the opening of the *Phänomenologie* the exposition of the movement from sensible consciousness to perception, Understanding and Reason, and also *Enz.* par.445-468, which presents the movement from sensible intuition to thought through representation *(Vorstellung)*. From this, it can be understood in what meaning Being and Thought mediate reciprocally. For the relation between thought and being, see also Hyppolite J., *Logique et existence*, PUF, Paris 1953.

57. With regard to this issue, it is worth mentioning the last Wittgenstein who tries to explain how arithmetic can apply to reality. In Wittgenstein's *Lectures on the Foundation of Mathematics, Cambridge 1939* (Diamond ed., Cornell Univ.Press, Ithaca 1976) p.43, we can read: "All the calculi in mathematics have been invented to suit experience, and then made independent of experience". See also L. Wittgenstein, *Remarks on the Foundations of Mathematics* , ed. by Anscombe, Rhees and Wright, The MIT Press, Massachusetts 1967. In the context of this discussion I do not distinguish between mathematics and logic, although this distinction can be made.

58. By 'Panlogism' I understand a point of view which identifies being with thought univocally, and which considers empirical matter as not-being.

59. See *Logic*, vol.1 pp.116-125 (I, 141-149).

60. The double meaning appears here also: by 'dialectic' one may understand the totality (human reality *qua* perceived and thought) and the part (human reality *qua* thought). Dialectic truly realizes the Russellian paradox concerning that set which is 'a member of itself'. On this point see also above at the end of par.3.1.1.

61. Indeed, in his *A Study of Hegel's Logic*, pp.294 et seq., Mure shows that, in human experience, the *Aufhebung* of the spatio-temporal sensation is not complete.

62. See Sartre, *Critique de la raison dialectique*, p.158.

63. See *Logik*, vol.2 pp.23-62 (II, 35-70); also *Enz.* par.115-20. In this Part of the book I follow the movements of the *Logik* and not those of the *Enz.*, simpler perhaps but also poorer.

64. And not merely 'organic', for one must not forget the leaps breaking off the continuity of development.

65. This structure 'reflecting' the structure of reality. For such an expression see Becker W., *Hegels Begriff*, p.8.

66. The term 'solution' *(Auflösung)* has to be understood in its double meaning: chemical solution or solution of a problem. Thus Reason 'resolves' (or 'dissolves') the independent determinations of the Understanding and relates them to one another.

67. See *Logik* vol.2 pp.63-76 (II, 71-83).

68. According to the double meaning of the Hegelian *Aufhebung* ; on this point see Notes 29 and 37 above.

69. On the problem of the production of a 'new immediate', see Note 54 above.

70. On the one hand Hegel, following Kant, sees the human being as constituting his own experience. But, on the other hand, the subject-matter is now the entire man's experience, comprising the linguistic, juridical, moral, political, artistic, religious, scientific and even philosophical experiences. Now, the crucial question is, of course: does man really live a total and unified experience? Hegel's philosophy, at least, answers that question affirmatively.

NOTES TO PART TWO

1. Philosophy may also treat of probability and seek to ground it. But then this grounding will be either certain or nothing at all—to put it in Kantian language.

2. As for the hypothetical method, it rests on the principle of refutation that requires its own grounding. Of course, the principle of refutation itself cannot be regarded as a hypothesis.

3. A well-known criticism, concerning the *Ethics*, blames Spinoza for having decided quite arbitrarily which proposition will be considered as an axiom and which as a theorem. On this point see e.g. H.A.Wolfson, *The Philosophy of Spinoza* (Harvard Univ.Press, Cambridge 1934, 2 vol.), vol.1 p.58.

4. See I.Kant, in his *Kritik der reinen Vernunf* (R.Schmidt ed., F.Meiner, Hamburg 1976) 760B-762B, where he refuses any use of 'axioms' to philosophy.

5. The following discussions will show that the dialectical solution precisely points to such a way, in a sense still to be elucidated.

6. On this point see G.W.F.Hegel, *Wissenschaft der Logik* (Lasson ed., F.Meiner, Hamburg 1975, 2 vol.), vol.1 p.9; or *Science of Logic* (transl. Johnston & Struthers, G.Allen, London 1929, 2vol.),I.39 (= vol.1 p.39).

7. The Greek etymology of the word 'method'—met-hodos—reminds us that the discussion is about a way leading to some thing. I think it advisable to understand the 'dialectical method' in this manner, as a basic process through which a certain content develops.

8. Under these circumstances, one must not said that one experience may correct another, but that thought alone is able to opt between two contradictory sensations—as in the classical example of the straight straw half immersed in water, and that looks like broken. In this case, the described solution relies on the supposed permanence of the material objects, innerly inanimate and externally in rest.

9. Aristotle, in his *Physics*. 225b15, showed that "there cannot be motion of motion or becoming of becoming or in general change of change". In this sense, then, the movement is constant, and that from the standpoint of thought.

10. See *Logik* vol.1 p.51 (I 79).

11. How is this consideration *(Betrachtung)* to be realized? Dialectic alone can teach us this: philosophy, done in Hegel's way, really consists in a certain circular activity. On a similar point of view see M.Heidegger, *Die Frage nach dem Ding* (M.Niemeyer, Tübingen 1969) p.187, and *Der Satz vom Grund* (Neske, Pfulligen 1957) chap.2.

12. See A.Trendelenburg in his *Logische Untersuchungen* (Hirzel, Leipzig 1970) pp.39-42,47-56, 69-75, and W.Becker in his *Hegels Begriff der Dialektik und das Prinzip des Idealismus* (Kohlhammer, Stuttgart 1969) pp.41-42,52-60, who have discovered the empirical presuppositions of the logical 'pure thought'. In my opinion, their respective criticism may be interpreted as an explication of the dialectical movement and its problems.

13. On Hegel's philosophy as subjective idealism and as realism, see *Logik*, vol.2 p.444 (II, 429). Beforehand, Kant had already exposed his philosophy as transcendental idealism and empirical realism.

14. Here, of course, the matter is about dialectical negation and about expression as a self-particularization of the universal Concept. Besides, the relevant critical study points to the unavoidable leaps and unmoved residue.

15. Hoffmeister, in the Introduction to his edition of the *Phänomenologie des Geistes* (F.Meiner, Hamburg ed.1952) p.XVI, rejects this thesis: seemingly, he is still attached to the formalistic thought of a single beginning. Hegel, for his part, in his *Logik* vol.1 p.137 (I, 161), mentions the possibility of two different beginnings, the one finite and the other infinite.

16. For that threefold syllogism see *Enzyklopädie der philosophischen Wissenschaften 1830* (F.Meiner, Hamburg ed.1975) par.575-577, and my doctoral Dissertation, 'La dialectique de Hegel en tant que logique philosophique: exposition et critique de ses principales caractéristiques' (Hebrew Univ. of Jerusalem 1983), 2° Part, chap.5 par.3 (in Hebrew, summary in French).

17. By 'the power of the Hegelian thinking' I understand the demand of 'thinking the opposites together'; hence the double movement and double meaning proper to this mode of philosophizing and, here, the double beginning... For the problems attached to that activity of 'thinking together', see the criticism of W.Becker, pp.62-65.

18. By 'modern scepticism' I mean an attitude like that expressed by P.K.Feyerabend in his *Against Method* (NLB, London 1975) or the relativism in vogue in some circles of contemporaneous philosophy.

19. In English, *'aufheben'* is often translated by 'to transcend', 'to overcome' or 'to sublate'. Yet 'to comprehend' with its double meaning ('to understand' and 'to contain') seems equally appropriate. In this twofold meaning, the Hegelian system claims 'to comprehend' the different philosophical principles discovered in the past.

20. See *Kritik*, XXXVI B-XXXVII B.

21. See *Enz*. par.10 note.

22. In this connection, see *Prolegomena zu einer jeden Künftigen Metaphysik* (Kants Werke IV, De Gruyter, Berlin 1968) p.282, where Kant says that the properties of things cannot move and pass to our faculty of representation.

23. On the problem of the beginning according to the Introduction to the *Phenomenology*, see Groll M., 'On the beginning of philosophy according to Hegel' in *IYYUN*, 9 (1958), in Hebrew.

24. By this expression, I understand the *a priori* way of sinking and restraining Hegel deals with in the *Phän*. pp.48-49, or *Phenomenology of Spirit* (transl.A.V.Miller, Oxford Univ.Press 1977) 35-6 (=pp.35-36).

25. In this way, in the place of Kant's *Kritik der reinen Vernunft*, Hegel proposes his *Phenomenology* as a criticism of knowledge which is also a criticism of the perceived object (Hegel does not separate epistemology from ontology).

26. In his *Phän*. p.67 (49), Hegel plays on the common root of the words *Zweifel* (doubt) and *Verzweiflung* (despair). As for the Socratic method, E.Dupréel, in his *La légende socratique et les sources de Platon* (R.Sand, Bruxelles 1922) pp.325-34, distinguishes between the Delphic Socrates who examines, the Numb Socrates who paralyses and the Midwife Socrates who gives birth to Ideas.

27. On this twofold relation see *Phän*. p.70 (52) and *Enz*. par.412. As concerns the Hegelian contradiction, see *Logik*, vol.2 pp.26-62 (II, 37-70) and the first Part of the present book, especially chap.4.

28. The immanence of the Hegelian method has to be pointed out: here knowledge is appraising itself, without relying on any external criterion—a method that may be called *'criterium sui'*.

29. The path of knowledge really progresses together with the path of being, as in the image of the divided line in Plato's *Republic* 509d.

30. That point of unification is announced in *Phän.* p.140 (110-11). Given the double movement which builds dialectic, however, it must be noticed that, in principle, a movement of internalization is prior to any exposed movement of externalization—and conversely.

31. On the relation of the novice to the Hegelian philosophy, see Y.Yovel, 'Reason, reality and philosophical discourse according to Hegel' in *IYYUN*, **26** (1975), 63-64 and 85-87 (in Hebrew).

32. That strategy of conversion is proposed by Kierkegaard as an indirect communication in his *The Point of View for My Work as An Author: A report to history* (transl.W.Lowrie, Harper & Row, N-Y. 1962) Part I and Part II chap.I.

33. Let me remind it: our domain of discussion is the thinking of being and knowledge (as human reality), while the thinker himself is and **knows**, and not the formal system beginning with arbitrary axioms, freely chosen by the researcher.

34. See *Phän.* p.74 (56). The very essence of dialectic lies in this activity of comprehending without setting aside, of distinguishing without excluding: the traversed way is thus 'comprehended' in the attained result.

35. In this way, the *a posteriori* path of systematic reconstruction complements the *a priori* path of sinking and restraining, mentioned in the Note 24 above. Dialectical development forms itself through this double movement, following the two dimensions of the *a priori* and the *a posteriori.*

36. And this without falling into relativism. At every level, truth is relative to the considered stage. However, the traversed way may be called 'absolute' in so far as it is ordered from abstract to concrete, from partial to total.

37. The dialectical movement may thus be presented as the explicit position of the different 'conditions of possibility' of human thought. In this sense, it reminds us of Kant's transcendental deduction.

38. This reciprocal mediation of opposites implies absolute idealism, which is simultaneously subjective idealism and realism; cf. Note 13 above. In other words, the discussion is about (ideel) identity and (real) difference of opposites, grasped together.

39. And no absolute Nothing. As we shall see later, here 'Nothing' has to be understood not only as negation, but also as privation.

40. Like Aristotle's primary matter.

41. In dialectic, as in formal thinking, the contradiction is understood as unbearable—hence the necessity of a 'solution'. Instead of a univocal cancelling, however, we find there a producing movement.

42. A certain kinship between dialectical development and transcendental deduction having been mentioned in Note 37 above, we may talk either of development or of 'deduction'.

43. See *Logic*, vol.1 p.52 (I, 80).

44. Hence, dialectically speaking, the double meaning of the concept of end which, in Hegel, signifies an end as well as a new beginning.

45. In a similar meaning Kant holds, in his *Kritik* 758B, that "the incomplete exposition must precede the complete".

46. For the presentation of dialectic as a development according to the moments of the in-itself, for-itself and in-itself-for-itself, see *Phän.* p.24 (14) and *Logik* vol.2 p.499 (II, 480). In place of the Kantian division of reality into things 'in themselves' and phenomena 'for us', Hegel proposes their dialectical identity.

47. When he talks about a 'possible experience', even Kant refers to an experience that can become real if the researcher wants it; in contrast to the fiction, unrealizable for us.

48. And this without being question of relativism; cf.Note 36 above. It is in the *Logik*, vol.1 p.54 (I, 82), that Hegel mentions an 'absolute beginning' (whose meaning is 'abstract beginning').

49. For the absolute as a result and the truth as the whole,see *Phän.* p.21 (11-12).

50. For the presentation of dialectic in three similar levels (of the Understanding, the dialectical and the speculative), see *Logik* vol.1 p.6 (I, 36) and *Enz.* par.79-82.

51. See *Logik* vol.1 p.58 (I, 85). W.Becker, in his *Hegels Begriff* p.35, says that Hegel writes 'nichts' (not anything) as 'Nichts' (Nothingness), hence the dialectic of Being and Nothingness. For my part, it seems to me that things are not so simple.

52. Aristotle too mentions a beginning out of non-being, understood as potential being; see his *Metaphysics*, 1069b15-20.

53. Since Kant's Copernican Revolution, man may be regarded as the constructing focus of his world—hence the expressions of 'human reality' and 'human thought'. As for nature as such, it constitutes a certain 'residue', irreducible to human theoretical activity. In my opinion, there is no dialectic of nature.

54. It seems to me that, thereby, Hegel also exposes the problem of the movement, answering thus the paradox of Zeno's arrow (see Aristotle's *Physics*, 239b5-10). Indeed, according to the Hegelian dialectic, the flying arrow is and is not at the same place, i.e. it moves.

55. In Hegel, by 'syllogism' it must be understood a dialectical movement interrelating three judgements to one another. For example:
the one positive: the beginning is;
the other negative: the beginning is not;
and the third uniting the two former: the beginning is and is not, it becomes
(we can see that the result truly contains the traversed path).

56. For the determinations of Coming To Be *(Entstehen)* and Passing Away *(Vergehen)*, see *Logik* vol.1 p.92 (I, 118). As regards the ambiguity specific to Hegelian dialectic, see Note 17 above.

57. See *Logik* vol.1 p.59 (I, 86), and *Differenz des Fichte'schen und Schelling'schen Systems der Philosophie* (F.Meiner, Hamburg ed.1962) p.77, also pp.27 and 42. In this expression, seemingly the first identity signifies 'thinking together', whereas the second signifies 'formal identity'.

58. This expression is famous since Stirling's book: *Das Geheimnis Hegels.*

59. On the dialectical contradiction, see Note 27 above. For the importance of the opposites 'Same' and 'Other', see Plato's *Sophistes* 254e and *Timaeus* 35a-b-39c.

60. Thus, dialectic may also be understood as a movement of the immediate revealing itself as mediated and becoming a new immediate at a higher level. J.v.D.Meulen, in his *Hegel; die gebrochene Mitte* (F.Meiner, Hamburg 1958) pp.45-53, criticizes this leap to the new immediate that seems to him as not valid. Here *'aufgehoben'* means 'resolved'.

61. G.R.G. Mure too, in his *An Introduction to Hegel* (Clarendon Press, Oxford ed.1966) pp.63-66, sees dialectic as a movement from experience to the thought of that experience—thought 'comprehending' experience. Now, the question is to think that thought.

62. For Hegel, any determination constitutes a relation to an other, to its other. For this issue see for ex.*Phän*. pp.45-46 (33). As for G.R.G.Mure, ibid. p.144, he does not separate quality from relation, the one implying the other and vice versa.

63. Hegel often uses the paradoxical language. Thus, for him isolation is a relation, for one can be isolated with regard to an other only. Concerning the incongruous examples of mud, hair and dirtiness, see Plato's *Parmenides*, 130c.

64. In the Transcendental Aesthetic of his *Critique*, by 'pure intuition' Kant understands the representations of space and time; for Hegel, in the *Logik* this expression means: empty, undeveloped thought—indeed, the programme of Hegel's logic is to abstract from space and time. Even the category of causality is not explained there with regard to time. Space and Time are examined in the *Encyclopaedia*, at the beginning of the 'Philosophy of Nature' which considers real being and not logical being any more.

65. All that, let it be underlined, to the extent that man lives a total and united experience.

66. This is only a new manner of expressing the identity of form (the method) and of content (the system). Thus, the method is said to consist in the consciousness of the form structuring the movement of the content. On this point, see *Logik* vol.1 p.35 (I, 64) and Note 7 above.

67. It is true that Iena's 'Logik' is prior to the *Phenomenology*. In place of the circular movement, therefore, the spiral movement could be mentioned.

68. See *Phän*. p.28 (18) and *Logik* vol.1 p.11(I, 41). In the *Encyclopaedia*, the par.1 recalls the point in question.

69. See *Enz*., par.387-412.

70. These opposite features (conservative/revolutionary) are pointed out by M.Riedel, *Studien zu Hegels Rechtsphilosophie* (Frankfurt 1969) p.100. No wonder then that right and left Hegelian movements developed. If ambiguity truly constitutes a basic law of dialectic, any one-sided reading will always lead to such a partition into adverse parties.

71. See *Logik* vol.1 p.9 (I, 39: for the 'external materials'). Culture in general and philosophy are conditioning one another in so far as, on the one hand, philosophy truly is the self-conciousness of its time and, on the other hand, in art, religion and philosophy, the logical Idea realizes itself in different forms—this movement being necessary to the constitution of self-consciousness.

72. For that assertion see *Phän*. 12 (3-4).

73. On this point see *Logik* vol.1 pp.7 and 53 (I, 37 and 80-1).

74. For this new point of view see *Enz.* par.13-14.

75. In his development, a thinker does not necessarily follow a linear progression. In particular, maybe that, on certain points, the young Hegel was less dogmatic and more opened to the questions of reality than the mature Hegel was.

76. It can be observed that, for his part, Aristotle was opposed to any notion of a 'science of the sciences'. He thought that every science does treat of a particular genus of being, and that being itself is not a genus.

77. In the present work, the Hegelian system is understood as essentially composed of three dialectical subsystems: *Phenomenology*, *Logic* and *Encyclopaedia*. I have especially studied that topic in my Dissertation ; on that point see Note 16 above.

78. From that Part of the work, however, an ambivalent and moderate answer comes out indirectly, in terms of 'yes and no', 'to a certain extent', 'it depends on the context'. That is what I call 'the third way'.

79. See H-F.Fulda, *Das Problem einer Einleitung in Hegels Wissenschaft der Logik* (Klostermann, Frankfurt am Main 1975) pp.273-301.

80. For this expression see H-F.Fulda, ibid., pp.296 and 299-300.

81. On the 'Epistemological Anarchy', see P-K.Feyerabend, *Against Method.*

82. This expression is to be found in Kant's *Kritik der praktischen Vernunft* (Akademie Textausgabe V, De Gruyter, Berlin 1968) p.87.

NOTES TO PART THREE

1. See Part Two above.

2. By 'human reality' I mean the social, historical and cultural reality in which we are living, acting and thinking; to distinguish from natural reality, characterized as physical, chemical or biological. Human reality alone is supposed to be dialectical. In other words, I reject any idea of a dialectic of nature.

3. The previous Part was concerned with the problem of the 'beginning of Science'. Here I am concerned with the general problem of the end, regarding both the history of philosophy and the history of human reality. This change in perspective originates in the thought of Hegel himself, who studies these two questions (on beginning and end) in these very terms.

4. By 'central works of the System' I understand Hegel's dialectical works published by him, on which this Part is mainly based : *Phenomenology of Spirit*, *Science of Logic* and *Encyclopaedia of Philosophical Sciences.*

5. On these different assertions, see Hegel's *Enzyklopädie der Philosophischen Wissenschaften 1830* (F.Meiner, Hamburg ed.1975) par.13-14. See also the beginning of the Preface to the *Phänomenologie des Geistes*, F.Meiner, Hamburg ed.1952 (or *Phenomenology of Spirit*, transl.A.V.Miller, Oxford Univ.Press 1977). At the end of *Phän.*, pp.558-559 (in English 487-8= pp.487-8), similar ideas can be found, expressed in terms of 'the end of Time'. Of course, the various introductions to *Einleitung in die Geschichte der Philosophie* (F.Meiner, Hamburg ed.1966) recall these diverse pretensions.

6. Concerning the general study of the principles of Hegelian dialectic, see my Dissertation: *La dialectique de Hegel en tant que logique philosophique: examen et critique de ses principales caractéristiques* (Hebrew Univ. of Jerusalem 1983) 1° Part (in Hebrew, summary in French).

7. The demand for philosophy to be presented in the form of an 'organic system' appears already in Kant's *Kritik der reinen Vernunft* (F.Meiner, Hamburg ed.1976) 860B. In this work a 'pretension concerning the end' is also to be found, particularly at the end of the Preface to the 2° ed. and on the last page. On these different topics, see Y.Yovel, *Kant and the renewal of Metaphysics*, Bialik Inst., Jerusalem 1973 (in Hebrew).

8. See e.g. *Wissenschaft der Logik* (F.Meiner, Hamburg ed.1975, 2 vol.), vol.1 p.16; or *Science of Logic* (transl.Johnston & Struthers, G.Allen, London 1929, 2vol.): I,46 (=vol.1 p.46).

9. In so far as philosophy 'reflects' human reality, talking of 'reality recognizing itself in Hegel's philosophy' seems legitimate. In concrete terms, the discussion is about a certain universal self-consciousness, attained by the individual involved in his time and studying the System.

10. On these different affirmations, see esp. *Enz.* par.13 and the diverse introductions to the *Einleitung* . Hegel thinks that different peoples reveal different principles, and then disappear from the scene of history. See also *Enz.* par.548-552 and 573.

11. Hegel classifies the world history according to the level of realization of the principle of freedom: in the Orient one alone is free, in ancient Greece a few ones and in modern Europe everyone is free in principle. On this formal scheme, see *Die Vernunft in der Geschichte* (F.Meiner, Hamburg ed.1970) pp.62 and 155-6; on pp.244-254 (also 154), adding the Roman and the Muslim worlds, Hegel divides history into five worlds.

12. In some contexts *'aufheben'*, an ambiguous term in Hegel, may be translated by 'to comprehend', ambiguous as well, since this term signifies 'to grasp' or 'to contain'.

13. On this point see S.Kierkegaard, *Concluding Unscientic Postscript to the Philosophical Fragments* (transl. D.F.Swenson, Princeton Univ.Press 1968) p.134 note.

14. We find there again the classical problem of the 'unity of the System'. On this topic see my Dissertation , 2° Part. chap.5.

15. It seems to me that the dialectical approach, as far as it is a concrete thinking, cannot and should not attain this ideal. As was underlined by the existentialist's revolt, human life shows as creative, ever-surprising and not to be grasped by us in the form of a closed system. On this point see Kierkegaard's *Postscript*, pp.107-113.

16. In Hegel's language, *"ein Sohn seiner Zeit"*. On this subject see the Preface to Hegel's *Grundlinien der Philosophie des Rechts* (F.Meiner, Hamburg ed.1967) pp.16-17.

17. In particular Hegel, without hesitating, accepts the idea that there is 'one' history, and even one 'world history'. Nowadays, it is becoming clear that this idea has to be re-examined—a task going beyond the boundaries of this work.

18. Obviously, the same applies to me, the author of these lines...

19. This Part deals with the problem of the end from the standpoint of human thinking, and does not raise the question of faith.

20. On the distinction between high and low level, see *Enz.* par.380. On the idea that, in the exposition of Hegel's philosophy, the middle level is the most adequate, see G.R.G.Mure, *A Study of Hegel's Logic* (Clarendon Press, Oxford ed.1967) pp.79-91, and M.Clark, *Logic and System* (Martinus Nijhoff, The Hague 1971) pp.68-71—both refer, in the *Logik*, to the doctrine of Essence.

21. On dialectic as an a *priori* movement of sinking and restraining and an *a posteriori* movement of systematic reconstruction, see below Part Six, chap.2-3.

22. At the beginning of the *Phenomenology* we find the sensible 'this, here and now' as ineffable; and, at the beginning of the *Logic*, Hegel holds that the difference existing between the determinations of Being and Nothing (which form Becoming) cannot be expressed.

23. As Aristotle asserts it in his doctrine of the right mean.

24. On the dialectic of Something and an Other, see *Logik*, vol.1 pp.104-110 (I, 129-35) and Part Six below, chap.5.

25. For Hegel, the determined is negative because it is tainted with a lack: it is not its Other. Of course, Hegel conceives non-A, the other of A, as positive as well. On this issue, see above Part One.

26. As far as no essential dimension is lost in the movement, as the Hegelian *Aufhebung* claims it to be.

27. On the dialectic of the Finite and the Infinite, see *Logic* vol.1 pp.103-146 (I, 129-169), and below Part Six, chap.5.

28. In his *Logic*, Hegel talks of categories recognizing themselves in their Other. In truth, the discussion is about the movement of thought of the thinker considering these categories.

29. A similar idea is proposed by M.Clark, in his *Logic and System* pp.107 and 187; he sees the 'final identity' as still tainted with dualism.

30. The double meaning of 'reflection' is also to be noticed. In Hegel, that word means at once human thinking and mediation of the determinations between themselves.

31. On this point see *Logik*, vol.1 p.100 (I, 125) and vol.2 p.164 (II, 167).

32. Thereby it may be undestood either a cosmic, external crash, or an inner accident, nature getting the upper hand over culture and destroying it, e.g. in the form of an accelerated increase of delinquency or of an ecological accident.

33. On this point see A.Kojève (*Introduction à la lecture de Hegel*, Gallimard, Paris 1947, pp.395, 432, 434 note 1 and 443) who foresees an "anéantissement définitif de l'Homme proprement dit", of course not a biological annihilation; and H-F.Fulda (*Das Problem einer Einleitung in Hegels Wissenschaft der Logik*, Klostermann, Frankfurt am Main 1975, pp.243-260) who perceives the end in the form of a *'Geschichte der Bildung'* realizing an entire correspondence between the System and its history.

34. Naturally, the discussion is not about the Hegelian opposition between accomplished essential history and indefinite, accidental movement of reality. This opposition will be examined in chap.5 below.

35. The relevant double meaning of the words 'to achieve' and 'end' cannot be missed, 'to achieve' signifying to reach by effort or to complete, and 'end' signifying extreme limit or purpose. Can we not discern here an example of the 'speculative spirit in the language' Hegel talks about in his *Logic*, vol.1 p.10 (I, 40)?

36. See *Logic*, vol.2 pp.383-406 (II,374-94), and the corresponding studies by E.Fleischmann, *La science universelle ou la logique de Hegel* (Plon, Paris 1968) pp.296-309; G.R.G.Mure, *A Study of Hegel's Logic*, pp.249-259; and Ch.Taylor, *Hegel* (Cambridge Univ.Press, ed.1977) pp.321-328. The *Encyclopaedia* too exposes that movement. Nonetheless I understand that the *Logic*, more complex but richer, is to be preferred.

37. Let me recall that here human reality is concerned, grasped as a product of theoretical and practical activity of man.

38. The source of that criticism seems to be Aristotle's *Posterior Analytics*, book I chap.19-23, which is dealing with the various issues concerning demonstrative syllogism.

39. As in formal logic, the contradiction has to be 'resolved'. But here it is supposed that it brings about a development and not about a univocal cancelling.

40. In this connection Hegel talks of the 'cunning of Reason'. This metaphor means that, in order to carry out its ends, the universal makes use of the particular who fancies satisfying his own needs alone. In this way, the Idea of Good is said to accomplish itself through the activity of the particular individual, in conflict with himself and with the others.

41. As for the dialectic of the Idea of Good, see *Logik*, vol.2 pp.477-483 (II, 460-5). Mure's *Study of Hegel's Logic* p.286, shows that, likewise, the Idea of Good is a contradiction: the Good is eternally realized and yet ever-realizing itself—an expression that constitutes another formulation of the thesis I vindicate here. See also Ch.Taylor, *Hegel*, p.325.

42. On the Hegelian *Aufhebung*, see *Logic*, vol.1 p.93 (I, 119-20) and *Phän.* p.90 (68, as supersession). M.Clark, in his *Logic and System*, pp.108, 114, particularly underlines that ambiguity. In the *Logic*, vol.2 p 404 (II, 393), the following paradoxical expression is to be noticed, concerning the *'ursprüngliche innere Äußerlichkeit des Begriffs'* (the 'original inner externality of the Concept').

43. Although Hegel says, at one time, that external being is 'non essential' (see *Logic*, vol.2 pp.402-403 or II 393-4) and, at other times, that exteriority is a determination of the Concept (p.405 or II, 394).

44. As expressed by Kierkegaard in his revolt against the System. See S.Kierkegaard, *Postscript*, e.g. pp.267-282.

45. See *Phän.* pp.255-282 (211-35)

46. It seems to me that here Hegel exploits the double meaning of the word 'individual'. If in dialectic one should not separate, yet it is advisable to differentiate: the intelligible is not the sensible... It is even possible to talk about the double meaning of the term 'meaning', the latter signifying either conceptual signification (referring to an ideel individual) or material denotation (referring to the real individual). 'Universal individual' also is ambivalent: it signifies either a concrete totality or a 'Sage' (see below). Here I mean a 'Sage' only.

47. On this point see Kierkegaard's *Postscript*, e.g. pp.282-288. Indeed, even a philosopher differs from the Concept.

48. The discussion being held from the standpoint of human thought, *stricto sensu* it may be said that the Idea is not only unrealizable but, furthermore, it is even unknown to us, since the opaque matter conceals it in part. If so, man is in quest of an end which, in principle, represents to him a certain mystery—immanent however.

49. For the double character of time, see M.Clark, *Logic and System*, p.47.

50. By '*Realphilosophie* ' it must be understood, in the *Encyclopaedia*, Philosophy of Nature and Philosophy of Spirit.

51. On the Concept as thought or comprehension, see also E.Fleischmann, *Science universelle*, p.241.

52. See *Enz.* par.257-259.

53. If at the beginning of the *Logic* Hegel identifies Being and intuition of Being, here he seems to identify Becoming and intuition of Becoming—indeed, Hegel grasps intuition as the seed of all thought to come. On the 'psychological' development from intuition to thought, see *Enz.* par.445-68.

54. The Concept, or thought determining the finite, is infinite in so far as it is reflective and self-determining.

55. On the dialectic of the Finite, see also chap.2 above.

56. See *Phän.* pp.38 (27, as existent) and 558 (487).

57. On the relation existing between time and Concept, see also A.Kojève, *Introduction*, pp.336-380.

58. On this question, see M.Clark, *Logic and System*, p.48. Aristotle, in his *Physics* 209a25 and 210b24, says that "place has not a place". Kant, for his part, talks about time as "immutable and fixed" in his *Kritik*, 183B and 58B.

59. The Hegelian dialectic constitutes itself through a double movement. Therefore, in the realm of pure thought or pure intuition, from 'A is B' one may always develop 'B is A'. In effect, dialectical identity is symmetrical and mediated. In the next chapter, it will be spoken about the 'essential which is accidental' and the 'accidental which is essential'.

60. At the beginning of the *Logic* the dialectic of Being, Nothing and Becoming is to be found, even though Being is 'empty intuition', 'empty thought'.

61. See *Phän.* p.39 (27). Aristotle, in his *Physics* 225b15, says likewise that there cannot be motion of motion or becoming of becoming or in general change of change. Consequently, no wonder that Hegel admires Aristotle's 'speculative spirit'; on this point see *Logik*, vol.1 p.192 (I, 212) and vol.2 p.433 (II, 420).

62. For this expression see *Phän.* p.74 (56).

63. Cf. chap.3.2. above.

64. Ambiguity comes again to light, and dialectic (here the abstract and/or concrete synoptic) proves to be a moment of itself...

65. However, there is no question of relativism or historicism, since any intelligible discussion is held in a general context to respect, which constitutes our culture here understood as Greek, Hebrew and Christian.

66. See Part Two above, chap.3.

67. The expression *'wirkliche Geschichte'* appears in *Phän.* p.559 (488, as actual history).

68. See chap.1.2. above.

69. Without overlooking, of course, the other fundamental questions raised by the second Heidegger, such as: what is Thought? what is Being, Truth, Philosophy, Language?

70. See respectively Mure, *A Study of Hegel's Logic*, pp.332-356, and J-P.Sartre, *Critique de la raison dialectique*, Gallimard, Paris 1960.

71. See Kant, *Kritik*, pref. to the 2°ed., and Th.S.Kuhn, *The Structure of Scientific Revolutions*, Univ. of Chicago, 1962. Talking about the 'sudden revolution in the way of thinking', Kant recalls the works of Thales, Bacon of Verulam, Galileo, Torricelli and De Stahl. Kuhn, for his part, deals with revolution as a global change in the vision of things, and quotes, *inter alia*, the paradigms of Copernic, Newton, Lavoisier and Einstein. Hegel, in his *Phänomenologie* p.15 (6), conveys a similar idea in terms of 'qualitative leap'.

72. Indeed, Continuous and Discrete Magnitudes are categories in the *Logic*, applying thus to the movement of reality; see *Logic* vol.1 p.193 (I, 213). It is the same for External Reflection (vol.2 p.17 or II, 29), a moment of 'real' determination breaking the continuity of the movement.

73. The point at issue is the passage from mediation of opposites to the new immediate, or from negation of negation to the new positive. As typical illustrations, mention can be made of the transition from the movement of Reason to the movement of Spirit in the *Phenomenology*, from objective to subjective logic in the *Logic*, and from absolute Idea to Nature in the *Encyclopaedia*.

74. What implies the entire realization of the different discovered principles. On this point see e.g. *Phän.* p.16 (7) and Ch.Taylor, *Hegel*, pp.426-427.

75. See *Logik* vol.1 pp.207, 333 (I, 227-8,343-4) and *Enz.* par.259 note.

76. See chap.1.2. above.

77. For the term *'plastisch'* , see *Phän.* p.52 (39, as 'plasticity') and *Logik* vol.1 pp.19,21 (I, 48,51 as 'adaptable').

78. To think a determination 'in itself' means to think this very determination and not something else; to think it 'for itself' means to think this determination not with an extraneous aim in view, not in order to solve a different problem.

79. Indeed, to human thought the essence of reality appears as particularized, for instance in the form of the spirit of a people, or of the spirit of the times, so that attaining some original essence directly is entirely out of reach. Is not the human vision simultaneously determining and reflective, that is determining and dividing (or positing essence as against appearance)?

80. This is the attitude adopted by W.Becker in his *Hegels Begriff der Dialektik und das Prinzip des Idealismus* , Kohlhammer, Stuttgart 1969; see esp. pp.40,101 and 105.

81. Maybe that the last Hegel, grown more cautious, got nearer to such a thesis. Indeed, he closes his *Vorlesungen über die Geschichte der Philosophie*, ed.Glockner, Frommans, Stuttgart 1928 (3vols.), in talking of the 'standpoint of the present time', 'at the moment'; see vol.3 p.690. Or still, at the end of his Philosophy of History we can read: "That is as far as consciousness has reached".

82. In chap.4 of this Part of the work, I mentioned the supratemporal thought and the temporal experience as present to each other too.

83. Indeed, duality is not dualism. If by 'dualism' one generally means a doctrine which, in a certain domain, admits two different principles, separated and essentially irreducible to each other, by 'duality' I understand the characteristic of this entity which, being twofold, contains two elements related to each other.

84. It seems to me that man, as a limited being, in principle is not able to have a concrete total viewpoint on totality; hence the set of the different partial points of view mediating one another.

85. At least bidimensional. However, to man other dimensions may be added, such as the aesthetic, mythical, emotional, playing dimensions. Human life being creative and unpredictable, there is no reason to limit and count...

86. And reciprocally, of course, in so far as every cultured being is the product of his sociohistorical surroundings, which he contributes to produce.

87. Meulen J.v.D in his *Hegel; die gebrochene Mitte* (F.Meiner, Hamburg 1958) uses such an expression to assert that every dialectical syllogism contains a 'broken' middle term.

88. Namely, if no 'last struggle' is to be found, there is yet an unceasing fight, strewn with victories and defeats.

NOTES TO PART FOUR

1. See G.W.F.Hegel, *Wissenschaft der Logik* (F.Meiner, Hamburg ed.1975, 2vol.), vol.2 pp.486 et seq.; or *Science of Logic* (transl.Johnston & Struthers, G.Allen, London 1929, 2vol.): II, 468ff. (=vol.2 pp.468ff.).

2. This constitutes the classical problem of the unity of Hegel's System. On this topic see especially the works of M.Clark, *Logic and System*, Martinus Nijhoff, The Hague 1971; H-F.Fulda, *Das Problem einer Einleitung in Hegels Wissenschaft der Logik*, Klostermann, Frankfurt am Main 1975; B.L.Puntel, *Darstellung, Methode und Struktur*, Bouvier Verlag, H.Grundmann 1973; and my Dissertation, *La dialectique de Hegel en tant que logique philosophique: examen et critique de ses principales caractéristiques*, Hebrew Univ. of Jerusalem 1983 (in Hebrew, summary in French), 2°Part., chap.5. See also Part Two above, chap.4, where the problem is examined from another point of view.

3. Other sources of difficulties may be proposed, such as the richness of the living object-subject or the problem of the exposition *(Darstellung)* of dialectical thinking through a discursive language—this problem is studied in the following Part Five.

4. See *Logik*, vol.2 pp.66-76 (II, 74-83).

5. That is why Fleischmann's interesting inquiry on this question has to be appreciated; see E.Fleischmann, *La science universelle ou la logique de Hegel* (Plon, Paris 1968) pp.148-152.

6. Such an expression as 'Essence posits the non-essential Show', in fact signifies that the thinker 'sinks into' the thought content and 'follows' its proper movement. However, it must be added that this following look is also the moving one. On the corresponding movement of 'sinking and restraining', see Part Six below, chap.3.

7. At the level of Understanding, reflection is reifying; therefore the movement appears as a thing. But, for Reason, reflection is a driving power; from this point of view, conversely the thing is understood as a movement. Now, in so far as man is both analytical Understanding and unifying Reason, the real shows itself to him as a movement as well as a thing.

8. Whereas in Kant, according to Hegel, essence is in-itself only; that is, it does not realize itself through the mediation of human activity.

9. This paradoxical expression is to be found in Hegel's *Differenz des Fichte'schen und Schelling'schen System der Philosophie* (F.Meiner, Hamburg 1962) p.77; it is recalled in *Logik*, vol.1 p.59 (I, 86).

10. Hegel says that, with this movement of 'stepping back', thought posits its own content as an other to itself which is facing it, as a *Gegen-stand* .

11. Alienation, generally regarded as the 'destiny' of man, is here understood as the result of his free decision: to separate in order to ignore the actual self-contradiction...

12. Paradoxically, Hegel presents the 'non-determination' of that other as the proper 'determination' of the Form. Likewise, at the beginning of the *Logic* Nothing was apprehended as Being.

13. Thus, to be determined is to have a form, a structure. E.Fleischmann, in his *science universelle*, p.126, says that this substratum is not Aristotle's primary matter, mentioned in the *Logic* vol.1 pp.386-387 (I, 392-3).

14. I understand that spirit implies matter which it negates, as well as matter implies spirit which determines it *qua* matter.

15. This term appears in *Logic*, vol.2 p.71 (II, 79 as 'susceptibility'). Kant, in his *Kritik der reinen Vernunft* (F.Meiner, Hamburg ed.1976) 317B, presents Matter as 'the determinable' *(das Bestimmbare)* and Form as 'the determination' *(die Bestimmung).*

16. Hegel says: Form gives thus to itself a *Bestehen*, a subsistence.

17. This contradiction may also be understood in the following way: it is within itself that Form posits itself as external Matter.

18. I understand Hegel's *Aufhebung* in its double meaning: it both negates and preserves—hence the arising of the mentioned tension. On this ambiguity, see for ex. J.D'Hondt, 'Le moment de la destruction dans la dialectique historique de Hegel' in *Revue Internationale de Philosophie*, **139-40** (1982), 125-137.

19. This is the case since, for Hegel, the determination is a relation to an other. Now, a relation is always a relation both of unification and of exclusion, the terms in relation (as identical) having also to exist by themselves (as different)—hence the paradox. This dialectical principle is mentioned by E. Fleischmann in his *science universelle* p.148. See also above Part One chap.3.

20. I have followed the movement of the *Logic* and not that of the *Encyclopaedia*, which is different and rather simplified. For ex., the note to par.133 presents the identity and difference of Form and Content in terms of '*umschlagen*' (mutual reversal), an expression which does not seem very informative. See in *Enzyklopädie der Philosophischen Wissenschaften 1830* (F.Meiner, Hamburg ed.1975) par.126-134.

21. This doctrine is to be found in his *Metaphysics*, Books Z, H and Theta. Concerning the interpretation of the doctrine, I rely especially on the studies by G.R.G.Mure, *Aristotle*, Oxford Univ.Press, N-Y. 1964; J.Owens, *The Doctrine of Being in Aristotelian Metaphysics*, Toronto 1951; and D.Ross, *Aristotle*, Methuen, London 1964.

22. The linear mode of reasoning starts from terms which have to be accepted straightaway, these being unable to ground themselves. On the other hand, we shall meet below the 'circular' mode of reasoning allowing the ground to ground itself, according to the double movement of the Hegelian mediation.

23. I understand that the god can be regarded as a transcendent cause, with regard to the Cosmos *qua* a whole only. As for the efficient cause, eventually, as regards the natural thing it proves to be identifiable to the two mentioned causes.

24. Different authors, as M.Heidegger in *Die Frage nach dem Ding* (Niemeyer, Tübingen 1969) p.187, and J.Piaget in *les formes élémentaires de la dialectique* (Gallimard, Paris 1980) chap.1, vindicate this circular mode of explanation. The metaphor of the circle comes to give expression to the movement of reciprocal mediation.

25. In Kant, this problem is known under the name of 'The problem of schematism'. The Kantian doctrine of the schematism must explain how, in human knowledge, its spontaneous and its receptive dimensions are united. While here, the discussion is about the problem of unification of the active with the passive principles.

26. Here the question is about an external other 'within', since thought is regarded as determining. As for the sensible given, it is to be grasped as an external other 'outside', independent of thought and irreducible.

27. Hegel, in his *Logik* vol.1 p.108 (I, 133-4) and in his *Enz.* par.44 note, rejects the Kantian concept of the thing-in-itself. As concerns Kant, in his *Kritik* he talks of matter, sometimes as that "which corresponds to sensation" (34B), and sometimes as "sensation" itself (74B, 209B). With this latter definition, Kant paves the way for Hegel's idealism.

28. According to Hegel, there is in spirit a drive to self-realization and self-knowledge. For this purpose, formally speaking spirit posits its own content as an other, at first alien from it (at the moment of self-differentiation). Through its movement it recovers its original unity (at the moment of self-identification), but this time as self-realized and self-conscious. Thus the *Enzyklopädie*, par.457, talks of spirit as *'sich selbst zum Sein, zur Sache zu machen'* (giving itself Being as the objectively real).

29. Indeed, it can be noticed that, in the Introduction to the Logic of the *Enzyklopädie*, par.19 et seq., metaphysics in general, empiricism and Kant's philosophy constitute the main stages in the production of objectivity by thought.

30. If Kant, in his *Kritik* 44B and 52B, entitles his philosophy 'transcendental idealism' and 'empirical realism', Hegel, in his *Logik* vol.2 p.444 (II, 429), presents this very ambiguity in terms of 'subjective idealism' and 'realism'. However, this time the discussion is about the total experience of man, in all its social, historical and cultural aspects—hence the paradoxes and inner contradictions, in this case also creative and not just destructive.

31. It can be observed that, if in Aristotle 'form' and 'matter' are relative terms, it is the same in Hegel with the terms 'universal' and 'particular', for the species may be viewed either as universal (with regard to the individual) or as particular (with regard to the genus).

32. See *Phänomenologie des Geistes* (F.Meiner, Hamburg ed.1952) p.88, or *Phenomenology of Spirit* (transl.A.V.Miller, Oxford Univ.Press 1977) 66 (=p.66), where the translation is slightly different.

33. On this point see *Logik*, vol.1 pp.77-78 (I, 104-5).

34. This idea is explicitly stated in *Enz.* par.85.

35. On the presence of an empirical dimension in 'pure' thought, see A.Trendelenburg, *Logische Untersuchungen* (Hirzel, Leipzig 1870) pp.39-42,47-56,69-75, and W.Becker, *Hegels Begriff der Dialektik und das Prinzip des Idealismus* (Kohlhammer, Stuttgart 1969) pp.41-42 and 52-60.

36. One may also talk of matter according to the different sciences, for ex. in terms of physical, biological or social matter. Hegel, following Kant on this point, reduces physical matter to forces of attraction and repulsion. Biological matter has been studied by Leibniz in his letters, under the name of 'second matter'. As regards social matter we can encounter it in Sartre, as the practico-inert of the *Critique de la raison dialectique*, Gallimard, Paris 1960. For my part, I study matter as encountered in everyday experience.

37. At one time the discussion is about experience, at another time about thought of experience. Now, in truth these two dimensions are one, to the extent that thought *aufhebt* (comprehends) human experience. To this very extent, form and content prove to be both essential.

38. The inevitability of leaps may also be accounted for in the following way. For instance, it may be asked: how is a dialectic of finite and infinite possible? If by 'finite' a category of thought is meant, then an abstract, uninteresting dialectic is formed. And if by 'finite' the sensible particular is meant, then it cannot be understood how that finite can develop into infinite, whatever the sense of 'infinite' may be. In truth, I understand that the task is 'to apprehend together' the thought of the finite and its empirical intuition; hence the leap, from inner thought to external intuition. So is obtained the (ideel) identity and

the (real) difference of finite and infinite—I see dialectic as consisting in the grasping of this tension, apprehended as creative.

39. What recalls the ambiguity of the Aristotelian substance, understood either as a thing existing separately (here the concrete object), or as the essential nature of the thing (here the abstract totality). Cf. D.Ross, *Aristotle*, p.166.

40. On the passage from the Idea to Nature, see *Enz.* par.244, the last lines of the *Logic*, and Bourgeois B., 'Dialectique et structure dans la philosophie de Hegel' in *Revue Internationale de Philosophie*, **139-140** (1982), 163-182. On logic realized as 'the spiritual', see *Enz.* par.574.

41. On the nature of the representation *(Vorstellung)*, see the 'psychological' movement in *Enz.* par.451-464. See also G.R.G.Mure (*A Study of Hegel's Logic*, Clarendon Press, Oxford ed.1967) pp.10 et seq., and M.Clark, *Logic and System*, pp.21-67. Attention must be paid to the fact that the presence, in human experience, of an *a priori* dimension—that is, of a conceptual one—constitutes a *sine qua non* of the Hegelian dialectic.

42. On the *a priori* movement of sinking and restraining, and the *a posteriori* movement of reconstruction, see farther Part Six, chap.2-3.

43. On this point see *Enz.* par.250 note. In a Kantian language, the discussion is also about 'derived' categories, the fundamental categories being exposed in the *Logic.*

44. On the movement of the solar system, see *Enz.* par.275-80. On the development of sensibility, see *Enz.* par.354, and also *Phän.* p.200 (160-1) and *Logik*, vol.2 pp.421-422 (II, 408-9).

45. On the relation between thought and representation, see *Enz.* par.20 and 20 note, and M.Clark, *Logic and System*, pp.21-39.

46. In my opinion, thought is unable to conceive an 'absolute' other, if not by reducing it to itself. And yet, the sensible given can be grasped (by the senses), reproduced and described from the outside—in other words, here a certain way 'back to Kant' is advocated. The question at stake is discussed in chap.4 below.

47. To put it another way, for concrete philosophy 'the best' is either the universal (pure thinking), or a synthesis of universal and particular (empirical representation). Concerning "the night in which...all cows are black" *(die Nacht...worin...alle Kühe schwartz sind)*, see *Phän.* p.19.

48. For Kant's original proposition regarding thought as against intuition, see *Kritik*, 75B.

49. Thus, the term 'concrete' has gained a double signification: at one time it means 'which perceives the empirical' (through representation), at another time 'which constitutes a totality including thought and representation'.

50. Of course, it all depends on the question what you mean by the word 'to think'. Here, this point is not treated directly.

51. On this point see *Enz.* par.246 note.

52. To a certain extent the Hegelian dialectic, when concretely grasped and exposed, *aufhebt* (comprehends) the Kantian thought, in this sense that the external moment is included in it.

53. In *Enz.* par.248 note, Hegel presents nature as the 'unresolved contradiction', that cannot be *aufgehoben* (transcended). Furthermore, in par. 250 note, he recognizes the fact that nature sets limits to

philosophy. Then, rather quickly he mentions the 'impotence of nature' when, in truth, impotence of philosophy is at stake...

54. See *Kritik* 197B and also 111A.

55. Obviously, what is meant here is nature as phenomenon. As for Hegel, he thinks that, in its essence, nature is but a form taken by spirit in the course of its self-realization.

56. See *Prolegomena zu einer jeden künftigen Metaphysik* (Kants Werke IV, De Gruyter, Berlin 1968) par.36; see also *Kritik*, 163B and 165B.

57. Or, more precisely: in Kant, man can know *a priori* the fundamental principles of physics alone; as for the particular empirical laws, the knowledge of them calls for experience. Hegel, for his part, does not separate the *a priori* from the *a posteriori* dimension, but regards dialectic as a double movement unfolding from experience to thought and vice versa.

58. In chap.4, we shall find this very assertion under the name of 'Vico's thesis'. In Kant, the ambiguous status of time and space is noteworthy: they belong to receptive sensibility (as intuitions), and yet organize the multiplicity of given impressions (as forms).

59. E.Fleischmann too establishes a parallel between matter and form of *Enz.* par.129 and the two meanings of nature in Kant. See *science universelle*, p.172.

60. Tradition used to separate the intelligible *a priori* from the sensible *a posteriori*.

61. It is in the 'psychological' movement, exposed in *Enz.* par.445-468, that Hegel explains, from the standpoint of the knowing individual, what he means by 'identity of universal thought and particular perceived'. Is it not possible to see in that exposition a sort of Kantian 'subjective deduction of categories'?

62. Besides, it may be noticed that the *Logic* exposes also a logical nature through the mediation of which the Concept rises up to the Idea. I mean thereby the modes of thought concerning nature (Mechanism, Teleology), exposed in the *Encyclopaedia* as Mechanics, Physics and Organic.

63. A thesis I already proposed in chap.1 above. This idea will appear again in chap.4: for dialectic, what matters is to give an account of concrete human experience as faithful as possible.

64. The *Phenomenology* and the *Lectures on Philosophy of Religion* also deal with nature as a moment of spirit's self-realisation, appearing at first as external and alien from itself, and bringing about the alienation of natural consciousness. I understand that the examination of these passages, basically uncritical, would not enrich the present study essentially.

65. The *Encyclopaedia*, par.254 note, explains why, in contrast with the *Logic*, the philosophy of nature starts precisely with the determination of quantity and not with that of quality: the reason lies in the fact that nature is the proper domain of otherness and externality—both determinations caracterizing quantity. In this connection, in the next chapter a certain Hegelian formalism will be mentioned.

66. See *A Study of Hegel's Logic*, p.305. Mure prefers the expression: 'That is a house', but for my discussion 'I see a house' will do.

67. As is admitted by Mure himself in his *Study of Hegel's Logic*, pp.295 and 304. The reader may reject the idealistic viewpoint, but then he must show that no *a priori* factor is working in the act of perception.

68. At this point, the ambiguity resides in the fact that 'external' is a term meaning 'outer', and also a thought-determination resulting from an activity of interiorization. Dialectic develops in this double movement, which equally applies to the determinations of space and time, in turn 'projected' outwards and reinteriorized. Kant, on the contrary, used to separate time, the form of inner sense, from space, the form of external sense in general.

69. What is at stake here is a (thought) experience about the thinking of experience.

70. See *A Study of Hegel's Logic*, p.309.

71. This gradation sends back to the 'psychological' movement, mentioned in Note 61 above.

72. What amounts to recognize, in human experience, an empirical and an *a priori* dimensions acting together.

73. On this paradoxical expression see *A Study*, p.310.

74. In other words, the external *qua* external is really 'within' thought.

74a. On the 'irresoluble antinomy' in Mure, see *A Study*, p.339.

75. See *A Study*, pp.314 et seq.

76. In this way, the here examined experience contains Hume's 'impression' and Leibniz' 'expression'.

77. Likewise, in chap.1 the form showed itself as depending on the matter to be structured.

78. Dialectically speaking, there is no direct self-knowledge.

79. On the side of the object, the residue is material or sensible. On the side of the subject, the point at issue is the imaginative and emotional private life, out of reach of universal thought.

80. It must be observed that, by definition, the 'sensible residue' is not an exteriorization of spirit. On Hegel's mentioned ideas, see *Enz.* par.447-448.

81. See *A Study*, pp.321-322. Accordingly, the nature of the sensible given is not exhausted by the determinations of particular, external, accidental, etc.

82. Therefore, thinking is not always sufficient. There are cases where self-experience is also required... Can legitimately a philosophy abstract from the rich, lived immediate experience, and call itself 'concrete'?

83. That is, the dialectical movement will always open out on various alternatives, the final selection depending on the relevant context.

84. See e.g. *Phän.* p.45 (33), *Enz.* par.28 and 465, and *Logik* vol.1 pp.30-31 and 42 (I, 60 and 71). In the *Logik*, vol.2 p.221 (II, 218), Hegel recalls Kant saying that "the object is that in the concept of which the manifold of a given intuition is united" (*Kritik* 137B). From this, Hegel concludes that the unity of the concept is what constitutes the object in its objectivity, namely in its being—in that perspective, the 'given intuition' has become superfluous...

85. A similar viewpoint was advocated by Parmenides (Being is Thought), Plato (Being is Idea), Aristotle (Being is God, a pure form), and Spinoza (Being is Substance).

86. On the above-mentioned expression, see *Phän.* p.9 (1) and *Logik* vol.1 p.6 (I, 36). On page 245 vol.2 (II, 239-40), the *Logic* seems to confirm my thesis.

87. On this double meaning, see the terms characterizing the sensible given as 'external', 'particular', 'accidental', indicated at the end of the precedent chapter.

88. See *Grundlinien der Philosophie des Rechts* (F.Meiner, Hamburg ed.1967) par.121, 124 note, 162, and *Phän.* p.25 (14-5); see also *Enz.* par.38 note. The *Phänomenologie*, pp.255-282 (211-35), deals also with the particular individual, but as a moment to be transcended.

89. See the discussion about Sense-Certainty in *Phän.* pp.79-89 (58-66); see also *Enz.* par.20 note.

90. In this case, art could constitute the means of expression peculiar to the sensible given, to emotion and imagination.

91. See especially 'Der Ursprung des Kunstwerkes' and 'Wozu Dichter?' in *Holzwege* (Klostermann, Frankfurt am Main 1980) respectively pp.1-72 and 265-316.

92. In this context, by 'direct action' I understand a social action, and by 'indirect action' an antisocial, nay a vandal behaviour.

93. By reducing quality to quantity, Hegel calls to mind the works of the first atomists... A superficial reading of the dialectic of Quantity, in the *Logic* as well as in the *Encyclopaedia*, is enough to testify to that description.

94. For these expressions ('The impotence of Nature' and the 'impotence of Reason') see *Logik* vol.2, respectively pp.247 and 251 (II, 241 and 245). See also *Enz.* par.250 and 250 note.

95. On this criticism levelled at formalistic thinking, see *Logik* vol.1 p.132 (I, 156) and vol.2 p.233 (II, 228).

96. In Kant, this topic is expressed in the form of the Copernican Revolution.

97. Cf. for ex. the Cartesian vision of things.,

98. On Hegel both as an idealist and as a realist, see *Logik* vol.2 p.444 (II, 429). I understand Hegel's 'absolute idealism' in these very terms.

99. On this point see J-B.Vico, *Principes de la philosophie de l'histoire* (trad. J.Michelet, Colin A., Paris 1963) p.79.

100. On nature as an opportunity and an obstacle at once, see Sartre's *Critique*, p.158.

101. On the gap within the Idea, see *Logik* vol.2 pp.412 and 484 (II, 399 and 466).

102. See *The Journals of Kierkegaard 1834-1854* (transl. and ed. A.Dru, Collins, London ed.1967) p.98 (1846).

103. See the Preface of *Phän.* pp.45 (32) and 58-59 (45). See also *Logik*, vol.1 p.20 (I, 49).

104. It is to be observed that in so far as 'inner' (immanent) and 'necessity' (necessary) are dialectical determinations, each of them 'contains' its other within itself.

105. On 'The instinct of Reason' *(der Vernunftinstinkt)*, see *Phän.* pp.187 et seq. (149ff.).

NOTES TO PART FIVE

1. On the problem of exposition, see B.L.Puntel, *Darstellung, Methode und Struktur* (Bouvier Verl., H.Grundmann, Bonn 1973) pp.29-47. If classical philosophy used to ask 'What is truth?', Hegel enriches the problematique at issue by raising the question of language explicitly: 'Can philosophical truth be adequately exposed?', he asks.

2. Western culture may be regarded as showing formalistic trends, in so far as Aristotelian logic prevails in it.

3. In this Part of the work, I do not distinguish between 'dialectical' and 'speculative', although Hegel does it sometimes, as in the *Enzyklopädie der philosophischen Wissenschaften 1830* (F.Meiner, Hamburg ed.1969) par.81-82. Owing to the fact that Hegel is not always consistent in his use of basic terms, lexical decisions have to be taken. Hegel talks particularly of *'spekulative Wahrheit'* or *'spekulative Satz'* when problems of language are at stake.

4. In dialectic, language is not regarded as a formal system of signs describing an external reality, but it is itself regarded as real.

5. On this problem of the sensible 'other', here as a medium of expression of the intelligible, see above Part Four, chap.3.

6. Dealing especially with the **logical** problem of the antinomy of language, and not with the nature and function of language in Hegel's philosophy, I mainly refer to the *Logic*, to the corresponding passages in the *Encyclopaedia* and to the Preface to the *Phenomenology*; and only incidentally to the *Phenomenology* itself and to the different *Lectures*, greatly examined from that latter point of view. For a general investigation into the problem of language in Hegel, see Th.Bodammer, *Hegels Deutung des Sprache*, F.Meiner, Hamburg 1969; D-J.Cook, *Language in the Philosophy of Hegel*, Kohlhammer, Stuttgart 1966; J.Simon, *Das Problem der Sprache bei Hegel*, Kohlhammer, Stuttgart 1966; and G.Wohlfart, *Der spekulative Satz*, De Gruyter, Berlin 1981. On the antinomy of language, see also Y.Yovel, 'Reason, reality and philosophical discourse according to Hegel' in *IYYUN* **26** (1975), 59-115 (in Hebrew).

7. Indeed, for the nominalist the universal is but a word, and the real, which is given to sense, constituted of particular elements.

8. It may be noticed that to the nominalist 'my body' raises a problem, in contradistinction to the bodies of the external world; since my body is, for me, at the same time inner and external, myself and an other. When I am pointing towards it, so to speak, it is pointing towards itself...

9. For more details, see Ch.Taylor, *Hegel* (Cambridge Univ.Press ed.1977) pp.3-124.

10. A similar point of view may be attributed to Hamann, as well as to Humboldt. In his work Taylor refers to *Herders Sämmtliche Werke*, ed.B.Suphan, Berlin 1891.

11. Referring to historical dialectic, one may observe the following movement: from immediate identity of sense and being with the Aristotelians, to the modern external difference, then to a certain mediated identity with the Romantics.

12. Let me recall it: here, not only is natural language said 'to reflect' reality, but it is itself regarded as real—hence the unavoidable circle of a language speaking about itself. On language (and work) as moment of reality, see especially Hegel's *Phänomenologie des Geistes* (F.Meiner, Hamburg ed.1952) pp.229 et seq., or *Phenomenology of Spirit* (transl.A.V.Miller, Oxford Univ.Press 1977) 187ff.(=pp.187 et seq.).

13. As it appears at the end of the *Enz.* par.553-577, and of the *Wissenschaft der Logik* (F.Meiner, Hamburg ed.1975, 2vol.), vol.2 p.484, or *Science of Logic* (trans. Johnston & Struthers, G.Allen, London 1929, 2vol.), II 466 (=vol.2 p.466). Hence the relevance, for a general doctrine of language, of studying the Hegelian philosophy of art and religion, which deals with different modes of expression of absolute spirit.

14. This is the case, of course, in so far as thought is truly the essence of human being, as Hegel asserts it again and again.

15. Should we not essentially distinguish between two different questions: 'What are human beings?' (the others grasped from the outside) and 'Who am I?' (I, from an inner standpoint)?

16. When we are saying 'Human language expresses thought', then a certain ambiguity takes form. Indeed, in this expression language turns out to be differentiated from thought and, from that viewpoint, it constitutes a formal system of signs. The antinomy of language, on the contrary, implies that the term 'language' has to be understood both as form and content, as comprising at once the sign and its meaning. A double meaning seems then inescapable, nay basic, even within a clarifying discussion about the problem of language.

17. The distinction is of importance, given the basic ambiguity already indicated: if in the Preface to his *Phenomenology* Hegel is dealing with the *spekulative Satz*, in his *Logic* he develops the movement of the divided *Urteil*. In order to avoid any risk of confusion, I have distinguished between judgement as *Satz* and judgement as *Urteil*. Given the fact that, here, just movements of thought are in question, the word 'judgement' seems appropriate; I shall use the term 'proposition' later on, when dealing with subsequent linguistic problems. As concerns Hegel, let me recall that, in his use of language, he sometimes shows some lack of coherence. For instance, if *'Urteil'* generally means 'division of the Concept', in his *Logic* Hegel sometimes talks of *'Satz'* as expressing the *Urteil* (*Logik*, vol.2 p.274 or II 267), or the reverse (vol.1 p.76 or I 103), at other times as a law in the form of 'All is A' (vol.2 pp.24-25, 335 or II 35-6,326), or still as a judgement connecting a particular predicate to a particular subject (vol.2 p.267 or II 261). Accordingly, I could not preserve the generally admitted parallel: *Urteil* = judgement and *Satz* = proposition.

18. A contemporaneous thinker might ask: in a meaningful language, are all the judgements reducible to the form 'S is P'? It seems to me that, in the context of a dialectical logic, in principle the answer is affirmative; for here 'quality' (or 'attribute') is understood as a relation to an other, and the relation as a category. Moreover, it must be observed that 'being' has here the sense of a movement and not of a fixed determination. On the nature of the Hegelian judgement, see also G.Jarczyk, *Système et liberté dans la logique de Hegel* (Aubier, Paris 1980) pp.54-88.

19. On the movement of the Judgement, see *Logik* , vol.2 pp.264 ff. (II, 258 et seq.).

20. It seems that Hegel has translated, in his own language, the Kantian doctrine of the empirical judgement exposed in the *Kritik der reinen Vernunft* (F.Meiner, Hamburg ed.1976) 92B-94B. In effect, the *Logik*, vol.2 p.267 (II, 260), talks of the subject of the empirical judgement in terms of an external object, and of its predicate in terms of a representation 'in the head', each one existing independently of the other. That vision of things is called by Hegel the 'subjective consideration' of the judgement.

21. See *Logik*, vol.2 pp.265-266 (II, 259), *Enz.* par.169 note and par.172 addition, and *Phän.* pp.51, 52 and 54 (38,39 and 40-1).

22. Cf. *Phän.* pp.53-54 (40).

23. The objective and universal Kantian judgement may also be grasped as identical, in that sense that here the predicate is grasped as belonging necessarily to the subject. Cf. *Kritik* 141B-143B.

24. See Leibniz, *Discourse on Metaphysics* (transl. P.C.Lucas & L.Grint, Manchester Univ.Press 1953) par.VIII, IX and XIII.

25. The representation *(Vorstellung)* may be said to be dialectical, in so far only as it 'contains' the Concept. By *'Realphilosophie'* I mean the Philosophy of Nature and the Philosophy of Spirit presented in the *Encyclopaedia.*

26. For ex. W.Becker, in his *Hegels Begriff der Dialektik und das Prinzip des Idealismus* (Kohlhammer, Stuttgart 1969) pp.96-97, asks if, from 'The eagle is brown' it is possible to construct 'The brown is eagle'... Yet, in a certain sense, from 'This eagle is brown' can it not be deduced 'This brown is (an) eagle'?

27. On the Hegelian conversion see also, in the *Phenomenology*, the different double movements constituting Consciousness and Self-consciousness, and the conversion of the judgement 'The Self is absolute Essence' *(das Selbst ist das absolute Wesen)* made there pp.521-522 (453).

28. Conversely, is somebody thinking seriously about a 'dialectization of formal logic'? Although some semantic paradoxes of the type 'I am a liar' might give rise to dialectical reflection. On the development of content even in mathematics and formal logic, see J.Hintikka, 'Are Logical Truths Analytic?' in *Knowledge and the Known* (D.Riedel, Dordrecht 1974) pp.135-159, and I.Lakatos, *Proofs and Refutations* (Cambridge Univ.press 1976) chap.6-7.

29. On this point see the many notes in the *Logic*, rejecting any use of mathematical signs—which are quantitative and external—for the expression of the movements of the spiritual (cultural) world: *Logik* vol.1 pp.334-335, 352-354 (I, 343-4,360-2) and vol.2 pp.226, 253 and 259 (II, 222,247,252-3).

30. For a similar view, see *Enz.* par.237 addition.

31. The expression 'movement of the judgement' is thus to be understood in its double meaning: if so far I have presented the ideel movement of the Hegelian *Satz*, now the discussion turns on the real movement of the empirical, divided *Urteil*, towards the dialectical *Satz* . On this movement of the *Urteil*, see *Logik* vol.2 pp.264-308 (II, 258-300).

32. With the addition of that fourth moment, Kant's Table of Judgements is entirely reconstructed dialectically. In this connection, Hegel remarks that the relation of the judgement to the concept must also appear.

33. Or, in other words, I want to say that the term 'table' denotes an external thing and has a meaning given by its lexical definition, the two constituting together its signification. Cf. also J-P.Sartre, *Critique de la raison dialectique* (Gallimard, Paris 1960) p.126.

34. No wonder that a 'value judgement' is mentioned, since the Idea of the Good is one of the categories developed in the *Logic* .

35. By analogy with the model of Chomsky (in *Cartesian linguistics*), it may be talked of 'surface structure', i.e. the judgement as divided *Urteil*, and 'deep structure', i.e. the judgement as dialectical *Satz*. Here, however, the reciprocal mediation replaces the linearity of Chomsky's formal 'universal grammar'.

36. When Hegel talks of language as the 'Being-there of Spirit' *(das Dasein des Geistes)*, he seems to be referring to its expressive nature; and, on the other hand, when he mentions it as 'the work of Understanding', he is discerning in it the inner syntactic division. On these two approaches, see respectively *Phän.* pp.458, 468 (395,405) and *Logik*, vol.1 p.104 (I, 130) or *Enz.* par.459 note.

37. And, of course, also through work and fight, topics which are not examined in the present book. In this connection, let me notice the equivocation due to the fact that Hegel, when speaking of *Satz* or *Urteil*, refers to a relation sometimes ontological, sometimes existing between representations, and sometimes existing between talked or written signs. Thus, in 'S is P' 'S' may be a grammatical subject, a logical subject or an external thing; as for 'P', it is a predicate 'in the head' or an external property. It is the same for the copula 'is', which functions as a sign or an ontological relation. In a philosophy which considers reality as a double movement of interiorization (towards thought) and of exteriorization (towards the phenomenon), the distinction between the different levels can be made, but certainly not their separation from one another.

38. See G.R.G.Mure, *A Study of Hegel's Logic* (Clarendon Press, Oxford ed.1967) chap.1 pp.1-27.

39. See especially *Phän.* p.29 (18).

40. On this point see *Phän.* pp.51-54. Hegel opposes also *Urteil oder Satz überhaupt* (*Phän.* p.51 (or 38)) to *philosophische Satz* (51, 52 (or 38,39)), form to content (52-53 (or 39)), or subjective to objective approach (cf. *Logik* vol.2 pp.267, 269 (or II 260,262)).

41. In his *Language*, pp.122-135, D.Cook asserts that Hegel prefers the uttered word, dynamic as time is, to the written word fixed in space: in this perspective, Hegel translates *'Logos'* by uttered 'Verb' or 'Word'. And yet, without the written word there is no founded memory, no inscribed tablets nor preserved archives, thus no States and no history either. It seems that, in a certain sense, the sensible sign also, besides the intelligible Concept, constitutes the permanent, the non transient dimension of the thing... It is to be noticed that, for De Saussure too, language is first of all a talked language; hence the importance of 'acoustic images' (the significant) and the danger of the 'illusions of writing'. See F.de Saussure, *Cours de Linguistique générale* (Payot, Paris 1973) respectively pp.32, 99, 155 and 56.

42. Indeed, when we are reading the philosophical text the first time, we tend to understand the judgement as divided *Urteil*, that is, as connecting various representations. Later on, the to and fro movement developing between subject and predicate (through our reading) allows to rise from empirical representation up to philosophical conception.

43. For the formalist, language 'correspond' to reality in so far as he sees the two as 'atomic', that is, as consisting of independent elements combining with one another. For the dialectician, language, as a moment of reality, is supposed to constitute a microcosm—hence the possible expression of reality through its agency, and the paradox of a reality expressed by a moment of itself.

44. So, a doubly double meaning arises: firstly, the double meaning of the proposition grasped as immediate and as speculative; and, secondly, the **ideel** speculative proposition as a movement of the 'immanent and necessary' Concept, and the **real** speculative proposition as a tension between ideel speculative proposition and immediate one. In a certain sense, the speculative proposition appears as a moment of itself...

45. Thought and language are identical as mediating one another: dialectical identity is meant. But they also differ for, in regard to the moving unifying thought, language appears as split up, static and sensible.

46. See Note 5 above. Everything depends also on what is meant by *'Aufhebung'* (here 'overcoming'). If the dimension of cancelling is emphasized, then a flowing but abstract dialectic is obtained; and, if the dimension of preserving is emphasized, there is no more dialectic but an empirical aggregate. When paying attention to the double meaning of the term *'Aufhebung'*, then an only partial dialectic is obtained, but at least a real and concrete one.

47. In this sense, the problem of language seems to be but a particular instance of the general problem of matter, sensible intuition and empirical representation conditioning the dialectical movement of thought, that is of 'speculative truth'.

48. See *Logik* vol.1 pp.75-79 (I, 102-5), and also *Enz.* par.88 note.

49. See *Logik* vol.1 p.67 (I, 94). The disagreement with formalistic thought lies, in my opinion, not in the very possibility of a significant discourse about Nothing, but in the ambiguity of this determination here understood negatively and also positively.

50. Writing again the proposition as 'Being is Nothing' will not modify the problematique at stake, since dialectical judgement has to be understood as identical. On this problematique see also *Logik* vol.1 pp.138, 141 (I, 161-2,164-5) and vol.2 p.54 (II, 63).

51. In this study, Hegel seems to recognize the essential role of the sensible given, in the form of a linguistic sign or an empirical representation, whereby the difference comes to be realized: the sensible given, thus, said to be "not true, not rational", finds its own truth, its own ground (see also Note 41 above). Likewise Parmenides, with the help of the spoken or written language, tried to demonstrate the non-being of the sensible, divisible, moving world...

52. See *Logic*, vol.2 p.252 (II, 246). Studying totalities, dialectic inevitably comes up against paradoxes, as Plato's *Parmenides* has clearly shown it. As for the formalistic thought, it bypasses the problem by abstracting. Yet, the relevant question is: in so far as the experienced reality constitutes a whole, is it possible to renounce a 'logic of totality', whose law is not identity but contradiction?

53. This statement is taken up again and developed at the beginning of the final discussion; see 6.1 below.

54. In particular, the sensible dimension persists in the form of metaphors at once deceiving and necessary which, like Plato's Mythos, present the spiritual (cultural) content in a vivid language. That is especially flagrant at the ends of Hegel's dialectical works, which present the Idea as 'imperishable life' (in the *Logic)* or as 'eternally alive, self-creating and enjoying itself' (in the *Encyclopaedia)* ; on this point, see also the last poetical lines of the *Phenomenology.*

55. In this discussion, I propose to render concrete the dialectical relation of the 'identity of the differentiated terms', which is also a 'difference of the identified terms', the opposites being at once identical (in reciprocal mediation) and different (independent). On this paradoxical relation, see Part One above. In Th.F.Geraetz, that double relation appears in the form of a 'double processus de scission et de réunion'; see Th.F.Geraetz, 'Les trois lectures philosophiques de l'Encyclopédie ou la réalisation du concept de la philosophie chez Hegel', in *Hegel-Studien*, **10** (1975), 231-254.

56. See *Logik* vol.2 pp.278-284 (II, 271-7), and *Enz.* par.172, where 'The individual is universal' becomes 'The individual is not abstractly universal'.

57. Likewise, can we affirm that any negative judgement implies a corresponding positive judgement? I think we can, in so far as the negation connects various terms of the 'same' genus, identical in that sense; in contradistinction to the meaningless 'infinite judgement', composed of terms wholly alien from each other, such as 'Spirit is not red' or 'The rose is not an elephant'; see *Logik* vol.2 p.284 (II, 277) and *Enz.* par.173 note.

58. See *Logik* vol.1 p.76 (I, 103).

59. When analysed, 'Un-ruhe' gives the idea of a movement. As we shall see later, Hegel is brought to manipulate language.

60. Indeed, in Hegel the logical categories are regarded as living determinations, as quasi-subjects, self-moving 'spiritual essentialities' according to *Phän.* p.31.

61. Thus, dialectic presupposes the double meaning of its determinations, identical as concepts, different as representations. Notice must also be taken of the Hegelian paradox of the 'indeterminate determination', explained by the fact that we are at the beginning only of the logical development.

62. To the extent that human thought is positing and determining, any object grasped by it must necessarily show itself as limited and related to its other. This assertion applies also to totality, as a particular determination opposed to the part. But it may happen that the totality remains vague, complex and underdetermined, leaving the category itself as somewhat underdetermined.

63. De Saussure too regards meaning as contextual; see De Saussure, *Cours de Linguistique*, p.126.

64. See *Logik* vol.1 p.79 (I, 109); also pp.141 and 145 (164-5 and 168).

65. J-L.Nancy, in his *La remarque spéculative* (Galilée, Auvers-sur-Oise 1973) p.51, makes mention of "la loi de l'insuffisance spéculative de toutes les copules grammaticales, du 'est' et du 'et'".

66. See G.Wohlfart, *Der spekulative Satz*, pp.240-241 and 290-291; on p.242, the author examines the speculative relation unfolding in the proposition between its syntax and its semantics.

67. What Hegel himself admits in the *Logik*, vol.2 pp.240-241 (II, 236), when he presents the universal beginning as *'ein nur Gemeintes '* (as only intended).

68. The dialectic of Sense-Certainty, at the beginning of the *Phenomenology*, seems to suggest that the transient, particular sensible, has no real existence relatively to the permanent intelligible universal, which really is. However, it may be asked: what would that 'high' universal become without its embodiment through the 'low' written marks? In this sense is it not possible to see, paradoxically, the sensible as a certain 'truth' of the intelligible? For a similar view, see Notes 41 and 51 above.

69. See *Enz.* par.85, where Hegel objects to the inadequate form affecting any definition.

70. For a similar vision of things, see M.Clark, *Logic and System* (Martinus Nijhoff, The Hague 1971) pp.68-71, 84-86, 146-150 and 189-193. This can be explained in the following way: here philosophy, the self-consciousness of human reality, is itself regarded as real, hence the paradox and the dividing reflection. We saw that the same holds for language, an expression of reality and itself real.

71. Thus, avoiding the extreme I remain truthful to classical tradition. By this I mean, for ex., Plato who does not think that all is at rest or all is in motion, Aristotle who proposes the rule of the right mean, and Kant working between naive dogmatism and destructive scepticism.

72. On the movement of the Syllogism, see *Logik* vol.2 pp.308-352 (II, 301-42), and *Enz.* par.181-193. See also Jarszyk, *Système et liberté,* pp.105-148.

73. Indeed, by 'dialectical syllogism' we cannot understand the exposed movement of the still formal and abstract 'syllogism of Understanding'. As for the 'rational syllogism', a movement of the concrete subject, it is briefly recalled and not expounded. Should we see in it the 'Syllogism of the three syllogisms' of the *Enz.* par.575-577? The discussion to come will specify. For the expressions 'syllogism of Understanding' *(Verstandesschluß)* and 'rational syllogism' *(vernünftiger Schluß)*, see *Logik* vol.2 pp.308-312, and *Enz.* par.181 and 182.

74. On the movement from transition to reflection in the other and to development, see *Logik* vol.1 pp.79, 109 (I, 106,134) and vol.2 pp.170, 240, 269 (II, 174,235,262-3), and *Enz.* par.84, 142 note, 161, 240 and 244.

75. Effectively, a relation can develop only between terms independent from a certain point of view. On this topic see also Note 55 above.

76. According to the double meaning of *'zugrunde gehen'*, which means 'to return to the ground' (to find a basis) and 'to go to the bottom' (to sink and vanish).

77. This syllogism can be explained as a development of the proposition 'The Individual is Universal' (I is U), whose copula has, at first, become the determination of the particular. In an Aristotelian perspective, in order to realize the basic syllogism I-P-U, the relations P-U and I-P have still to be explicated; hence the two following propositions: 'P is U' (the Particular is Universal) and 'I is P' (the Individual is Particular).

78. At the same time, thus, the pattern of the 'Syllogism of the three syllogisms' is rejected, that which was the assumed 'Syllogism of Reason' concluding the *Encyclopaedia* with the par.575-577. On this pattern see also *Enz.* par.198, 201, 207, 217, 269 note and 571. Philosophy is 'its own time apprehended in thought', Hegel says. However, at the end of his central works he passes far beyond this determination, what every critical mind can observe.

79. On a similar point of view see G.R.G.Mure, *A Study*, pp.79-91 and 342, and M.Clark, *Logic and System*, pp.70 and 90.

80. Anyway, in the *Logic* the Syllogism constitutes but a moment, necessary and insufficient as such, since it refers to a still abstract Subjective Reason opposed to Objective Reason.

81. See J-P.Sartre, *Critique*, pp.92-94. It is here possible to distinguish between microdialectic, the movement of a single proposition, and macrodialectic, the movement of large ensembles, problematic in the sense that, in the System, various gaps come and break their rhythm of development—as it can be seen, for ex., in the transitions from *Phenomenologie* to *Logic*, from objective to subjective logic in the *Logic*, and from logic to Realphilosophie in the *Encyclopaedia.*

82. This demand for 'letting the thing be itself', orginating with Hegel, has been emphasized by the second Heidegger.

83. However, it must be pointed out that, in his *Logik* vol.1 p.36 (I,65), Hegel presents also the different titles, found in the book, as a 'historical' product of external reflection which already knows the whole and suggests a preliminary division: even from this point of view 'the external is within'.

84. His philosophy of nature is sharply criticized on this point. Anyhow, the *Encyclopaedia* seems to be written from the standpoint of the 'transition', proper to the doctrine of Being; on Notes expressing such an idea, see *Enz.* pp.3, 20 and par.18 note.

85. Right from the start of the work, that problematique shows up. We find there two titles, 'System of the Science, first part: the Phenomenology of Spirit' and 'First part: Science of the Experience of Consciousness', and two inner divisions, one in capital letters (A, B, C, AA, BB, CC), and the other in Roman numerals from I to VIII.

86. Indeed, generally speaking the spatio-temporal *Darstellung* (exposition) of the Hegelian Concept, in the medium of discursive language, brings to separate that which in truth should be apprehended together—this constitutes the problem of exposition, recalled at the beginning of the present Part. For this issue, see my Dissertation: *La dialectique de Hegel en tant que logique philosophique: exposition et critique de ses principales carctéristiques,* Hebrew Univ. of Jerusalem 1983, 2°Part chap.b (in Hebrew, summary in French).

87. See *Logik* vol.1 pp.23-24, 29-30, 41 and 64 (I, 53,59,70 and 90). See also Note 83 above.

88. A conclusion which can be confirmed by the reading of the various 'Additions' to the *Encyclopaedia,* and of the *Lectures* on Philosophy of History, Aesthetics, Philosophy of Religion and History of Philosophy.

89. In this final discussion, I confine myself to the problem of the speculative proposition and of the antinomy of language, without referring anymore to the different Introductions and Notes written by Hegel in common language.

90. We meet thus again the essential relation holding between essence and phenomenon, truth and example—a relation in which the first term realizes itself through the mediation of the second.

91. For ex., firstly the tautological, formal Ground repeats the grounded in other words; and secondly, the real Ground does not ground, being different from the grounded. Or still, formal Causality is identical with the effect; while determinate, finite real Causality develops into an alternated infinite progression, of effects which are themselves causes, and so on. On these diverse movements, see *Logik* respectively vol.2 pp.76-88 and pp.189-198 (II, 84-94 and 191-200); also pp.365-371 (355-361).

92. For a similar approach, see Note 44 above.

93. See in the *Encyclopaedia* the well-known par.459 on language. There, not only images, symbols and signs are examined, but also the metonymy in the form of various associations between images and representations, recalled at the par.455 and 456.

94. See *Enz.* par.12 note; also *Phän.* pp.87-88 (65) which mentions the animals as realizing the unity of subject and object by the consumption of food. 'At least in part' for, as Part Four (on Matter and Nature) has shown, a residue remains, 'indigestible'.

95. On the representation as of service to the concept, see also *Logik* vol.2 p.357 (II, 347 on images and notions) and *Enz.* par.1-5 and 20.

96. On the Hegelian proposition as analytic and synthetic, see *Logik* vol.2 pp.491 and 32 (II, 472-3 and 42).

97. In contradistinction to formalistic criticism, I understand that the ambiguity characterizing the natural language precisely constitutes the source of its many-sided richness and of its fluidity. On the 'Wisdom embodied in the language' see A.Koyré, 'Note sur la langue et la terminologie hégélienne' in *Revue phil.de la France et de l'étranger,* **CXII** (1931), 415.

98. As in Kant, dealing with the Transcendental Deduction; see his *Kritik* 754B-764B. Science is not regarded here as an instrument designed for the solution of some theoretical or practical problems, but as a mode of expression of reality, itself real, in the same way as language, art, religion and philosophy.

99. As shown in the first Part of this work.

100. On the importance of poetical language and its limits according to Hegel, see D.Cook, ***Language*** pp.101-124.

101. However, let me precise: the discussion is about 'the concrete universal', an individual both universal and particular. For natural language, when spoken in a 'plastic' way, can be characterized as an 'individualized' language: not only does it express the individual but, moreover, its very terms are themselves individualized, their meaning unceasingly varying with the surrounding context.

102. As observed by J-L.Nancy in his *La remarque*, p.106: the articulated proposition seems to be 'la seule syntaxe possible du discours'. In this framework, from the inadequate mathematical language to the organic and religious languages, often used by Hegel, Hegel has at his disposal a range of signs and symbols whereby he can move up and down the scale, according to its various goals.

103. And that synchronically, the context giving the meaning ; but diachronically too, the achieved result comprehending the way leading to it. That is, the dialectician has to appropriate, as much as possible, the culture of his time and its history.

104. In other words, concrete thought is temporal: it forms and explains itself with the help of a series of representations succeeding one another—in this way, a certain return to the Kantian way of philosophizing is indeed accomplished. A certain intellectual intuition is perhaps possible 'one instant', but the need for self-explanation and concreteness will ever oblige 'to go back to the cavern': in principle, the *Vorstellung* (representation) proves unconquerable.

105. That is, significant human (historical, social and cultural) reality and opaque natural (physico-chemical and biological) reality form, together, the world in which we, as 'reflective' men, are living, thinking and acting, reconciled in part only.

106. In Hegelian logic, notice must be taken of the important function fulfilled by the metaphors 'to contain', 'within', 'inside' and 'outside', although the logical category is not a box with an inner and an outer.

107. It all depends whether the matter is about predicative (sensible, external, divided and static) proposition, or about dialectical (conceptual, inner, identical and dynamic) judgement. Thus, we encounter again the double meaning of 'dialectic', which signifies either abstract dialectic as against the predicative dimension, or concrete dialectic uniting both abstract dialectic and predicative dimension. On this double meaning see Notes 16 and 44 above.

108. These sentences define the dialectical relation of 'self-reflection'. In this sense every determination is a totality and a moment of itself; on this point see Note 44 above.

109. I understand the discussion as being about B.Russell's and the young Wittgenstein's logical atomism. On the contrary, the last Wittgenstein asserts that in mathematics every renewed proof introduces a new concept; see Wittgenstein, *Remarks on the Foundations of Mathematics* (ed.Wright, Rhees and Anscombe, The MIT Press, Massachusetts, ed.1967) Part II prop.31, Part III prop.30-31, Part.IV prop.45. In this concrete vision of things, can we not discern a certain Hegelian idea of the 'movement of the Concept'?

110. On such a reconstruction see Carnap, *Der Logische Aufbau der Welt*, F.Meiner, Hamburg 1966.

111. For a development of content even in mathematics and in formal logic, see Note 28 above.

112. Thus the predicative proposition, a moment of the dialectical movement, constitutes an independent determination too. On this paradoxical relation, see Notes 55 and 75 above.

113. On this pattern of explanation, see *Enz.* par.380. Therefore, in spite of Hegel's opposed assertion, we cannot think 'by names' alone: in dialectic, for ex., the conventional sign makes sense when relying on the sensible symbol; rendering concrete by empirical representation is always essential.

114. See *Logik* vol.1 p.10 (I, 40); also p.94 (120). See also A.Koyré, 'Note sur la langue',.425-426.

115. As possible translations of Hegel's *'aufheben'*, in English I suggest such ambivalent terms as 'to comprehend' (B 'comprehends' the opposition between A and not-A) and 'to resolve' (B 'resolves' the opposition between A and not-A, or 'dissolves it')—see also 'to comprise'. Of course, the usual 'to transcend' and 'to overcome' (I sometimes use them, depending on the context), or 'to sublate' and 'to supersede', are fitting too. In French, *'dépasser'* gives a good idea of Hegel's meaning (in English, 'to pass' with its various suffixes may also help). Anyway, *'aufheben'* should rarely be translated by 'to suppress' or any similar univocal, negative term. On this issue, see the interesting remarks of J-L.Nancy in *La remarque*, pp.17-18.

Finally Hegel, instead of the positive *'Einheit'*, sometimes writes *'Ungetrenntheit'*, stressing hence the negative side of the relation at stake; see p.108 above.

116. See *Logik* vol.1, respectively pp.96, 104 and 26 (I, 122,130 and 55).

117. Therefore, from time to time we may ask ourselves: do we really read Hegel, or only our own prejudices? If in principle this hermeneutic question cannot receive a univocal answer, nevertheless a permanent re-reading and re-thinking may enable to move any postulate 'laid down as a firm basis', and thus to rise from the standpoint of a rigid and formalistic representation to the fluid and concrete dialectical standpoint.

118. Thus, in his *Logik* vol.1 pp.94-95 (I, 120), Hegel suggests having recourse to expressions of Latin origin when the mother tongue is not rich enough, or still when 'reflected' determinations are concerned; for the foreign language, less familiar, proves more suitable for a pertinent consideration of the thing in question.

119. In so doing Hegel finds again the *'to ti en einai'*, the Aristotelian quiddity also expressed with the help of a past time *(en)*. On this 'family connection' in German, see *Logik* vol.2 p.3 (II, 15); also *Phän.* p.85 (63).

120. On the use of 'Unterschied' or 'Difference', see P-J.Labarrière & G.Jarczyk in their work of translation: *Hegel, science de la logique* (Aubier-Montaigne, Paris 1972, 3 vols.), vol.1 book 1 p.XXIX.

121. On this point see Note 101 above. D.Cook in his *Language*, p.67 Note 22, asserts that Leibniz also suggested to philosophize in German.

NOTES TO PART SIX

1. See G.W.F.Hegel, *Wissenschaft der Logik* (F.Meiner, Hamburg ed.1975, 2vol.), vol.2 pp.485-504; or *Science of Logic* (transl. Johnston & Struthers, G.Allen, London 1929, 2vol.), II, 467-84 (= vol.2 pp.467-84).

2. See above respectively Part One and Part Five.

3. As for the mathematical being, its problematic status is well-known: must it be understood as a sign, a meaning or an independent reality? Until now, realists and non-realists are still debating about that question. The young Plato, for his part, seems to have regarded the number as a 'third', a middle term between sensible experience and pure thought.

4. In this way, the problem of the relation between soul and body can be posed again.

5. Any idea of a dialectic of nature is then rejected.

6. Although some researchers, working in the field of experimental sciences, think the opposite. On this issue see E.Nagel, *The Structure of Science* (Routledge & Kegan, London ed.1968) pp.336-397, and K.Lambert & G.G.Brittan, *An Introduction to the Philosophy of Science* (Prentice-Hall, N-J. 1970) pp.60-65, 100-106. However, we shall see that, in Hegel, it is possible to talk of an explanation which 'cancels dialectically', that is, which *aufhebt* or develops.

7. Among the Greeks, dialectic was still busy with any problem and any subject, let it be formal, empirical or cultural. Therefore, it could always degenerate into sophistry.

8. In mathematics, there are of course other sorts of proofs, such as mathematical induction, proof by counter-example or by reducing to the absurd, but this is not the point here.

9. On the object treated by mathematics, see for ex. W.Kneale & M.Kneale, *The development of Logic* (Clarendon Press, Oxford 1964) chap.VII pp.435 et seq., dealing with the works of Cantor and Frege.

10. As additional non formal ways of thinking, transcendental philosophy, phenomenology and hermeneutics may be recalled.

11. See I.Kant, *Kritik der reinen Vernunft* (F.Meiner, Hamburg ed.1976) 754B-766B.

12. See Hegel, *Enzyklopädie der philosophischen Wissenschaften 1830* (F.Meiner, Hamburg ed.1975) par.85; see also *Logik*, vol.1 p.125 (I, 150).

13. See *Logik*, vol.2 pp.442-450 (II, 428-35), and *Enz*. par.259 note.

14. On the indefinable see G.Frege, *Posthumous Writings*, (ed.Hermes, Kambartel & Kaulbach, Blackwell pub., Oxford 1979) pp.89, 174; and Whitehead & Russell, *Principia Mathematica* (Cambridge Univ.Press ed.1964, 3 vols.), vol.1 pp.6-13 and 91-97. For a non formalistic approach in mathematics, see I.Lakatos, *Proofs and Refutations*, Cambridge Univ.Press 1976.

15. For a criticism of the use of Kantian definitions in mathematics, see G.Frege, *The Foundations of Arithmetic* (transl. J-L.Austin, Oxford 1950) par.88.

16. In this approach, natural language is regarded as real and as constituting an essential dimension of human reality, in contradistinction to formal, artificial language.

17. Popper's 'conjecture' too is thereby rejected.

18. Yet, see on Newton's analytic and synthetic 'new beginning', rejecting any need of hypothesis, in J.Hintikka & U.Remes, *The Method of Analysis*, Reidel, Dordrecht 1974; as against Popper's linear method of falsification presented in his *Conjectures and Refutations*, Routledge & Kegan, London ed.1968.

19. In particular, Heidegger raises again the following classical questions: what is Being, Philosophy, Language, Thinking, the principle of Ground, a thing?

20. See Hegel, *Phänomenologie des Geistes* (F.Meiner, Hamburg ed.1952) pp.35-39, or *Phenomenology of Spirit* (transl.A.V.Miller, Oxford Univ.Press 1977) 24-27 (=pp.24-27); see also *Logik*, vol.1 p.35 (I, 63-4).

21. On 'deduction' out of the concept, see Kant's *Kritik*, 810B-822B. On the relation between way and result in Hegel, see the beginning of the Preface to the *Phänomenologie*, pp.9-12 (1-3). In Wittgenstein's notion of a 'formation of concepts', I seem to discover a similar point of view: every new proof is said to introduce a new concept; see L.Wittgenstein, *Remarks on the Foundations of Mathematic* (ed.Wright, Rhees & Anscombe, the MIT Press, Massachusetts 1967) App.II 2, II 24, 31, 41, III 29, 30, 45, IV 9. On 'Concept-Formation', see also Lakatos, *Proofs and Refutations*, chap.1 par.8.

22. See *Logik*, vol.1 pp.177-335 (I,198-344). It is true that, for Frege, in mathematics 'to deduce is to calculate'. Nevertheless, in his work he is mainly concerned with concepts, judgements and thoughts, and not with combinations of formal signs. On 'living' mathematics, neither mechanical nor authoritarian, see Lakatos, *Proofs and Refutations*, and H.Khalfon, 'Un trajet en mathématiques' dans *Pratique des mots*, **56** sept.86.

23. In this way, a reversal of positions takes place: from the primacy of the quantitative dimension to that of the qualitative one—hence the minor status given to classical mathematics by Hegel. For my part, in the realm of human studies I advocate the 'coexistence of the opposites', that is, the concrete dialectic uniting abstract (one-sided) dialectic and formal thinking.

24. Yet, a certain ambiguity persists, for the mathematical being seems to remain outside, neither empirical nor formal, as Kant has shown it. Cf. J.Hintikka, 'Kant on the Mathematical Method' in *Knowledge and the Known*, (Reidel, Dordrecht 1974) pp.160-183.

25. According to Kant, mathematics also grasps universal and particular together. However, they do so by linear construction (they construct a concept, that is, they exhibit *a priori* the intuition which corresponds to the concept) and not by reciprocal mediation. For an organic or synthetic vision of the deduction, see Frege, *The Foundations*, par.88 and 91, Hintikka, 'Are Logical Truths Analytic?' in *Knowledge*, pp.135-159, and Lakatos, *Proofs and Refutations*, chap.6-7. On transcendental logic or dialectic as an organon, see G.R.G.Mure, *An Introduction to Hegel* (Clarendon, Oxford ed.1966) pp.100, 114-115, 119-120, and Y.Yovel, *Kant and the renewal of Metaphysics* (Bialik Inst., Jerusalem 1973, in Hebrew), pp.17,23 and 296.

26. In his *Kritik* Kant separates the two sources of knowledge: the spontaneous (universal) **concept** and the given (particular) empirical **intuition**. As regards space and time, here considered as sensible intuitions *a priori*, the ambiguity of their status is to be noted: receptive as sensible, they are active as organizing the given manifold.

27. From this point of view, Hempel's linear model of explanation—characterized as 'deductive-nomological'—does really not satisfy. For it subsumes univocally the phenomenon to explain under the corresponding universal law; whereas in dialectic, as in transcendental deduction, the point of departure is not the explaining principle but the explained object: the cultural thing has primarily to be respected, and not the law. On this model see C.G.Hempel, *Aspects of Scientific Explanation* (The Free Press, N-Y. 1965) pp.331-496. For a criticism of this model in Physics, see W.C.Salmon, *Scientific Explanation and the Causal Structure of the World*, Princeton Univ.Press 1984.

28. Just as the hermeneutical circle discussed in H-G.Gadamer, *Wahrheit und Methode* (J.C.B.Mohr, Tübingen ed.1972) pp.164, 178, 250, 495.

29. On this point see the paradox of Meno in Plato's *Meno*, 80d-e, arising when Meno asks about the possibility of learning something new. See also K-R.Dove, 'Hegel's phenomenological method' in *Rev.of Met.*, June 1970, 617.

30. See W.Becker, *Hegels Begriff der Dialektik und das Prinzip des Idealismus* (Kohlhammer, Stuttgart 1969) pp.7 et seq.

31. The idea of an organic system was first exposed explicitly by Kant, who regarded Reason as a system; see Kant's *Kritik*, 766B, 860B-866B. As for Hegel, he considers the whole reality, the natural as well as the spiritual, as forming a system.

32. Let me recall that, from a dialectical point of view, human spirit is at once analytic Understanding *(Verstand)* and synthetic Reason *(Vernunft)*. Of course, this Hegelian distinction constitutes a reinterpretation of the corresponding Kantian distinction.

33. In Spinoza, a similar idea is to be found; see his *Ethics*, 2°Part. prop.XXXV.

34. See *Phän.* pp.33-34 (22-3).

35. In other words, 'non A' is understood as 'non-A', the negative as positive. If non A as formal contradictory of A is indeterminate, signifying the disparate aggregation of all the things that are not A (for ex., 'not being a subject' can denote a comma, the boiling point, the lightning, etc.), now, 'Non-A' as contrary of A is B, its opposite within some definite genus (for ex., within the domain of human reality, non-subject is an object, non-means is an end, etc.). Dialectical contradiction uniting contrary opposition and formal contradiction, B can be written 'non A and non-A', a 'tensed' dialectical category as an 'identity of differentiated terms'. On different problems attached to dialectical contradiction, see above Part One chap.4.

36. The idea of a systematic history of philosophy appears already in Kant, although as sketched out only; see Kant's *Kritik*, 880B et seq.

37. This example is discussed by Hegel in his *Logik*, vol.2 pp.216-218 (II, 214-215).

38. Since *aufheben* is to negate dialectically, that is to develop from within.

39. See *Logik*, vol.2 pp.497-502 (II, 478-82).

40. The ambiguity of the procedure is to be remarked: it implies an 'immanent' development, with the help of an external leading strand.

41. As Kant himself, who has written the *Prolegomena zu einer jeden künftigen Metaphysik*. On Kant's certitude to be never contradicted essentially, see his *Kritik*, pref.to 2°ed., XLIII B; see also the last words of that work.

42. On such a reconstruction see G.R.G.Mure, *An Introduction to Hegel*, Clarendon Press, Oxford ed.1966, and *A Study of Hegel's Logic*, Clarendon Press, Oxford ed.1967.

43. See Note 40 above, which already mentioned the ambiguity of a movement that is at once inner and outer. Moreover, external reflection proves constitutive, as is shown in *Logik*, vol.2 pp.17-18 (II, 29-30).

44. I have thus studied, *inter alia*, the fruitful criticisms of Kierkegaard, Trendelenburg, Meulen J.v.D. (see Note 60 below) and W.Becker. These studies have greatly helped me in my understanding of Hegel's dialectic, on its deficiencies and relative validity.

45. On this point, see above Part Four chap.3.

46. Consult, for ex., *Enz.* par.13-14.

47. On this point see Cl.Lévi-Strauss, *La pensée sauvage* (Plon, Paris 1962) pp.338-357, and *anthropologie structurale deux* (Plon, Paris 1973) pp.377-422.

48. On the movements of reflection, see *Logik* vol.2 pp.13-23 (II, 25-34).

49. See Kant's *Kritik der Urteilskraf* (F.Meiner, Hamburg 1974) 'Einleitung', par.IV pp.15-17.

50. Similarly, Hegel makes mention of a 'Reason which understands' *(verständige Vernunft)* and of an 'Understanding which reasons' *(vernünftiger Verstand)*; see *Logik*, vol.1 p.6 (I, 36). Indeed, unlike Kant, within the knowing subject he does not separate empirical consciousness (or experience) from philosophical consciousness (or the thinking of experience).

51. See also Note 26 above. The apprehension of reflection as determining and of logic as organon, seems to justify the use of organic explanation in the realm of human reality, in so far as human reality really forms a system; in contradistinction to Kant, who refers to the organic model as applying to nature alone, and as constituting but a heuristic exposition; cf. *Kritik der Urteilskraft*, par.78.

52. See *Phän.* pp.48-51 (35-8), and also p.27 (17).

53. For a similar idea, see the beginning of the *Logik*, vol.1 p.20 (I, 49).

54. It is no use asking for an optimal point, 'neither too close nor too far', from where one could see better; for the discussion is about dialectic, a logic of ambiguity, and not about an empirical thinking.

55. In *Phän.* p.50 (37), Hegel deals with the 'two subjects' of the speculative proposition, the grammatical and the knowing subjects. By adding the movement of knowledge itself, which constitutes 'the Concept', we are led to realize that in fact they are three. On the Hegelian Concept as a subject, see *Logik* vol.2 pp.211ff. (II, 209 et seq.). On the two subjects of the speculative proposition, see above Part Five at the end of chap.5.

56. See *Logik*, vol.2 p.222 (II, 219).

57. In contradistinction to the Aristotelian God, an 'unmoved mover' that is an immanent cause of the world, and that is also external to it (see Aristotle's *Metaphysics*, 1075a11-25)... From that angle the Concept, as the totality of the movement, also reveals itself as a moment only, being a movement formed by the activity of the thinker.

58. To distinguish from the 'cunning of Reason'. On the cunning of knowledge, see *Phän.* p.46 (33) and p.64 (47) as well.

59. On the Concept as 'divine', see *Logik* vol.2 pp.356 and 505 (II, 345 and 485). On the thinker's *a priori* movement of sinking and restraining, see also K-R.Dove, 'Hegel's phenomenological method'.

60. On a similar criticism see A.Trendelenburg, *Logische Untersuchungen* (Hirzel, Leipzig 1870) pp.68-79; W.Becker, *Hegels Begriff*, pp.63-65, and Meulen J.v.D., *Hegel; die gebrochene Mitte* (F.Meiner, Hamburg 1958) pp.45-53.

61. Kant's problem of the 'a priori synthetic judgment' has turned, in Hegel, into the problem of the 'identity of opposites'—an expression whose meaning is: a double movement of the Concept through the mediation of the thinker who develops *a priori* and reconstructs *a posteriori*. From this point of view, dialectic is really comprehensible as a productive and critical organon. On the scientific method as a double movement of *resolutio* (analysis) and *compositio* (synthesis), see Hintikka & Remes, *Method of Analysis*.

62. To put it another way, the point at issue is to justify the act of predicating A of the subject B. See also W.Becker, *Hegels Begriff*, pp.44 et seq.

63. On these different ideas, see Plato's *Parmenides*, 1° Part.

64. See Plato's *Republic*, 509b-c.

65. See *Parmenides*, 2° Part.

66. See *Sophistes*, 256a-257b.

67. See Aristotle's *Metaphysics*, 998b24-27; see also *Logik*, vol.2 p.235 lines 19-22 (the 5 last lines of II,230).

68. The problem of the beginning arises again and again...

69. Antonymies and paradoxes are the price to pay for an all-embracing thought that grasps human reality in its multidimensionality without abstracting nor reducing, and recognizes each of its dimensions as relatively autonomous.

70. For such an attitude of Hegel towards Kant see, for ex., *Logik* vol.1 pp.15, 108 (I 44,133-4), vol.2 pp.221-233 (II, 218-28). On philosophy as a science of self-consciousness in this sense, see Y.Yovel, *Kant*, pp.177-184.

71. On the 'inverted world', see *Phän.* pp.121-124 (96-8) and p.25 (15), *Logik* vol.2 pp.129-134 (II,135-9); S.Rosen, *G.W.F.Hegel* (Yale Univ.Press, ed.1976) pp.140-150, and J.C.Flay, 'Hegel's "Inverted world"' in *Rev.of Met.*, **92**, June 1970, 662-678; on the '*Spiegelbild*', see G. Wohlfart, *Der spekulative Satz* (De Gruyter, Berlin 1981) pp.60, 81 and 97.

72. In Hegel, the linear mode of explanation is examined in particular in the dialectic of Ground (see *Logik*, vol.2 pp.76-88 or II, 84-94): as 'real Ground' the ground explains badly, since it is different from the grounded; and with 'formal Ground' there is no explanation at all, but only a repetition of the grounded in other words.

73. Furthermore, the logical explanation has to be understood in its context, that is, among other things, in relation to the relevant Phenomenological way and Real (natural and spiritual) development exposed in the '*Realphilosophie*' .

74. In his philosophy, Hegel reinterprets the Kantian doctrine of constitution. Dialectic is intelligible, indeed, in so far only as an *a priori* dimension is actively present in our experience.

75. On the problem of exposition *(Darstellung)*, see above at the beginning of Part Five. Dialectical thought requires the 'grasping together' of that which, unavoidably, is exhibited separately in space when the philosophical discourse is written down.

76. On the movements of Being from the standpoint of Essence, see *Logik* vol.1 pp.64 and 66 (I, 91 and 93). On Essence as the reconstruction of the categories of Being, see *Logik* vol.1 p.86 (I, 112) and vol.2 pp.7-9, 97-99 (II, 20-1,103-5).

77. The Idea itself has to be understood dialectically, here as a double movement unfolding from subjectivity to objectivity and vice versa.

78. In the quoted example, the matter is about the transition from the necessity of Substance to the freedom of the Concept; see *Logik* vol.2 pp.202ff. (II, 203 et seq.). The mentioned leap is also mitigated by the ambiguity of the concerned terms (necessity is contingent and freedom proves also to be the rule of necessity), and by the converse movement of the subjective Concept to Objectivity, exposed in *Logik*, vol.2 pp.351-359 (II, 342-349).

79. See *Logik*, vol.1 p.74 (I,101); also vol.2 p.226 (II, 222-3).

80. Th.Haering asserts that Hegel organizes the philosophical content according to three different principles: firstly, that of the traditional metaphysics (classifying from particular to general), then a certain principle of dialectical organisation and, to crown it all, the principle of 'spiritual monism' arranging according to a trinitarian model (Idea, Nature, Spirit). See Th. Haering, *Hegel, sein Wollen und sein Werk*, Leipzig 1929, vol.2.

81. See *Logik*, vol.1 pp.101-146 (I, 127-69), especially pp.101-110 (127-35); see also *Enz.* par.89-95. The dialectic of Finite and Infinite, more abstract in my view, will be summarized briefly.

82. Cf.*Logik*, vol.1 p.108 (I, 133-4), and also pp.15, 32, 45 (44,62,73), vol.2 pp.111-112, 232 (II, 117-8,228); and *Kritik* 249A, 307B, 312B, 333B, etc.

83. Here, as in formal or empirical thinking, the aim is not 'to justify everything'. In the dialectical mode of explanation, beginning of elucidation and end of development are enlightening one another.

84. Indeed the Hegelian logic, a non formal one, is also an ontology and, in principle, it deals with things in their intelligible aspects.

85. Cf.*Enz.*, add.to par.94. In Hegel the various *(Verscheiden)*, merely external, is not the different *(Unterschieden)* which is differentiated inwardly. On this point see *Logik*, vol.2 pp.32-48 (II, 43-58).

86. For such a thesis, see above Part Five chap.5.

87. The Hegelian logic is also a 'thought of thought', the *Phenomenology* thinking the human experience, the *Logic* considering that very thought. On Hegelian logic as a thought of thought, see *Enz.* par.17 and 19 note, *Logik* vol.1 pp.23-47 (I, 53-75) and its last words on 'The pure Concept which forms a Concept of itself'.

88. Although imagination and memory are moments of 'pure' thought in Hegel, as it can be noted in the 'psychological' way presented in *Enz.* par.445-468.

89. For ex., G.R.G.Mure in *A Study of Hegel's Logic*, pp.45-52, or E.Fleischmann in *La science universelle ou la logique de Hegel* (Plon, Paris 1968) pp.71-81.

90. Parmenides also, in Plato's *Parmenides* 135e, asks the young Socrates to consider what, in every case, follows from the hypothesis concerning the non-existence of the object, after having considered what follows from the hypothesis concerning its existence. See also *Logik* vol.1 p.137 (I, 161).

91. The paradox is manifest: the task is to think 'in general' something that is 'determined', hence particular... Yet, even in Aristotle the 'general' species is said to be 'specific' with regard to another species and under the same genus.

92. See *Kritik*, 15B-17B; also 205B.

93. The thinking of Something and an Other will not constitute Kant's '*a priori* synthetic', but Hegel's '*a priori* and *a posteriori* identity of the opposites'. Purely *a priori*, the Hegelian logic could not be an organon producing a certain truth. However, with the help of the leading strand of the historical and cultural experience, a certain developing 'identity of opposites' seems possible, at least from the standpoint of the Age. Cf. Note 61 above.

94. Hegel explicitly refers to the Platonic category *'to eteron'*, the other according to *Sophistes*, 254e et seq. On this point see *Logik*, vol.1 p.105 (I, 130-131).

95. In the *Logik*, vol.1 p.109 (I, 134), Hegel asserts that, at the level of Being we are dealing with, the thought category seems to be complete without its Other. It seems so *(scheint)*, but in fact it is not so.

96. I choose a sensible example just as a help to thinking. The so-called 'pure' thought realizing itself through the mediation of empirical thought, this attempt seems to me legitimate. See also *Enz.*, addition to par.92.

97. Indeed, just as according to the Gestalt psychology a pattern is always perceived as against a certain background, so the meaning of a term has to be grasped here with regard to its surroundings, its determining opposite or context.

98. Even in this sense, the Hegelian dialectic proves realistic and idealistic at once.

99. Every thinker being 'a child of his time' and, moreover, being influenced by his own experience.

100. That is, the constitutive relation is rightly defined as 'inner'. On the other hand, the excluding relation is grasped as 'external': everyone knows that, in arithmetic, 'one' is not 'two'. On the determination as a relation to an other, see *Logik* vol.1 pp.108-110, 112 and 57 (I, 134-5,137 and 84); see also *Phän.* p.47 (34). Conversely, in *Logik* vol.2 p.13 (II, 25), the relation, in the form of 'Reflection', is grasped as a category.

101. In this case too the matter is about a 'thought of thought', the dialectical thought examining the results proposed by the physicist **afterwards** only—**post-factum**. For a detailed discussion of atomistic thinking by Hegel, see esp. *Logik*, vol.1 pp.156-157 (I, 179-80).

102. For ex., depending on the context, in physics light is said to be grasped as a corpuscular or as a wave phenomenon.

103. On this ambivalent movement, see above chap.3.

104. From an objective point of view. Now, from a subjective point of view, in chap.3 above the question was to examine the act of consideration as moving the dialectic. In other words, the contradiction lies in the cultural things in so far as man is looking at them and determines them.

105. For Hegel, the physical thing is a Concept 'in itself and for us': devoid of any consciousness, it does not develop itself but undergoes modifications. On the other hand, with respect to the thinking man the Concept is 'for him', hence his movement of self-development. On this point see *Logik* , vol.1 pp.122-123 (I, 147-8).

106. This category whose movement we are following, we are producing it by this act itself of following! In fact, there is a movement of thought *(Sache)* and not a movement of thing *(Ding)*.

107. Without forgetting, of course, the ambiguity of a necessary detaching 'stepping back' *(recul)* and of the (partly external) leading strand of the history of philosophy.

108. It must be said that Something **is** a quality, and not that it **has** a quality. For here the discussion is about inner, constitutive relation, and not about external, predicative relation.

109. As against Kant, Hegel affirms explicitly that where a limit is, there is also a passing beyond it; see *Logik*, vol.1 pp.115, 121 and 122 or I, 140-1 and 146-7 (on *die Schranke*, the Barrier as a posited limit), and *Enz.* par.60 note.

110. Here it is about 'dialectical' necessity, which means a necessity at once necessary and contingent, that is, a relative necessity.

111. At this level (of the analytic thinking), the principle of the excluded middle seems to apply—but only as a moment to be overcome. We must not misunderstand: essentially, dialectic does not deal with empirical propositions of the form 'it is raining' or 'it is not raining', but with the movement of living and creative cultural things.

112. Hegel talks about *Veränderung* (becoming an other); see *Logik*, vol.1 p.103 (I, 125 as 'change'). Strictly speaking, in dialectic 'becoming' is not 'developing': if a becoming may be recurrent and cyclic, as an alternated movement between being and nothing, a development is always creative.

113. Which is a term containing this double meaning: determination and destiny. As it is well known, Hegel likes to manipulate language and to play on words.

114. Is not this a discussion about the Sartrian Being-for-itself of *Being and Nothingness*, which is an individual unable to realize himself, the Being-for-others proving to be alienating and not mediating—hence his ineluctable failure? From a Hegelian point of view, the matter at stake is about the world of tools, a world in which everyone is a means for an other, with no determined end to be aimed at.

115. See above Note 105.

116. Hegel is particularly fond of this paradoxical mode of language, originating from his 'negation of negation'—a transition to a higher stage. For ex., see *Logik*, vol.1 pp.69, 93, 118, 119, 124, 131, and 397 (I, 96,119,143,144,149,155 and 403). In German, the effect is more visible.

117. Dialectically speaking, any human development implies a practice (an experience) and a theory (a train of reasoning) supporting each other, so that history may advance.

118. The ambivalence lies there: the difference is at once real and ideel; see *Logik*, vol.1 p.113 (I, 139). At this level, the dialectic in question can be interpreted as the dialectic of the Same and the Other, an opposition especially dear to the Greeks.

119. We meet again the *'für uns'* (for us) of the *Phenomenology*. Indeed, there is some similarity between the logical dialectic of Something and an Other and the dialectic of Sense-Certainty in the

Phenomenology. Thus, in the *Logik* vol.1 p.104 (I, 130), Something is presented as a This *(Dieses)* that is Designated.

120. On this image see Note 71 above.

121. The Finite and the Infinite being, each one, finite and infinite at once; see *Logik*, vol.1 pp.137-138 (I, 161); see also pp.134 and 141 (158,164-5). Hegel's strategy at this point is the following: if the finite is at once finite and infinite, it is then both real and ideel; moreover, in *Logik*, vol.2 pp.443-444 (II, 429), Hegel talks of knowledge as uniting subjective idealism (the finite as ideel) and realism (the finite as real)—his absolute idealism may be expounded in these very terms. Cf. also Note 98 above.

122. See *Logik*, vol.1 p.141 (I, 164-5) and *Enz*. par.95 note. Likewise, I propose 'The identity and difference of the identified and differentiated terms', despite Hegel's other formulation in his *Logik*, vol.1 p.59 (I, 86). Speculative truth being contextual, to a new examination corresponds a new formula.

123. As far as human reality is truly dialectical, of course, and develops through reciprocal mediations on the impulse of inner contradictions.

124. For ex., between Hegel's doctrine of Essence in the *Logic* and that in the *Encyclopaedia*, important differences are to be noticed in the movement of the categories. Cf. respectively *Logik*, vol.2 pp.26-155 (II, 37-159) and *Enz*. par.115-141.

125. On such a critical standpoint overcoming the hackneyed antinomies, see P.Bourdieu, *Le sens pratique* (Ed.de minuit, Paris 1980) esp. pp.7-50.

126. Thus, the *Phenomenology* is supposed to show how the *Logik*, as a viewpoint of pure thought, is possible.

127. While explaining experience, therefore, dialectic is also developing itself... In other words, two centuries after Hegel, in a contemporaneous world turned into a sophisticated technical civilization, which sees such tremendous cultural and political changes, strictly speaking dialectic can no more be the same.

128. See *Kritik*, 763B-764B.

129. I am here referring to a dialectic of the Whole and the Moment (the Part), such as its reconstruction is made possible by *Logik*, vol.2 p.33 (II, 44)—therefrom it may be understood that Moment and Whole are, each one, moment and whole in turn. According to the **principle of ambivalence** I advocate, the true constitutes itself ambiguously from a standpoint which is total and partial as well.

130. Concretely speaking, the logical explanation should acquire its meaning in the context of the System (cf. Note 73 above). Unfortunately, the researchers fail to agree about the nature of this System. On that topic see my Dissertation, *La dialectique en tant que logique philosophique*, 2° Part. chap.5.

131. In my opinion, the Hegelian synthesis is at once **homogeneous and heterogeneous**, reconstructing thus the different Kantian syntheses. This double nature is asserted, for ex., in the principle of 'The identity (the homogeneous moment) and difference (the heterogeneous moment) of the identified and differentiated terms'.

132. Here, the question is only to know whether the conceptual being is grasped as real, whether one is a realist in this sense—the strict empiricist is not such a realist. On dialectic as analytic and synthetic see *Logik*, vol.2 pp.32, 491 and 499-500 (II, 42,472-3 and 480).

133. Now the ambiguity consists in that human reality comprehends within itself its own explanation; or still, conversely, that the explanation, a 'reflection' of reality, is itself a moment of that reality.

134. In Kant's transcendental deduction too a double movement can be detected: a progressive one (out of the 'I think') and a regressive one (out of the fact that mathematical and physical sciences do exist). These movements are here separated, however, and the progressive deduction plays a prevailing role. Furthermore, to Kant's *a priori* deduction Hegel opposes his own bidimensional 'deduction'—as *a priori* development and *a posteriori* reconstruction (cf.*Logik* vol.1 p.47 or I, 75 and Note 93 above). Again, if Kant thinks that Reason cannot really contradict itself, Hegel, on the contrary, holds that Reason is dialectical, that is, a 'unity in division', which is not only self-explaining but also self-creating. On the relation between Kant's and Hegel's thoughts, see A.Stanguennec, *Hegel critique de Kant*, PUF, Paris 1985.

135. On the inner contradiction, see *Logik*, vol.2 pp.58-59 (II, 66-8); on the category as 'containing its own opposite', see *Logik* for ex. vol.2 pp.56 and 496 (II, 65 and 477).

136. The Concept truly is a 'tensed' unity of Being (here: simple being) and Essence (here: ground).

137. See *Logik*, vol.1 p.117; also vol.2 pp.62 and 252 (II, 70 and 246). Cf. above Part Three at the beginning of chap.4.

138. According to Hegel, from a logical point of view the finite must live and die so that the infinite might live—that is its fate. In this sense, the opaque matter is not a cause but only a means. Besides, could one not say that Spinoza's **Deus sive natura** has become Hegel's **Deus sive cultura**?

139. See above Part One.

140. We must be careful: the discussion is not about the dialectic of the Inner and the Outer exposed in *Logik*, vol.2 pp.150-155 (II, 154-9). In the *Logic* it looks as if, at the level of Essential Relation (*Logik*, vol.2 pp.136-155 or II, 142-159), Hegel himself tends towards the organic-biological interpretation of dialectic. On such an explicit interpretation, cf. *Logik*, for ex. vol.1 pp.43, 122-123 (I, 72,147-8), and vol.2 pp.153, 224 (II, 157,221). On the problematique linked with the couple 'inner-outer', cf.Kant's *Kritik* 373A, 378A, and above Part Three Note 42 and Part Four Notes 17 and 26.

141. See above Notes 57 and 129. It is possible to consider a dialectic of 'the Whole and the Part (the moment)', differing from that of 'The Whole and the Parts' exposed in the *Logik*, vol.2 pp.138-144 (II, 143-8). In this connection, see the ambiguity of 'Accidentality' which becomes 'the Accidents' (*Logik*, vol.2 pp.186-187 or II, 189), and of the 'Many Ones' which becomes 'Multiplicity' (vol.1 pp.157-160 or I, 180-2). Cf. above Part Five Notes 12, 43, 44, 70 and 108.

142. It is to be pointed out that these basic categories (identity and difference, inner and outer and, in the following paragraph, form and content) remind of Kant's concepts of reflection *(Reflexionsbegriffe)*, except for the pair 'whole and moment'; cf. *Kritik* 316B-324B.

143. These few words briefly sum up the study made above in Part Five, chap.4.

144. Ambiguously, the real is at once the fact *(Sein)* and the opposition between intention (or pretension) and the fact (between *Sollen* and *Sein*). On this ambiguity see *Enz.*, add. to par.234. The opposition *Sein/Sollen* itself seems to be basic.

145. This vision of things calls to mind the phenomenological movement, presented by Hegel in the Introduction to his *Phenomenology*, pp.63-75 (46-57). The phenomenological dialectic, in its many concrete aspects, will be examined in a following work.

146. On that tearing present in the Idea, see *Logik* vol.2 pp.412 and 484 (II, 399 and 466). In this sense too, Hegel's thought may be said to contain Kant's thought.

147. Only some points relevant to this final discussion will be mentioned here.

148. It is true that Hegel protests mainly against the use, in philosophy, of the 'low' mathematical categories; cf.*Logik*, vol.1 pp.207-212, 334-335 (I, 227-32,343-4) and vol.2 pp.253, 497-498 (II, 247,478-9). As for the use, in the logical explanation, of categories not explained yet, see *Logik*, vol.1 pp.19-20 (I, 49).

149. On a non vicious circular reasoning, see D.N.Walton, 'Are circular arguments necessarily vicious?' in *American Phil.Quaterly*, **22** (Oct.85), 263-274. The spiral movement is formed by the self-developing circular explanation.

150. See W.Becker, *Hegels Begriff*, pp.86-95. The author refers especially to the determinations of identity, difference and opposition.

151. On this point see *Logik*, vol.1 pp.13, 15 (I, 43,44), and the dialectic of Means and End exposed above in Part Three, chap.3.1.

152. Here we meet again the paradoxical relation of the Whole (the concrete dialectic) grasped also as a Moment (as abstract dialectic), or of a relation at once of repulsion and attraction. See also Note 24 above.

153. See W.Becker, *Hegels Begriff*, pp.63-64. For his part, the author comes to the conclusion that speculative thought is categorically impossible.

154. In the proposed examples, the difficulty increases because of the fact that knowing *a priori* what it means to think 'apart' Nothing or the Infinite , is not clear at all. Kierkegaard is the philosopher who has especially developed the 'indirect method'; see his *Concluding Unscientific Postscript to the philosophical Fragments* (transl. D.F.Swenson, Princeton Univ.Press 1968) pp.67-74 and 225-266. For an indirect knowledge, see above Part Four, p.103.

155. See D.Marconi, 'Logique et dialectique—Sur la justification de certaines argumentations hegeliennes', in *Revue Phil.de Louvain*, Nov.1983, 563-579; in particular 569.

156. Dialectic lives on this contradiction and does not die of it, to the extent that it realizes itself as 'infinite', that is, as all-encompassing, open, fluid and self-correcting. However, when it becomes 'finite', that is rigid, one-sided and dogmatic, then it passes away (disappears) like every finite being.

157. The biological-organic interpretation of dialectic, sometimes put forward by Hegel, seems to indicate that the time was not 'ripe', notwithstanding Hegel's youth optimism displayed in the *Phenomenology*, p.58 (44).

Besides, this misunderstanding may also be due to the organic system underlying Hegel's philosophy. Yet, on the one hand the System is organic but in a weak sense (leaps are breaking it); and, on the other hand, the organic form can facilitate, without hindering univocally, the dialectical movement of the content. Namely, I understand that, at this point, organic system and dialectical thought both oppose and unite. On this problematique, see above Notes 129, 140, 141, 152, and also 31, 51.

158. Which is also a determining one, transforming every mediated term into a new immediate, and vice versa—in contradistinction to formal negation which is univocally universal, abstract and excluding.

ANALYTICAL TABLE OF CONTENTS

EXPLANATORY GLOSSARY

1. PRESENTED AND DISCUSSED CATEGORIES

Part One
Being: *Sein*
Essence: *Wesen*
the Concept: *der Begriff*
Identity: *Identität*
Difference: *Unterschied*
Variety: *Verschiedenheit*
Opposition: *Gegensatz*
Contradiction: *Widerspruch*

Part Two
Being: *Sein*
Nothing: *Nichts*
Becoming: *Werden*
Coming to be: *Entstehen*
Passing away: *Vergehen*

Part Three
End: *Zweck*
Means: *Mittel*

Part Four
Form: *Form*
Matter: *Materie*
Content: *Inhalt*
Ground: *Grund*

Part Five
Judgement: *Urteil*
Proposition: *Satz*
Syllogism: *Schluß*
Middle Term: *Mitte*

Part Six
Something: *Etwas*
An Other: *ein Anderes*
Modification: *Veränderung*
Finitude: *Endlichkeit*
Infinity: *Unendlichkeit*
Limit: *Grenze*
Barrier: *Schranke*

2. MAIN USED CATEGORIES

The One: *das Eins*
Many: *Vieles*
Whole: *Ganze*
Part: *Teil*
Inner: *Inner*
Outer: *Äußer*
Universal: *Allgemeine*
Particular: *Besondere*
The Individual: *das Einzelne*

3. CURRENTLY USED OPPOSITES :

Psychological:
representation/concept : *Vorstellung/Begriff*
Understanding/Reason : *Verstand/Vernunft*

Ontological :
matter/spirit : *Materie/Geist*
conceptual event/empirical thing : *Sache/Ding*

Linguistic :
subjective intention/objective signification : *Meinen/Bedeutung*
only intended/said : *nur gemeint/gesagt*

Methodological :
correct/true : *richtig/wahr*
justification/proof: *Rechtfertigung/Beweis*
reconstruction/sinking/restraining: *Wiederherstellung/zu versenken/Enthaltsamkeit*
Immediate/mediated: *unmittelbar/vermittelt*
in itself/posited: *an sich/gesetz*

Categorial
Being/Ought to be: *Sein/Sollen*
Unity/Separation: *Einheit/Trennung*
Contingency/Possibility/Reality/Necessity: *Zufälligkeit/Möglichkeit/Wirklichkeit/Notwendigkeit*

4. SOME HEGELIAN DOUBLE MEANINGS AND LANGUAGE MANIPULATIONS

(see Part Five, Consequences 3 and 4)

aufheben, Aufhebung: Hegel keeps the meanings of 'making to cease' and 'preserving'; hence the double meaning between 'to cancel' and 'to preserve'; has sometimes the meaning of 'to move', 'to develop'; in English, may be translated by the ambiguous terms 'to comprehend' or 'to comprise' (meaning: to grasp and to contain) and 'to resolve' (meaning: to dissolve a thing and to solve a problem)—see Part Five Note 115.

auflösen, Auflösung: A term whose meaning is close to that of 'aufheben'; has also the double meaning between 'to resolve' a problem and 'to dissolve' the rigid, isolated determination, that is, to move it (see Part One Note 66).

Bestimmung: has the double meaning between 'determination' and 'destiny'. Thus, self-contradiction is said to be the 'Bestimmung' of man.

darstellen, Darstellung: has the double meaning between 'to embody the latent content by the unveiling of the presuppositions' and 'to present the different developments as a set as coherent as possible'.

erinnern: Hegel plays on the morphology of the word 'er-innern', hence the double meaning between to remind and to interiorize.

Einheit (Unity) and **Untrennbarkeit** (inseparability): depending on the context, Hegel can write alternately 'Unity of the opposites' or 'Separation of the inseparable' (see Part Five, 4.5.).

Reflection: Hegel exploits the physical meaning of the term (as optical reverberation) and its philosophical meaning (as mode of thinking). Hence the **Reflexionsbestimmungen** (Identity, Difference and Contradiction), 'reflection in itself which is also reflection in the other and vice versa'.

Unruhe: that Hegel reads as Un-ruhe or rest-lessness, which gives a certain notion of movement. Whence the characterization of Time as unrestlessness (anxiety).

Urteil: read by Hegel as Ur-teil, a judgement understood as a division. So that Hegel may assert that 'All is Urteil'; that is, reality appears as an atomic empirical world which, through the mediation of Judgement and Syllogism, returns to the original conceptual unity.

Zugrunde gehen: means, according to the German double meaning, 'to go back to the basis' or 'to go under'.

BIBLIOGRAPHY

Aristotle, *The basic Works of Aristotle*, ed.McKeon, Random House, N-Y 1941.

Bar-On A.Z., *Foundations of Reality and Knowledge*, Bialik. Inst., Jerusalem 1967 (in Hebrew).

Becker W., *Hegels Begriff der Dialektik und das Prinzip des Idealismus*, Kohlhammer, Stuttgart 1969.

Bodammer Th., *Hegels Deutung des Sprache*, F.Meiner, Hamburg 1969.

Bourdieu P., *Le sens pratique*, Les éditions de minuit, Paris 1980.

Bourgeois B., 'Dialectique et structure dans la philosophie de Hegel' in *Revue Intern. de Phil.*, **139-140** (1982), 163-182.

Clark M., *Logic and System*, Martinus Nijhoff, The Hague 1971.

Cook D-J., *Language in the Philosophy of Hegel*, Kohlhammer, Stuttgart 1966.

Dove K-R., 'Hegel's phenomenological method' in *Rev. of Met.*, June 1970, 615-641.

Dupréel E., *La légende socratique et les sources de Platon*, R.Sand, Bruxelles 1922.

Feyerabend P-K., *Against Method*, NLB, London 1975.

Fleischmann E., *La science universelle ou la logique de Hegel*, Plon, Paris 1968.

Frege G., *The Foundations of Arithmetic*, transl. J-L.Austin, Oxford 1950.
Ecrits logiques et philosophiques, trad. Imbert, Ed.du Seuil, Paris 1971.
Posthumous Writings, transl.Long, White, Blackwell, ed.Hermes, Kambartel, Kaulbach 1979.

Fulda H-F., *Das Problem einer Einleitung in Hegels Wissenschaft der Logic*, Klostermann, Frankfurt am Main 1965 (2°ed.1975).

Gadamer H-G., *Hegels Dialektik*, J.C.B.Mohr, Tübingen 1971.
Wahrheit und Methode, J.C.B.Mohr, Tübingen ed. 1972.

Geraetz Th.F., 'Les trois lectures philosophiques de l'Encyclopédie ou la réalisation du concept de la philosophie chez Hegel' in *Hegel-Studien* (1975), 231-254.

Grégoire F., 'Hegel et l'universelle contradiction' in *Rev. phil. de Louvain*, 44 (1946).

Groll M., 'On the beginning of Philosophy according to Hegel' in *IYYUN*, 9 (1958).

Haering Th., *Hegel, sein Wollen und sein Werk*, 2vol., Leipzig 1929.

Hegel G.W.F., *Differenz des Fichte'schen und Schelling'schen System der Philosophie*, F.Meiner, Hamburg 1962.
Phänomenologie des Geistes, F.Meiner, Hamburg ed.1952; or
Phenomenology of Spirit, transl. A.V.Miller, Oxford Univ.Press 1977.
Wissenschaft der Logik, F.Meiner, Hamburg ed.1975, 2 vol.; or
Science of Logic, transl. Johnston & Struthers, Allen & Unwin, London 1929, 2vol.
Enzyklopädie der Philosophischen Wissenschaften im Grundrisse (1830), F.Meiner, Hamburg 1975; or
Hegel's Logic (Encyclopaedia part I), transl. Wallace, Clarendon Press, Oxford 1975.
Grundlinien der Philosophie des Rechts, F.Meiner, Hamburg ed. 1967.
Einleitung in die Geschichte der Philosophie, F.Meiner, Hamburg ed. 1966.
Vorlesungen über die Geschichte der Philosophie, Frommans, Stuttgart 1928.
Die Vernunft in der Geschichte, F.Meiner, Hamburg ed. 1970.
Sämtliche Werke, ed. Glockner, Frommans, Stuttgart 1928.

Heidegger M., *Der Satz vom Grund*, Neske, Pfullingen 1957.
Die Frage nach dem Ding, Niemeyer, Tübingen 1969.
Holzwege, Klostermann, Frankfort am Main 1980.

Hempel C.G., *Aspects of Scientific Explanation*, The Free Press, N-Y 1965.

Hintikka J., *Knowledge and the Known*, Reidel Pub., Dordrecht 1974.
Logic, Language-games and Information, Clarendon Press,Oxford 1973 (ed. 1979).

Hintikka J. & Remes U., *The Method of Analysis*, Reidel Pub., Dordrecht 1974.

D'Hondt J., 'Le moment de la destruction dans la dialectique historique de Hegel' in *Rev. Intern. de Phil.* **139-140** (1982), 125-137.

Hyppolite J., *Logique et existence*, PUF, Paris 1953.

Jarczyk G., *Système et liberté dans la logique de Hegel*, Aubier, Paris 1980.

Kant I., *Kritik der reinen Vernunft*, F.Meiner, Hamburg ed.1976; or
Critique of Pure Reason, transl. Kemp Smith, Macmillan, London 1929 (ed. 1990).
Kritik der praktischen Vernunft, Kants Werke V, De Gruyter 1968.
Kritik der Urteilskraft, F.Meiner, Hamburg 1974.
Prolegomena zu einer jeden künftigen Metaphysik, Kants Werke IV, De Gruyter, Berlin 1968.
Kants Werke, Akademie Textausgabe, De Gruyter, Berlin 1968.

Khalfon H., 'Un trajet en mathématiques' dans *Pratique des mots*, **56** (sept. 1986).

Kierkegaard S., *Concluding Unscientific Postscript to the philosophical Fragments*, transl. D.F.Swenson, Princeton Univ. Press 1968.
The Point of view for My Work as An Author, Harper Torchbook, N-Y. 1962.
The Journals of Kierkegaard 1834-1854, ed. Dru, Collins, London ed. 1967.

Kneale W. & M., *The Development of Logic*, Clarendon Press, Oxford 1964.

Kojève A., *Introduction à la lecture de Hegel*, Gallimard, Paris 1947.

Koyré A., 'Note sur la langue et la terminologie hégélienne' in *Rev. phil. de la France et de l'étranger*, **CXII** (1931).

Kuhn Th.S., *The Structure of Scientific Revolutions*, Univ. of Chicago 1962.

Lakatos I., *Proofs and Refutations*, Cambridge Univ. Press 1976.

Lambert K.& Brittan G.C., *An Introduction to the Philosophy of Science*, Prentice-Hall 1970.

Lévi-Strauss Cl., *La pensée sauvage*, Plon, Paris 1962.
anthropologie structurale deux, Plon, Paris 1973.

Marconi D., 'Logique et dialectique—Sur la justification de certaines argumentations hégéliennes' in *Rev. Phil. de Louvain* (1983), 563-579.

Meulen J.v.D., *Hegel; die gebrochene Mitte*, F.Meiner, Hamburg 1958.

Mure G.R.G., *Aristotle*, Oxford Univ. Press, N-Y. 1964.
An Introduction to Hegel, Clarendon Press, Oxford ed. 1966.
A Study of Hegel's Logic, Clarendon Press, Oxford ed. 1967.

Nancy J-L., *La remarque spéculative*, Gallilée, Auvers-sur-Oise 1973.

Nagel E., *The Structure of Science*, Routledge & Kegan, London ed. 1968.

Nietzsche F., *Thus Spoke Zarathustra*, transl. R.J.Hollingdale, Penguin Books, London ed. 1969.

Owens J., *The Doctrine of Being in Aristotelian Metaphysics*, Toronto 1951.

Piaget J., *les formes élémentaires de la dialectique*, Gallimard, Paris 1980.

Plato, *Plato: Collected Dialogues*, eds. Hamilton & Cairns, Princeton, Univ.Press 1961.

Popper K.R., *Conjectures and Refutations*, Routledge & Kegan, London ed. 1969.

Puntel B.L., *Darstellung, Methode und Struktur*, Bouvier, Grundmann 1973.

Riedel M., *Studien zu Hegels Rechtsphilosophie*, Frankfurt 1969.
'Causal and Historical Explanation' in *Essays on Explanation and Understanding*, ed. Manninen & Tuomelu, Reidel, Dordrecht 1976.

Rosen Menahem, *La dialectique de Hegel en tant que logique philosophique; exposition et critique de ses principales caractéristiques*, Doctoral Dissertation, Univ.of Jerusalem 1983 (in Hebrew,summary in French).

Rosen S., *G.W.F.Hegel*, Yale Univ. Press, ed.1976.

Ross D., *Aristotle*, Methuen, London 1964.

Rotenstreich N., *From Substance to Subject*, Martinus Nijhoff, The Hague 1974.

Salmon W.C., *Scientific Explanation and the Causal Structure of the World*, Princeton 1984.

Sartre J-P., *Critique de la raison dialectique*, Gallimard, Paris 1960.

Saussure F. de, *Cours de linguistique générale*, Payot, Paris 1973.

Simon J., *Das Problem der Sprache bei Hegel*, Kohlhammer, Stuttgart 1966.

Spinoza B., *Ethics*, ed. J.Gutmann, N-Y. 1974.

Stanguennec A., *Hegel critique de Kant*, PUF, Paris 1985.

Taylor A.E., *Plato, The Man and His Work*, Methuen, London 1960 (ed. 1966).

Taylor Ch., *Hegel*, Cambridge Univ. Press ed. 1977.

Trendelenburg A., *Logische Untersuchungen*, Hirzel, Leipzig 1870.

Vico J-B., *Principes de la philosophie de l'histoire*, trad. Michelet, Colin, Paris 1963.

Walton D.N., 'Are circular arguments necessarily vicious?' in *Amer.Phil.Quat.* (oct.85), 263-274.

Whitehead & Russell,*Principia Mathematica*, Univ.Press Cambridge, 1964, 3 vol.

Wittgenstein L., *Tractatus logico-philosophicus*, transl.C.K.Ogden, Routledge & Kegan, London 1981.
Wittgenstein's Lectures on the Foundation of Mathematics, Cambridge 1939,ed. Diamond C., Cornell Univ. Press, Ithaca 1976.
Remarks on the Foundations of Mathematics, ed. Wright, Rhees, Anscombe, the MIT press,Massachusetts 1967.

Wohlfart G., *Der spekulative Satz*, De Gruyter, Berlin 1981.

Wolfson H-A., *The Philosophy of Spinoza*, Harvard Univ. Press, Cambridge 1934, 2vol.

Yovel Y. *Kant and the renewal of Metaphysics*, Bialik Inst., Jerusalem 1973.
'Reason, reality and philosophical discourse according to Hegel' in *IYYUN* **26** N°1-3 (1975), 59-115.

INDEX OF NAMES

a) Hegel, Kant and Aristotle are not indicated here, this book dealing essentially with their way of thinking

b) 134 = on p.134
0.12 = Note 12, Introduction
II.18 = Note 18, Part Two

INDEX OF SUBJECTS

134 = on p.134

V.82 = Part Five Note 82

106. K. Kosík, *Dialectics of the Concrete*. A Study on Problems of Man and World. [Boston Studies in the Philosophy of Science, Vol. LII] 1976
ISBN 90-277-0761-8; Pb 90-277-0764-2
107. N. Goodman, *The Structure of Appearance*. 3rd ed. with an Introduction by G. Hellman. [Boston Studies in the Philosophy of Science, Vol. LIII] 1977
ISBN 90-277-0773-1; Pb 90-277-0774-X
108. K. Ajdukiewicz, *The Scientific World-Perspective and Other Essays, 1931-1963*. Translated from Polish. Edited and with an Introduction by J. Giedymin. 1978
ISBN 90-277-0527-5
109. R. L. Causey, *Unity of Science*. 1977 ISBN 90-277-0779-0
110. R. E. Grandy, *Advanced Logic for Applications*. 1977 ISBN 90-277-0781-2
111. R. P. McArthur, *Tense Logic*. 1976 ISBN 90-277-0697-2
112. L. Lindahl, *Position and Change*. A Study in Law and Logic. Translated from Swedish by P. Needham. 1977 ISBN 90-277-0787-1
113. R. Tuomela, *Dispositions*. 1978 ISBN 90-277-0810-X
114. H. A. Simon, *Models of Discovery and Other Topics in the Methods of Science*. [Boston Studies in the Philosophy of Science, Vol. LIV] 1977
ISBN 90-277-0812-6; Pb 90-277-0858-4
115. R. D. Rosenkrantz, *Inference, Method and Decision*. Towards a Bayesian Philosophy of Science. 1977 ISBN 90-277-0817-7; Pb 90-277-0818-5
116. R. Tuomela, *Human Action and Its Explanation*. A Study on the Philosophical Foundations of Psychology. 1977 ISBN 90-277-0824-X
117. M. Lazerowitz, *The Language of Philosophy*. Freud and Wittgenstein. [Boston Studies in the Philosophy of Science, Vol. LV] 1977
ISBN 90-277-0826-6; Pb 90-277-0862-2
118. Not published
119. J. Pelc (ed.), *Semiotics in Poland, 1894–1969*. Translated from Polish. 1979
ISBN 90-277-0811-8
120. I. Pörn, *Action Theory and Social Science*. Some Formal Models. 1977
ISBN 90-277-0846-0
121. J. Margolis, *Persons and Mind*. The Prospects of Nonreductive Materialism. [Boston Studies in the Philosophy of Science, Vol. LVII] 1977
ISBN 90-277-0854-1; Pb 90-277-0863-0
122. J. Hintikka, I. Niiniluoto, and E. Saarinen (eds.), *Essays on Mathematical and Philosophical Logic*. 1979 ISBN 90-277-0879-7
123. T. A. F. Kuipers, *Studies in Inductive Probability and Rational Expectation*. 1978
ISBN 90-277-0882-7
124. E. Saarinen, R. Hilpinen, I. Niiniluoto and M. P. Hintikka (eds.), *Essays in Honour of Jaakko Hintikka on the Occasion of His 50th Birthday*. 1979
ISBN 90-277-0916-5
125. G. Radnitzky and G. Andersson (eds.), *Progress and Rationality in Science*. [Boston Studies in the Philosophy of Science, Vol. LVIII] 1978
ISBN 90-277-0921-1; Pb 90-277-0922-X
126. P. Mittelstaedt, *Quantum Logic*. 1978 ISBN 90-277-0925-4
127. K. A. Bowen, *Model Theory for Modal Logic*. Kripke Models for Modal Predicate Calculi. 1979 ISBN 90-277-0929-7
128. H. A. Bursen, *Dismantling the Memory Machine*. A Philosophical Investigation of Machine Theories of Memory. 1978 ISBN 90-277-0933-5

129. M. W. Wartofsky, *Models*. Representation and the Scientific Understanding. [Boston Studies in the Philosophy of Science, Vol. XLVIII] 1979
ISBN 90-277-0736-7; Pb 90-277-0947-5
130. D. Ihde, *Technics and Praxis*. A Philosophy of Technology. [Boston Studies in the Philosophy of Science, Vol. XXIV] 1979 ISBN 90-277-0953-X; Pb 90-277-0954-8
131. J. J. Wiatr (ed.), *Polish Essays in the Methodology of the Social Sciences*. [Boston Studies in the Philosophy of Science, Vol. XXIX] 1979
ISBN 90-277-0723-5; Pb 90-277-0956-4
132. W. C. Salmon (ed.), *Hans Reichenbach: Logical Empiricist*. 1979
ISBN 90-277-0958-0
133. P. Bieri, R.-P. Horstmann and L. Krüger (eds.), *Transcendental Arguments in Science*. Essays in Epistemology. 1979 ISBN 90-277-0963-7; Pb 90-277-0964-5
134. M. Marković and G. Petrović (eds.), *Praxis*. Yugoslav Essays in the Philosophy and Methodology of the Social Sciences. [Boston Studies in the Philosophy of Science, Vol. XXXVI] 1979 ISBN 90-277-0727-8; Pb 90-277-0968-8
135. R. Wójcicki, *Topics in the Formal Methodology of Empirical Sciences*. Translated from Polish. 1979 ISBN 90-277-1004-X
136. G. Radnitzky and G. Andersson (eds.), *The Structure and Development of Science*. [Boston Studies in the Philosophy of Science, Vol. LIX] 1979
ISBN 90-277-0994-7; Pb 90-277-0995-5
137. J. C. Webb, *Mechanism, Mentalism and Metamathematics*. An Essay on Finitism. 1980 ISBN 90-277-1046-5
138. D. F. Gustafson and B. L. Tapscott (eds.), *Body, Mind and Method*. Essays in Honor of Virgil C. Aldrich. 1979 ISBN 90-277-1013-9
139. L. Nowak, *The Structure of Idealization*. Towards a Systematic Interpretation of the Marxian Idea of Science. 1980 ISBN 90-277-1014-7
140. C. Perelman, *The New Rhetoric and the Humanities*. Essays on Rhetoric and Its Applications. Translated from French and German. With an Introduction by H. Zyskind. 1979 ISBN 90-277-1018-X; Pb 90-277-1019-8
141. W. Rabinowicz, *Universalizability*. A Study in Morals and Metaphysics. 1979
ISBN 90-277-1020-2
142. C. Perelman, *Justice, Law and Argument*. Essays on Moral and Legal Reasoning. Translated from French and German. With an Introduction by H.J. Berman. 1980
ISBN 90-277-1089-9; Pb 90-277-1090-2
143. S. Kanger and S. Öhman (eds.), *Philosophy and Grammar*. Papers on the Occasion of the Quincentennial of Uppsala University. 1981 ISBN 90-277-1091-0
144. T. Pawlowski, *Concept Formation in the Humanities and the Social Sciences*. 1980
ISBN 90-277-1096-1
145. J. Hintikka, D. Gruender and E. Agazzi (eds.), *Theory Change, Ancient Axiomatics and Galileo's Methodology*. Proceedings of the 1978 Pisa Conference on the History and Philosophy of Science, Volume I. 1981 ISBN 90-277-1126-7
146. J. Hintikka, D. Gruender and E. Agazzi (eds.), *Probabilistic Thinking, Thermodynamics, and the Interaction of the History and Philosophy of Science*. Proceedings of the 1978 Pisa Conference on the History and Philosophy of Science, Volume II. 1981 ISBN 90-277-1127-5
147. U. Mönnich (ed.), *Aspects of Philosophical Logic*. Some Logical Forays into Central Notions of Linguistics and Philosophy. 1981 ISBN 90-277-1201-8
148. D. M. Gabbay, *Semantical Investigations in Heyting's Intuitionistic Logic*. 1981
ISBN 90-277-1202-6

149. E. Agazzi (ed.), *Modern Logic – A Survey*. Historical, Philosophical, and Mathematical Aspects of Modern Logic and Its Applications. 1981 ISBN 90-277-1137-2
150. A. F. Parker-Rhodes, *The Theory of Indistinguishables*. A Search for Explanatory Principles below the Level of Physics. 1981 ISBN 90-277-1214-X
151. J. C. Pitt, *Pictures, Images, and Conceptual Change*. An Analysis of Wilfrid Sellars' Philosophy of Science. 1981 ISBN 90-277-1276-X; Pb 90-277-1277-8
152. R. Hilpinen (ed.), *New Studies in Deontic Logic*. Norms, Actions, and the Foundations of Ethics. 1981 ISBN 90-277-1278-6; Pb 90-277-1346-4
153. C. Dilworth, *Scientific Progress*. A Study Concerning the Nature of the Relation between Successive Scientific Theories. 2nd, rev. and augmented ed., 1986 ISBN 90-277-2215-3; Pb 90-277-2216-1
154. D. Woodruff Smith and R. McIntyre, *Husserl and Intentionality*. A Study of Mind, Meaning, and Language. 1982 ISBN 90-277-1392-8; Pb 90-277-1730-3
155. R. J. Nelson, *The Logic of Mind*. 2nd. ed., 1989 ISBN 90-277-2819-4; Pb 90-277-2822-4
156. J. F. A. K. van Benthem, *The Logic of Time*. A Model-Theoretic Investigation into the Varieties of Temporal Ontology, and Temporal Discourse. 1983; 2nd ed., 1991 ISBN 0-7923-1081-0
157. R. Swinburne (ed.), *Space, Time and Causality*. 1983 ISBN 90-277-1437-1
158. E. T. Jaynes, *Papers on Probability, Statistics and Statistical Physics*. Ed. by R. D. Rozenkrantz. 1983 ISBN 90-277-1448-7; Pb (1989) 0-7923-0213-3
159. T. Chapman, *Time: A Philosophical Analysis*. 1982 ISBN 90-277-1465-7
160. E. N. Zalta, *Abstract Objects*. An Introduction to Axiomatic Metaphysics. 1983 ISBN 90-277-1474-6
161. S. Harding and M. B. Hintikka (eds.), *Discovering Reality*. Feminist Perspectives on Epistemology, Metaphysics, Methodology, and Philosophy of Science. 1983 ISBN 90-277-1496-7; Pb 90-277-1538-6
162. M. A. Stewart (ed.), *Law, Morality and Rights*. 1983 ISBN 90-277-1519-X
163. D. Mayr and G. Süssmann (eds.), *Space, Time, and Mechanics*. Basic Structures of a Physical Theory. 1983 ISBN 90-277-1525-4
164. D. Gabbay and F. Guenthner (eds.), *Handbook of Philosophical Logic*. Vol. I: Elements of Classical Logic. 1983 ISBN 90-277-1542-4
165. D. Gabbay and F. Guenthner (eds.), *Handbook of Philosophical Logic*. Vol. II: Extensions of Classical Logic. 1984 ISBN 90-277-1604-8
166. D. Gabbay and F. Guenthner (eds.), *Handbook of Philosophical Logic*. Vol. III: Alternative to Classical Logic. 1986 ISBN 90-277-1605-6
167. D. Gabbay and F. Guenthner (eds.), *Handbook of Philosophical Logic*. Vol. IV: Topics in the Philosophy of Language. 1989 ISBN 90-277-1606-4
168. A. J. I. Jones, *Communication and Meaning*. An Essay in Applied Modal Logic. 1983 ISBN 90-277-1543-2
169. M. Fitting, *Proof Methods for Modal and Intuitionistic Logics*. 1983 ISBN 90-277-1573-4
170. J. Margolis, *Culture and Cultural Entities*. Toward a New Unity of Science. 1984 ISBN 90-277-1574-2
171. R. Tuomela, *A Theory of Social Action*. 1984 ISBN 90-277-1703-6
172. J. J. E. Gracia, E. Rabossi, E. Villanueva and M. Dascal (eds.), *Philosophical Analysis in Latin America*. 1984 ISBN 90-277-1749-4
173. P. Ziff, *Epistemic Analysis*. A Coherence Theory of Knowledge. 1984 ISBN 90-277-1751-7

174. P. Ziff, *Antiaesthetics*. An Appreciation of the Cow with the Subtile Nose. 1984 ISBN 90-277-1773-7
175. W. Balzer, D. A. Pearce, and H.-J. Schmidt (eds.), *Reduction in Science*. Structure, Examples, Philosophical Problems. 1984 ISBN 90-277-1811-3
176. A. Peczenik, L. Lindahl and B. van Roermund (eds.), *Theory of Legal Science*. Proceedings of the Conference on Legal Theory and Philosophy of Science (Lund, Sweden, December 1983). 1984 ISBN 90-277-1834-2
177. I. Niiniluoto, *Is Science Progressive?* 1984 ISBN 90-277-1835-0
178. B. K. Matilal and J. L. Shaw (eds.), *Analytical Philosophy in Comparative Perspective*. Exploratory Essays in Current Theories and Classical Indian Theories of Meaning and Reference. 1985 ISBN 90-277-1870-9
179. P. Kroes, *Time: Its Structure and Role in Physical Theories*. 1985 ISBN 90-277-1894-6
180. J. H. Fetzer, *Sociobiology and Epistemology*. 1985 ISBN 90-277-2005-3; Pb 90-277-2006-1
181. L. Haaparanta and J. Hintikka (eds.), *Frege Synthesized*. Essays on the Philosophical and Foundational Work of Gottlob Frege. 1986 ISBN 90-277-2126-2
182. M. Detlefsen, *Hilbert's Program*. An Essay on Mathematical Instrumentalism. 1986 ISBN 90-277-2151-3
183. J. L. Golden and J. J. Pilotta (eds.), *Practical Reasoning in Human Affairs*. Studies in Honor of Chaim Perelman. 1986 ISBN 90-277-2255-2
184. H. Zandvoort, *Models of Scientific Development and the Case of Nuclear Magnetic Resonance*. 1986 ISBN 90-277-2351-6
185. I. Niiniluoto, *Truthlikeness*. 1987 ISBN 90-277-2354-0
186. W. Balzer, C. U. Moulines and J. D. Sneed, *An Architectonic for Science*. The Structuralist Program. 1987 ISBN 90-277-2403-2
187. D. Pearce, *Roads to Commensurability*. 1987 ISBN 90-277-2414-8
188. L. M. Vaina (ed.), *Matters of Intelligence*. Conceptual Structures in Cognitive Neuroscience. 1987 ISBN 90-277-2460-1
189. H. Siegel, *Relativism Refuted*. A Critique of Contemporary Epistemological Relativism. 1987 ISBN 90-277-2469-5
190. W. Callebaut and R. Pinxten, *Evolutionary Epistemology*. A Multiparadigm Program, with a Complete Evolutionary Epistemology Bibliograph. 1987 ISBN 90-277-2582-9
191. J. Kmita, *Problems in Historical Epistemology*. 1988 ISBN 90-277-2199-8
192. J. H. Fetzer (ed.), *Probability and Causality*. Essays in Honor of Wesley C. Salmon, with an Annotated Bibliography. 1988 ISBN 90-277-2607-8; Pb 1-5560-8052-2
193. A. Donovan, L. Laudan and R. Laudan (eds.), *Scrutinizing Science*. Empirical Studies of Scientific Change. 1988 ISBN 90-277-2608-6
194. H.R. Otto and J.A. Tuedio (eds.), *Perspectives on Mind*. 1988 ISBN 90-277-2640-X
195. D. Batens and J.P. van Bendegem (eds.), *Theory and Experiment*. Recent Insights and New Perspectives on Their Relation. 1988 ISBN 90-277-2645-0
196. J. Österberg, *Self and Others*. A Study of Ethical Egoism. 1988 ISBN 90-277-2648-5
197. D.H. Helman (ed.), *Analogical Reasoning*. Perspectives of Artificial Intelligence, Cognitive Science, and Philosophy. 1988 ISBN 90-277-2711-2
198. J. Wolenski, *Logic and Philosophy in the Lvov-Warsaw School*. 1989 ISBN 90-277-2749-X

199. R. Wójcicki, *Theory of Logical Calculi*. Basic Theory of Consequence Operations. 1988 ISBN 90-277-2785-6
200. J. Hintikka and M.B. Hintikka, *The Logic of Epistemology and the Epistemology of Logic*. Selected Essays. 1989 ISBN 0-7923-0040-8; Pb 0-7923-0041-6
201. E. Agazzi (ed.), *Probability in the Sciences*. 1988 ISBN 90-277-2808-9
202. M. Meyer (ed.), *From Metaphysics to Rhetoric*. 1989 ISBN 90-277-2814-3
203. R.L. Tieszen, *Mathematical Intuition*. Phenomenology and Mathematical Knowledge. 1989 ISBN 0-7923-0131-5
204. A. Melnick, *Space, Time, and Thought in Kant*. 1989 ISBN 0-7923-0135-8
205. D.W. Smith, *The Circle of Acquaintance*. Perception, Consciousness, and Empathy. 1989 ISBN 0-7923-0252-4
206. M.H. Salmon (ed.), *The Philosophy of Logical Mechanism*. Essays in Honor of Arthur W. Burks. With his Responses, and with a Bibliography of Burk's Work. 1990 ISBN 0-7923-0325-3
207. M. Kusch, *Language as Calculus vs. Language as Universal Medium*. A Study in Husserl, Heidegger, and Gadamer. 1989 ISBN 0-7923-0333-4
208. T.C. Meyering, *Historical Roots of Cognitive Science*. The Rise of a Cognitive Theory of Perception from Antiquity to the Nineteenth Century. 1989 ISBN 0-7923-0349-0
209. P. Kosso, *Observability and Observation in Physical Science*. 1989 ISBN 0-7923-0389-X
210. J. Kmita, *Essays on the Theory of Scientific Cognition*. 1990 ISBN 0-7923-0441-1
211. W. Sieg (ed.), *Acting and Reflecting*. The Interdisciplinary Turn in Philosophy. 1990 ISBN 0-7923-0512-4
212. J. Karpiński, *Causality in Sociological Research*. 1990 ISBN 0-7923-0546-9
213. H.A. Lewis (ed.), *Peter Geach: Philosophical Encounters*. 1991 ISBN 0-7923-0823-9
214. M. Ter Hark, *Beyond the Inner and the Outer*. Wittgenstein's Philosophy of Psychology. 1990 ISBN 0-7923-0850-6
215. M. Gosselin, *Nominalism and Contemporary Nominalism*. Ontological and Epistemological Implications of the Work of W.V.O. Quine and of N. Goodman. 1990 ISBN 0-7923-0904-9
216. J.H. Fetzer, D. Shatz and G. Schlesinger (eds.), *Definitions and Definability*. Philosophical Perspectives. 1991 ISBN 0-7923-1046-2
217. E. Agazzi and A. Cordero (eds.), *Philosophy and the Origin and Evolution of the Universe*. 1991 ISBN 0-7923-1322-4
218. M. Kusch, *Foucault's Strata and Fields*. An Investigation into Archaeological and Genealogical Science Studies. 1991 ISBN 0-7923-1462-X
219. C.J. Posy, *Kant's Philosophy of Mathematics*. Modern Essays. 1992 ISBN 0-7923-1495-6
220. G. Van de Vijver, *New Perspectives on Cybernetics*. Self-Organization, Autonomy and Connectionism. 1992 ISBN 0-7923-1519-7
221. J.C. Nyíri, *Tradition and Individuality*. Essays. 1992 ISBN 0-7923-1566-9
222. R. Howell, *Kant's Transcendental Deduction*. An Analysis of Main Themes in His Critical Philosophy. 1992 ISBN 0-7923-1571-5

223. A. García de la Sienra, *The Logical Foundations of the Marxian Theory of Value.* 1992 ISBN 0-7923-1778-5
224. D.S. Shwayder, *Statement and Referent.* An Inquiry into the Foundations of our Conceptual Order. 1992 ISBN 0-7923-1803-X
225. M. Rosen, *Problems of the Hegelian Dialectic.* Dialectic Reconstructed as a Logic of Human Reality. 1993 ISBN 0-7923-2047-6

Previous volumes are still available.

KLUWER ACADEMIC PUBLISHERS – DORDRECHT / BOSTON / LONDON

FSC
www.fsc.org
MIX
Papier aus verantwortungsvollen Quellen
Paper from responsible sources
FSC® C105338

If you have any concerns about our products,
you can contact us on
ProductSafety@springernature.com

In case Publisher is established outside the EU,
the EU authorized representative is:
Springer Nature Customer Service Center GmbH
Europaplatz 3, 69115 Heidelberg, Germany

Printed by Libri Plureos GmbH
in Hamburg, Germany